Ecological Studies

Analysis and Synthesis

Edited by

W. D. Billings, Durham (USA) F. Golley, Athens (USA)
O. L. Lange, Würzburg (FRG) J. S. Olson, Oak Ridge (USA)
H. Remmert, Marburg (FRG)

Volume 90

Jill Baron
Editor

Biogeochemistry of a Subalpine Ecosystem

Loch Vale Watershed

Contributors

M. A. Arthur, A. S. Denning, M. A. Harris,
M. A. Mast, D. M. McKnight, P. McLaughlin,
B. D. Rosenlund, S. A. Spaulding, P. M. Walthall

With 62 Illustrations

Springer-Verlag
New York Berlin Heidelberg London Paris
Tokyo Hong Kong Barcelona Budapest

Jill Baron
National Park Service Water Resource Division
Natural Resource Ecology Laboratory
Colorado State University
Fort Collins, CO 80523, USA

Cover Illustration: Southwest view of Loch Vale Watershed, Rocky Mountain National Park, Colorado. Continental Divide is on the horizon; The Loch is the lake in the foreground. Pen and ink drawing by Stephanie Stern.

Library of Congress Cataloging-in-Publication Data
Biogeochemistry of a subalpine ecosystem: Loch Vale Watershed/
Jill Baron, editor.
p. cm.—(Ecological studies)
Includes bibliographical references and index.
ISBN-13: 978-1-4612-7670-8
1. Mountain ecology—Colorado—Loch Vale Watershed.
2. Biogeochemistry—Colorado—Loch Vale Watershed. 3. Loch Vale Watershed (Colo.) I. Baron, Jill. II. Series.
QH105.C6B56 1991
574.5′222—dc20 91-35098

Printed on acid-free paper.

Softcover reprint of the hardcover 1st edition 1992

Typeset by Thomson Press (I) Ltd., New Delhi, India.

9 8 7 6 5 4 3 2 1

ISBN-13: 978-1-4612-7670-8 e-ISBN-13: 978-1-4612-2788-5
DOI: 10.1007/978-1-4612-2788-5

Foreword

Rocky Mountain National Park was established in 1915, one year before the creation of the National Park Service. The mandate of the National Park Service is to preserve and protect areas of exquisite beauty and cultural value for the benefit and enjoyment of future generations. National parks mean many things to many people, and, in often stirring words, a National Parks and Conservation Association report states the National Park System is a magnificent and uniquely American gift to the American people and the world. In the early years of the Service, park superintendents actively promoted and developed parks to accommodate visitors. Then, as now, parks represented a democratic ideal, that even the greatest treasures should be available to all. Seventy five years ago, however, park managers saw little need for active management of natural resources, unless it was to enhance visitors' experience. And few managers saw the need for a stable and independent research program on which to base management decisions. Thus began a legacy of erratic, often passive, resource management based more on politics and in-house studies than on validated scientific information.

The world is a different place than it was 75 years ago. Human population growth, changes in land use, and ever more sophisticated technology affect the very fabric of life on Earth. As local-, regional-, and global-scale changes occur from human tampering with the environment, the integrity of natural ecosystems is threatened worldwide. In a world where natural lands are becoming rare, national parks serve an increasingly important role as baselines from which to measure and understand change.

So, times have changed and so must the basis for national park management. Issues have become much more complex biologically, and contentious socially. Interdisciplinary studies incorporating the best of science are essential, not just desirable. The balance of resource management efforts within national parks needs to embrace the concept of ecosystem management. Ecosystem management implies an understanding of the physical and biological processes underlying the ecosystems and landscapes which are the parks. Much of the necessary information on natural variability must come from long-term studies conducted on a time scale similar to that of the processes being studied. Long-term research places unusual events in perspective, enables us to distinguish natural variability from human-imposed change, and can empower park managers with the knowledge required to effectively manage and mitigate the many threats facing national park resources.

The status of research in the national parks—both in quality and quantity—is far below the needed levels. Fortunately, there is a small but growing movement within the National Park Service which is building a scientifically sound base of knowledge about natural resources. The Loch Vale Watershed study described in this volume is an example of the type of holistic, long-term program that is essential. I hope that it will serve as a model for greatly expanded efforts at basic ecosystem research within national parks, and, in doing so, provide the beginning of a new cooperation between scientists and resource managers—a cooperation founded on scientific understanding.

Jerry Franklin
University of Washington
Seattle, Washington

Contents

Contributors

Mary A. Arthur
Boyce Thompson Institute
Cornell University
Ithaca, NY 14853

A. Scott Denning
Department of Atmospheric Sciences
Colorado State University
Fort Collins, CO 80523

Mitchell A. Harris
Department of Entomology
Colorado State University
Fort Collins, CO 80523

M. Alisa Mast
U.S. Geological Survey
Water Resources Division,
Mail Stop 415
Denver, CO 80225

Diane M. McKnight
U.S. Geological Survey
Water Resources Division, MS 408
Denver, CO 80225

Paul McLaughlin	National Park Service Alaska Regional Office 2525 Gambell Street Anchorage, AL 99503
Bruce D. Rosenlund	U.S. Fish and Wildlife Service Golden, CO 80401
Sarah A. Spaulding	U.S. Geological Survey Water Resources Division, MS 408 Denver, CO 80225
P. Mark Walthall	Agronomy Department Louisiana State University Baton Rouge, LA 70803

Relief Map of North-Central Colorado

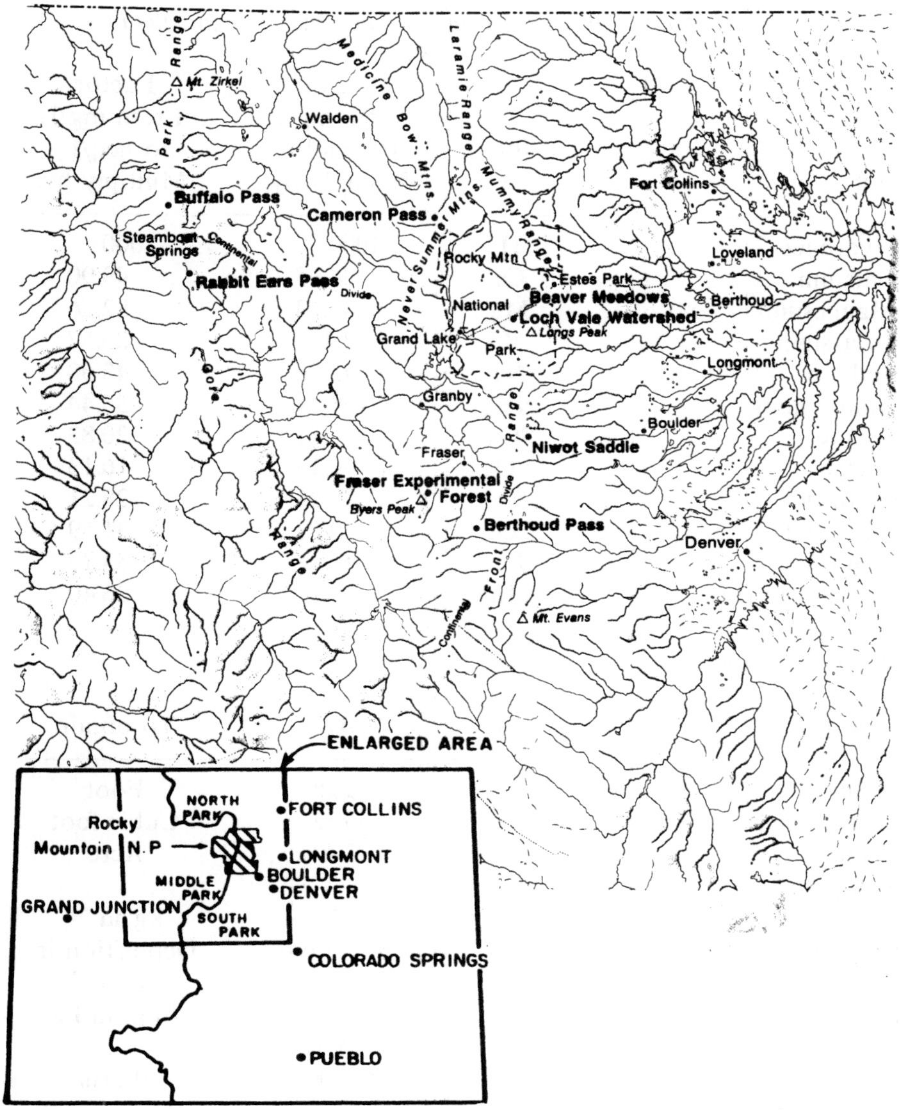

Conversion Factors

Chemical parameters: Conversion from mass to moles or equivalents

Ion	Ion as	Factor: From mg/L to μmol/L Multiply by:	Factor: From mg/L to μeq/L, Multiply by:
Hydrogen	H	1000	1000
Calcium	Ca	24.95	49.90
Magnesium	Mg	41.13	82.26
Potassium	K	25.57	25.57
Sodium	Na	43.50	43.50
Ammonium	NH_4	55.44	55.44
Sulfate	SO_4	10.41	20.83
Nitrate	NO_3	16.13	16.13
Chloride	Cl	28.21	28.21
Phosphate	PO_4	10.53	31.59
Iron (Fe^{3+})	Fe	17.91	53.72
Manganese (Mn^{2+})	Mn	18.20	36.40

Physical Parameters:

Multiply	As	By	To obtain
Meter	m	3.28	Foot
Cubic meter	m^3	35.32	Cubic foot
Hectare	ha	2.47	Acre
Milligrams per square meter	$mg\ m^{-2}$	10^{-2}	$kg\ ha^{-1}$
Milligrams per liter	$mg\ L^{-1}$	$10^{-1} \times$ cm precipitation	Deposition in $kg\ ha^{-1}$
Milligrams per liter	$mg\ L^{-1}$	$m^3\ 10^{-3}$ discharge	Flux in kg
Joules per day (12 hours)	$J\ day^{-1}$	43,200	Watts

1. Introduction

Jill Baron

Ecosystem Analysis

Natural systems have their own rhythms and progressions through time that strongly influence, if not dictate, the nature of individual components. Bedrock, soils, vegetation, surface waters, and their communities are all shaped by climate, natural disturbances such as fire, and interaction with each other. Each component, whether animate or inanimate, is part of the ecosystem. Each unit contributes to the cycling, transforming, or holding of energy and materials. To understand the dynamics of one part of a system, then, one must also know about the system as a whole.

Units can be surveyed from a static perspective, and this is necessary to obtain initial characterizing properties, but they must also be understood from a dynamic perspective (Schumm 1978). Systems change over time in response to external pressures such as changing climate or atmospheric deposition. Systems also change in response to internal processes; forest succession, soil development, and lake sedimentation are examples. Stochastic events are important in the shaping of many ecosystems. They are especially important in the alpine and subalpine landscapes that are the subject of this volume. Examination of ecosystem components from a holistic perspective increases the time required for study. It increases the information needed to be assimilated, but in the long run helps to unravel the complexity inherent in natural systems.

This is a book written for two audiences. First, it is a scientific document of the biogeochemical processes of Loch Vale Watershed, Rocky Mountain National Park, Colorado, by which I hope to share the understanding we have gained about subalpine and alpine processes with other ecologists. Second, it is a record of long-term ecological research; I hope to apprise managers of public lands of the power associated with foreknowledge of ecosystem structure and functions. Foreknowledge allows quantitative assessment of the effects of human activities upon natural systems. Most importantly, it allows *prediction* of the effects of development. Predictions based on sound scientific understanding can be used to prevent or mitigate loss of ecological integrity.

The Role of the National Park Service

The United States was the first nation on Earth to recognize the value of setting aside wilderness as part of the national heritage. In 1916 the U.S. Congress passed the National Park Service Organic Act, which mandated that the newly created Service preserve and protect areas of natural and cultural resources value for future enjoyment by the people of the United States. The U.S. National Park system is still the most extensive natural lands preservation system and serves as an important example to other nations who are establishing their own national preserves.

When parks were established, many of them were remote from human habitation and influence. Over time, however, encroaching populations and the ever-expanding influences of land use and pollution have created national parks that are natural islands surrounded by development. Natural areas within national parks have come to assume ever greater importance as preserves and refuges for dwindling wild populations. Parks today provide some of the few remaining undisturbed natural systems within which we can ask questions about ecosystem process and function.

Managers of national parks must assume an increasingly important responsibility for wise resource management in light of these considerations. The interpretation of resource management has changed over time with broadening understanding of ecological interactions (Agee and Johnson 1988). A contemporary management policy is evolving around the recognition that natural systems are dynamic, not static. Each ecosystem has its own unique responses, or resiliency, to both internal and external stress (Holling 1978). The more a manager understands about the natural system under his or her protection, the less chance there is for environmental surprise. Surprise is a human response elicited by unexpected system behavior; surprise can be politically and economically costly, socially embarrasing, and dangerous to both natural and human resource integrity.

The value of conducting ecosystem-level studies to minimize surprise about natural and disturbed systems cannot be too heavily emphasized. The ecosystem concept provides a framework for observing the interactions of biological communities and their abiotic environments and for studying the change of these relationships over time (Likens 1985). The better we understand ecosystem components in the context of the whole system, including watershed and airshed, the more appropriate will be our management actions regarding their husbandry. Long-term studies are essential to observe and understand natural system dynamics. This understanding, in turn, provides the necessary reference base from which questions can be asked about the effects of unnatural disturbance.

Management of natural systems is no longer a contradiction in terms. A combination of human population expansion with the products and by-products of technology has caused humanity to become a global-scale biogeochemical force (Baron and Galvin 1990). With our capability to alter the climate of Earth, no natural environment, no matter how seemingly remote, is immune from human influence. The state of the natural resources of the Earth is now dependent on human actions. And acknowledgment of the scope of human influence implies management, whether through neglect and inaction or through informed policy generation and implementation.

The Loch Vale Watershed Study

The Loch Vale Watershed study in Rocky Mountain National Park was initiated to minimize surprise with respect to a particular human-caused disturbance: acidic atmospheric deposition. Our objective was to increase our understanding of both natural and potentially acidified biogeochemical pathways in the alpine and subalpine environment of Rocky Mountain National Park. The study took an ecosystem approach, attempting to quantify major elemental and pollutant flux and identifying sources, sinks, and controls of the major ions. We defined the boundaries of our ecosystem to be the physical boundaries of Loch Vale Watershed and adopted the "small watershed technique" as the overall context for study.

This technique originated in Colorado, where the U.S. Forest Service applied it to questions of inputs and outputs at Wagon Wheel Gap in the early 1900s (Bates and Henry 1928). It has been widely used since then and has become well known as the technique applied in the Hubbard Brook ecosystem study (Likens et al. 1977). Instead of "black-boxing" internal processes while quantifying watershed inputs and outputs, we have attempted to understand and quantify inputs and outputs from each of the major ecosystem components of the landscape: bedrock, soils, vegetation, and surface waters (Figure 1.1). Because so little was known about these components of the alpine and subalpine zones, much of our effort went into documenting basic characteristics and processes. Because the study was based on the threat of acidic deposition, we also explored the airshed (most of

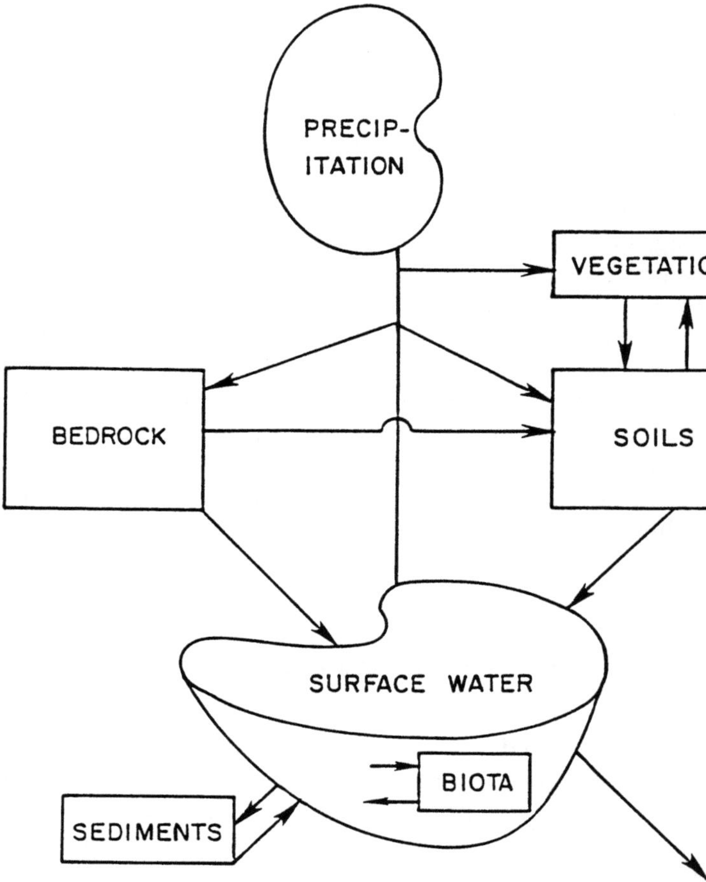

Figure 1.1. Conceptual model of Loch Vale Watershed study.

the western United States) that contributes both moisture and chemical compounds to Loch Vale Watershed.

Rocky Mountain National Park is sensitive to acidic deposition (Eilers et al. 1986; Turk and Spahr 1989). Much of the park area is exposed granitic rock with 147 alpine and subalpine lakes and more than 700 stream kilometers (Figure 1.2). Naturally reproducing populations of trout are present in 59 lakes (Rosenlund and Stevens 1990). It is well documented (Gibson 1986; Schindler 1988) that deposition of strong acid anions can rapidly exhaust the acid-neutralizing capacity (ANC) of systems underlain by slow-weathering bedrock types, allowing lakes and streams to acidify. The acidity itself, and more importantly aluminum, which becomes soluble at pH levels below 5.0, are extremely toxic to aquatic organisms.

We chose Loch Vale Watershed (LVWS) for study for a number of reasons. A reconnaissance in 1981 of more than 40 lakes and streams within four

Figure 1.2. Sky Pond (upper) and Glass lake (lower) in Loch Vale Watershed.

major drainages of Rocky Mountain National Park (Gibson et al. 1983) provided a chemical and geographical base of information that aided in the selection of a specific watershed study site. Of the drainages sampled, Loch Vale Watershed exhibited the lowest summertime alkalinities and showed an inverse gradient of ANC with elevation. There are four lakes within the watershed, three alpine and one located below treeline, allowing for comparison of alpine with subalpine lakes (Figure 1.3). There is sufficient soil development to allow characterization of different soil types and their influence on surface water chemistry. An old-growth Englemann spruce–subalpine fir forest is within the valley basin. A controlled surface water outlet from the entire basin allowed us to quantify stream discharge. One final reason for the selection of LVWS is that the watershed is accessible, although remote, throughout the year.

The LVWS study was organized around four working questions intended to focus our efforts for maximum understanding about alpine and subalpine

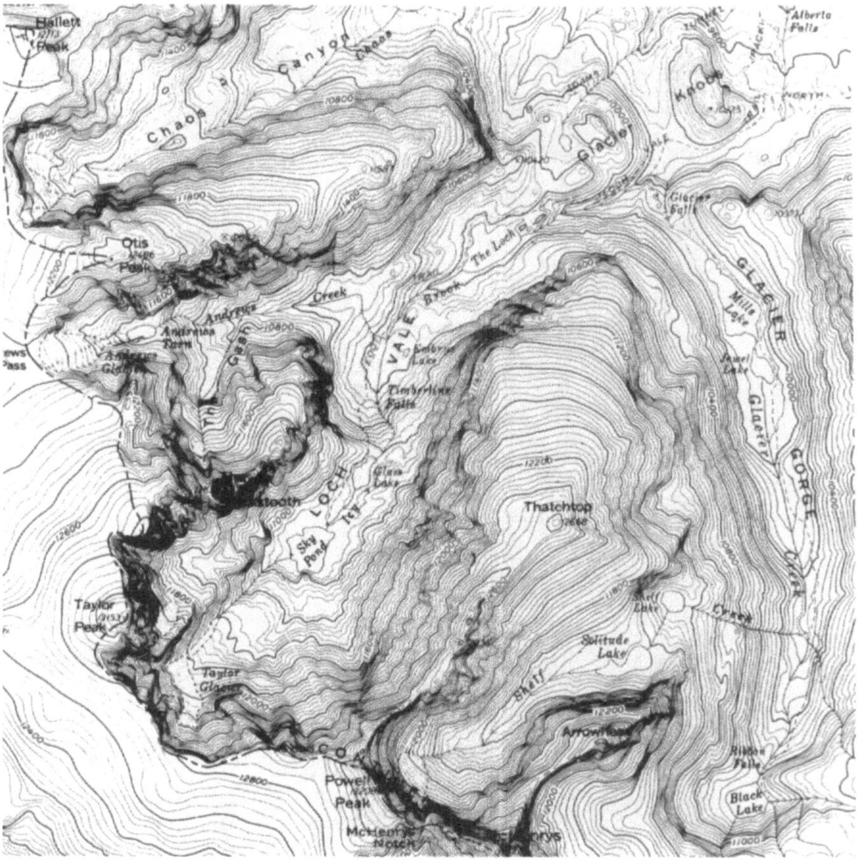

Figure 1.3. Topographic map of Loch Vale Watershed (contour intervals in feet).

responses to acidic deposition. The evolution of the LVWS project has been concurrent with the accumulation of much basic knowledge about atmospheric processes of transformation and transport, hydrologic flowpaths, chemical controls within many types of soils, and the nature of stream and lake dynamics (NAPAP 1989). We have been fortunate in being able to incorporate much of this new information into our interpretations.

Working Question 1. Are current levels of acidity in precipitation in Rocky Mountain National Park greater than historic levels? Information was collected to test this hypothesis both historically with the analysis of lake sediment cores and with regular analysis of precipitation samples (see Chapter 4, Deposition, this volume). By addressing this question, we established a frame of reference along a continuum (ranging from no industrially caused acidic deposition to very acidic deposition) for understanding LVWS processes and reactions to acidic deposition.

Working Question 2. Does buffering of surface waters come from soil processes such as weathering of inorganic soil materials and exchange reactions from soil organic matter? This hypothesis formed the basis for our soils investigations and appeared to be valid for a short period each spring. It was invalid on a yearly or greater time scale, as discussed in Chapters 7 and 8.

Working Question 3. What are the controls of surface water chemical composition? We postulated that the composition of lakes and streams fluctuated over time, influenced by hydrologic processes as well as products from bedrock weathering, soils, and vegetative and aquatic biota. With this large working question we attempted to define the major influences on lake and stream waters. A number of separate studies were conducted under the umbrella of this question, including characterizations of bedrock and geochemical weathering (Chapter 6), vegetation (Chapter 5), and the physical, chemical and hydrologic characterization of lakes and streams (Chapters 3 and 8).

Working Question 4. Can we define and predict surface water chemical composition on the basis of knowledge of precipitation, bedrock, soil, vegetation, and aquatic organisms? This, of course, is *The Big Question* that every manager of natural resources wants answered. We used this question to integrate knowledge of Loch Vale Watershed. We used this question to place our understanding of individual processes within a conceptual framework (Figure 1.1; see Chapter 10). And we used this question to test the validity of ecosystem analysis as a resource management tool (Chapter 11).

A Consideration of Error

Among the many assumptions that form the foundations of this study is one suggesting that numbers collected in the field accurately represent reality. The injection of sources of error can come at many points along the way from sample collection to final report, and those attempting to quantify natural systems must be aware of this. The effect of two or more sources of error on the final interpretation of data may be additive, although in other cases the effect may be multiplicative (Winter 1981; Bigelow 1986). If we are very lucky, they may cancel each other. We tried to assess the confidence with which numbers are presented in as many cases as possible in the following chapters.

The many sources of possible error in the Loch Vale Watershed study, as in any field study, include: (1) precision and accuracy of the field instrumentation; (2) spatial or temporal extrapolation of point data; (3) use of residuals in some cases for determining water and chemical budgets; (4) incorporation of qualitative data in instances where quantitative data were unavailable; (5) sample contamination in the field, in the field laboratory, or in the analytical laboratory; (6) mistakes made in data handling and manipulation, and, (7) at all stages, observer bias. Sources of uncertainty are discussed with the data presented and were taken into account in interpretation. Acknowledgment of uncertainty does not mean rejection of the data, although samples shown to be contaminated were not used. Estimates of uncertainty are used, instead, to place confidence limits on the interpretations made regarding processes within LVWS.

We concluded that the combination of ecological uncertainty and the variability introduced by mountainous terrain makes sweeping generalizations of many results impossible. This is not as unfortunate as it seems. It is evident that extrapolations of deposition patterns and chemical composition to other parts of the southern Rocky Mountains, let alone to the next watershed, are inappropriate. The hydrologic regime probably differs among watersheds of different aspects and between watersheds with and without snowfields. On the other hand, some basic characteristics of bedrock geochemistry, soils, vegetation, and surface waters, many of which were heretofore unknown, may be valid throughout granitic mountainous areas of the western United States. For instance, Mast's discovery (Mast et al. 1990) of calcitic microveins in the granitic bedrock of LVWS may have solved a question that has puzzled geochemists for some time. Future investigations are needed to determine whether similar microveins are responsible for observed levels of surface water calcium in other granitic regions of the world.

Conclusions

The attempt to acquire accurate information on the chemistry and hydrology of a steep mountain basin has been a humbling experience. In winters we prepared for severe cold, vicious winds, equipment malfunction, and potential avalanches. Typical days included a 10-km or longer round trip on skis with a heavy backpack or sled. In the summer we prepared for lightning storms, surprise blizzards and hail, equipment malfunction, and equipment tampering by animals (mostly marmots). All equipment in LVWS has been transported on the backs of pack animals (including human pack animals) or by helicopter. While the two major pieces of monitoring equipment, the weather station and the stream gauging station, are sophisticated in operation, their installation with handheld pickaxes and shovels could have taken place in any remote developing country in the world.

Great increases in the understanding of ecosystem processes have occurred over the last several decades. These increases have come about because of conditions that have been eloquently discussed by others (Likens 1983; Franklin 1987). First and foremost among them has been the ability to interpret long-term databases. Unfortunately, acquisition of long-term data depends on many tenuous ingredients, and the loss of any one ingredient can disrupt the line of continuity. Successful long-term data sets come from dedicated research sites with broad institutional and financial bases of support. They depend on continuous scientific rigor at all stages of analysis. They also often depend on inputs of luck.

Down on the working level of daily operations, perhaps the fundamental key to successful ecosystem-scale research was to liberally apply common sense to every phase of research, from personnel safety to data analysis. On a day-to-day basis, we were fortunate to have project personnel that were flexible, creative, and willing to live with uncertainty. Equipment was backed up by other equipment; plans were backed up with contingency plans. Experience taught that research and field staff who were given a variety of responsibilities contributed the most insights whereas those with limited responsibilities burned out and quit.

One other ingredient contributed to our long-term ecological research program. Early in the project it was decided to focus our efforts into developing and maintaining a high quality physical and chemical database. We then made these data freely available. Because of this, Loch Vale Watershed has become attractive to other scientists who need physical and chemical information as the bases from which to ask other questions. This "economy of science" led to more research than we could have paid for and a greater level of insight than would have occurred otherwise.

Acknowledgments

The LVWS project has been built on the contributions of many people and institutions, and I take great pleasure in acknowledging as many as possible. All of them, from the volunteer field assistants to those who contributed valuable scientific insights, have been vital to the success of this project. Their efforts are greatly appreciated. I especially want to thank Ray Herrmann, whose holistic view toward the management of national park lands has been valuable since 1976. My family has provided tremendous support: Dr. Dennis Ojima, in the nurturing of ecosystem concepts and insistence on scientific rigor; Dr. Alma Baron, who served as a role model by successfully combining career and family long before this was an accepted path for women; my father, Lee, and my children Claire and Kyle Ojima, who helped maintain a proper perspective on the whole business. Scott Denning and David Bigelow served as catalysts and critics.

Heartfelt thanks are offered to: David Abeles, Sara Abozeid, Göran Ågren, Steve Alexander, Joe Arnold, Mary Arthur, Mary Anne Atencio, Bruce Bandorick, Linda Bandhauer, David Beeson, Ad van den Berg, Dan Binkley, Elizabeth Binney, Arlene Boaman, Karen Bradley, Mark Brenner, Katherine Bricker, Emma Bricker, Owen Bricker, Chester Brooks, Aaron Brown, Phil Chapman, James Clayton, Dave Clow, Jack Cosby, Debi Cox, Michael Dawson, Diana Dear, Nolan Doesken, Tim Drever, Kristi Duncan, Rob Edwards, Tim Fahey, John Fitzgerald, Klaus Flach, Cindy Fudge, Jim Galloway, Jim Gibson, David Gluns, Douglas Hansen, Mitch Harris, Eugene

Hester, Ken Horton, Jon Jackson, Dave Jepsen, Corrine Johnson, Gwen Kittel, Bob Lautenslager, Willard Lindsay, Cliff Martinka, Alisa Mast, Sandy McCarley, Paul McLaughlin, John McFee, Kay McElwain, Diane McKnight, Marilyn Morrison, Steve Muehlhauser, Mari Nakada, Michele Nelson, Tica van Nes, Trung Nguyen, Stephen Norton, Jerry Olson, Sheryn Olson, Brian Olver, Cherry Payne, Bill Parton, Stan Ponce, James Rabold, Michael Reddy, John Reuss, Bruce Rosenlund, David Schimel, Keith Schoepflin, Alison Sheets, Sarah Spaulding, David Stevens, Nancy Stevens, David Swift, Kirk Thompson, Jim Thompson, Chuck Troendle, Marge Van Arsdale, Kristiina Vogt, Mark Walthall, James Ward, Kathy Warren, Mary Beth Watwood, David Weingartner, Al Williams, Thomas Winter, Bob Woodmansee, Steve Zary, and Duane Zavodil. The National Park Service, U.S. Geological Survey, U.S. Bureau of Reclamation, U.S. Forest Service, the Solar Energy Research Institute, Colorado State University, and the Swedish Agricultural University provided financial and logistical support for this project. Finally, I want to thank G. E. Likens, whose influence permeates this book. Had I realized how much he has shaped ecosystem science, I would not have slept through his lectures in 1975.

References

Agee JK, Johnson DR (1988) Ecosystem Management for Parks and Wilderness. University of Washington Press, Seattle.

Baron J, Galvin KA (1990) Future directions of ecosystem science: towards an understanding of the global biological environment. BioScience 40:640–642.

Bates CG, Henry AJ (1928) Forest and Stream-Flow Experiment at Wagon Wheel Gap, Colo. Mon. Weather Rev. Suppl. 30, U.S. Department of Agriculture Weather Bureau, Washington, D.C.

Bigelow DS (1986) Quality Assurance Report. NADP/NTN Deposition Monitoring, Field Operations 1978 through 1983. National Atmospheric Deposition Monitoring Program. Natural Resource Ecology Laboratory, Colorado State University, Fort Collins Colorado.

Eilers JM, Kanciruk P, McCord RA, Overton WS, Hook L, Blick DJ, Brakke DF, Kellar P, Silverstein ME, Landers DH (1986) Characteristics of Lakes in the Western United States—Vol. II: Data compendium for selected physical and chemical variables. EPA-600/3-86/054B, U.S. Environmental Production Agency, Washington, D.C.

Franklin JF (1987) Past and future of ecosystem research—contribution of dedicated experiment sites. In: Swank WT, Crossley DA, JR., eds. Forest Hydrology and Ecology at Coweeta, Vol. 66, pp. 415–424. Ecological Studies, Springer-Verlag, New York.

Gibson JH, ed. (1986) Acid Deposition: Long-Term Trends. National Academy Press, Washington, D.C.

Gibson JH, Galloway JN, Schofield C, McFee W, Johnson R, McCarley S, Dise N, Herzog D (1983) Rocky Mountain Acidification Study. FWS/OBS-80/40.17, Division of Biological Sciences, Eastern Energy and Land Use Team, U.S. Fish and Wildlife Service, Washington, D.C.

Holling CS (1978) Adaptive Environmental Assessment and Management. Wiley, Chichester, England.

Likens GE (1983) A priority for ecological research: address of the past president. Ecol. Soc. Am. Bull. 64:234–243.

Likens GE, ed. (1985) An Ecosystem Approach to Aquatic Ecology: Mirror Lake and Its Environment. Springer-Verlag, New York.

Likens GE, Bormann FH, Pierce RS, Eaton JS, Johnson NM (1977) Biogeochemistry of a Forested Ecosystem. Springer-Verlag, New York.

Mast MA, Drever JI, Baron J (1990) Chemical weathering in the Loch Vale Watershed, Rocky Mountain National Park, Colorado. Water Resour. Res. 26:2971–2978.

NAPAP (1989) National Acid Precipitation Assessment Program State-of-Science and State-of-Technology Reports. Office of the Director, NAPAP, 722 Jackson Place NW, Washington, D.C.

Rosenlund BR, Stevens DR (1990) Fisheries and Aquatic Management, Rocky Mountain National Park 1988–1989. Colorado Fish and Wildlife Assistance Office, U.S. Fish and Wildlife Service, Golden, Colorado.

Turk JT, Spahr NE (1989) Chemistry of Rocky Mountain Lakes. In: Adriano DC, Havas M, eds., Acid Precipitation, Vol. 1: Case Studies, pp. 181–208. Springer-Verlag, New York.

Schindler DS (1988) Effects of acid rain on freshwater ecosystems. Science 239:149–157.

Schumm SA (1978) Fluvial Geomorphology. Wiley, New York.

Winter TC (1981) Uncertainties in estimating the water balance of lakes. Water Resour. Bull. 17:82–115.

2. Regional Characterization and Setting for the Loch Vale Watershed Study

Jill Baron and M. Alisa Mast

Geology

The Colorado Front Range is a north-south trending massif located in the central region of Colorado (refer to relief map in frontmatter). It is bounded on the east by the Denver Basin, on the north by the Wyoming state line, and on the west by North, Middle, and South Parks. In the south it merges with a complex highland overlooking South Park (Ives 1980). The Front Range is a Laramide structure with a core of crystalline rocks made of predominantly Precambrian granite, schist and gneiss. The core is framed by steeply dipping Paleozoic sediments (Figure 2.1; Lovering and Goddard 1959). The metamorphic schist and gneiss, thought to be of sedimentary origin, are the oldest rocks in the range. Together with the Precambrian granites these rock types make up most of the terrain in Rocky Mountain National Park (Cole 1977). A series of Tertiary porphyritic intrusive rocks are concentrated along the Front Range mineral belt. This narrow belt, extending southwestward from Boulder, contains nearly all the productive mineral deposits of the Front Range. Another zone of intrusive rocks borders the northwestern corner of the national park and extends through the Cameron Pass area, but it has not produced ores of commercial value (Lovering and Goddard 1959). No valuable metal deposits have been found within the boundary of Rocky Mountain National Park.

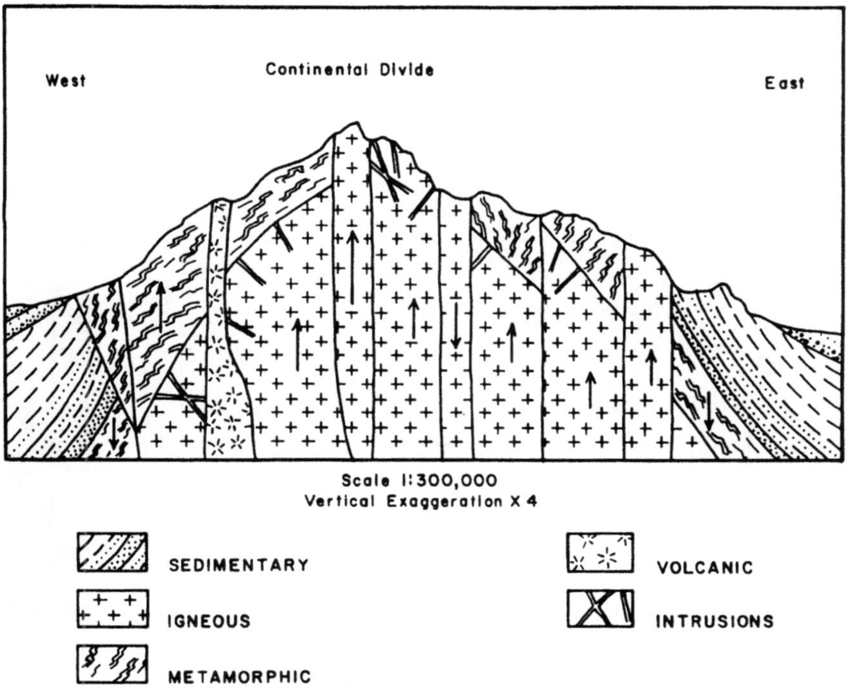

Figure 2.1. Generalized geologic cross section of the Colorado Front Range. [From Ives IE, ed. (1980) Geoecology of the Colorado Front Range: A Study of Alpine and Subalpine Environments. Westview Press, Boulder, Colorado.]

The maximum relief from the plains east of the mountains to the crest is about 2500 m. The Continental Divide forms a sharply defined crestline with peaks up to 4200 m along the Front Range. To the west, waters drain into the Pacific via the Colorado River drainage. Waters flow east to the Gulf of Mexico through the Mississippi River drainage. Viewed from a low angle (Figure 2.2) the Front Range forms a series of moderately smooth erosion surfaces that slope off gently toward the east (Lovering and Goddard 1959; Ives 1980). These benches are dissected at right angles to the crestline by deep trenches that formed as river valleys and were later altered by glaciation (Ives 1980).

Geologic History

The earliest recognizable event during Front Range history was the deposition of thick sequences of graywacke and shale. These may have been interbedded with basalts, conglomerates, and calcareous units during the

Figure 2.2. Aerial photograph of Front Range showing erosion surfaces dissected by deeply incised valleys. View is to southwest.

early Precambrian (Braddock 1969; Cole 1977). These sediments were subsequently deformed and recrystallized into gneiss and schist by as many as four deformational and two regional metamorphic events (Cole 1977). The most intense period of metamorphism is thought to have taken place around 1.7 billion years ago. This period of metamorphism produced a regional pattern of mineral assemblages ranging from greenschist to upper amphibolite facies (Braddock and Cole, 1979). The Boulder Creek Granite was emplaced during this time. Following this period of maximum metamorphism, about 1.4 billion years ago, the Silver Plume Granite and Sherman Granites were formed and located in extensive bodies throughout the Front Range. A less intense period of alteration followed this intrusive event but did not result in noticeable recrystallization of the rocks (Braddock and Cole 1979).

During the Paleozoic and Mesozoic Eras, the site of the present Front Range was a high area surrounded by sedimentary basins. It was submerged during the upper Cretaceous Period (Lovering and Goddard 1959). During the Laramide Orogeny, the mountain-building event that raised the Front Range, the region was faulted and uplifted. At this time the rocks were intruded by a series of porphyritic igneous rocks, causing vein deposits to be formed. Following uplift, the cover sediments were stripped to expose the Precambrian rocks that core the range today (see Figure 2.1).

Quaternary Geology

The spectacular scenery and rugged topography of the Front Range are the products of extensive Quaternary glaciation. Alpine cirques, glacial moraines and U-shaped valleys are typical geomorphic features at high elevations throughout the Front Range and Rocky Mountain national Park. Except for a small ice cap that developed over the northwest corner of Rocky Mountain National Park, the ice formed valley glaciers. The smooth erosion surfaces of the highest elevations remained unglaciated (Madole 1976).

At least three separate Pleistocene advances are recorded in morainal deposits along the flanks of the Front Range: Buffalo, Bull Lake, and Pinedale (Madole 1976). Remains of Buffalo advances are recorded from outside Rocky Mountain National Park. The Bull Lake advance occurred about 100,000 years B.P. Low valley and floor moraines from this advance are found within the subalpine zone of the park and are vegetated by subalpine forests and meadows. The deposits from the Bull Lake period are now covered by a grey-brown podzolic soil about 1.5 m thick (Richmond 1980).

The Pinedale glaciation was the last major ice advance, between 30,000 and 7,000 years B.P. The Pinedale advance left extensive till throughout the Front Range in a band between 2400 and 3200 m wide. The large lateral moraine that is prominent from the Bear Lake Road (Bierstadt Moraine) on the way to Loch Vale Watershed is of Pinedale age.

Three minor ice advances, which occurred during the Holocence, are termed Neoglacial. These advances left identifiable glacial debris at elevations between 3300 and 3750 m. Neoglacial deposits include moraines, rock glaciers, talus, and debris flows, and all these are evident above treeline in Loch Vale Watershed. The Triple Lakes advance, 3000–5000 years B.P., left moraines typically 5–10 m high in the valley heads above treeline. These moraines are old enough in some areas to have developed a turf covering, and the talus typically has abundant lichen cover. The Audubon advance, 950 to 1850 years B.P., left abundant talus at elevations above 3510 m. The small moraine above Sky Pond dates from this period.

The most recent glaciation, termed the Arapahoe Peak advance, includes deposits that have accumulated during the last few centuries. Both Taylor Glacier and Andrews Glacier in Loch Vale Watershed are remnants of the Arapahoe Peak advance. Historic records indicate Front Range glaciers have been receding since the mid-1800s (Madole 1976). Although the Sky Pond cirque was not reoccupied during the Arapahoe Advance, Andrews Tarn probably was, and the moraine at the outlet of the tarn is characteristic of this last glacial period.

Taylor and Andrews glaciers within Loch Vale Watershed are characterized as drift glaciers by Outcalt and MacPhail (1980). They persist because of annual additions of windblown snow from alpine surfaces west of Loch Vale Watershed. Shading from adjacent towering valley walls minimizes direct sunlight. Outcalt and MacPhail described Andrews Glacier

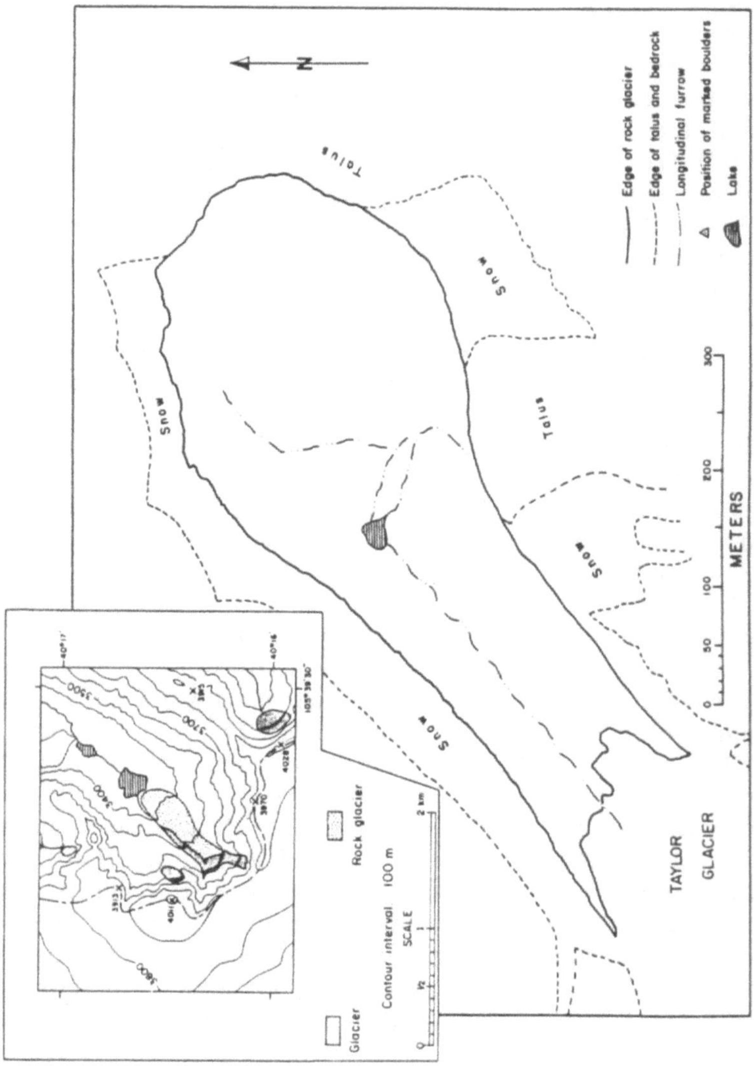

Figure 2.3. Taylor rock glacier. Inset map of rock glacier and Taylor Glacier adapted from U.S.G.S. McHenrys Peak Quadrangle, 7.5 min. ser., 1957. In inset, Continental Divide is marked with dashed line. Broad periglacial surface is west of Divide. (Reprinted with permission.)

as a col-fed glacier; it occurs on the lee side of a col (or gentle mountain pass) leading to the relatively flat erosion surface of Flattop Mountain. This relatively low area funnels snow down to maintain Andrews Glacier. Taylor Glacier is a gully-glacier, or an extremely narrow glacier. The steep gradient permits glacial movement, and shading during the summer prevents much loss of mass from ablation (Outcalt and MacPhail 1980).

Below Taylor Glacier is an active rock glacier that was one of several Front Range rock glaciers studied in the 1960s (White 1980). This tongue-shaped mixture of rock and ice extends almost to the upper shore of Sky Pond (Figure 2.3). It fills almost all the upper basin above the lake. Several couloirs northwest of the glacier allow a continuous stream of rockfall- and avalanche-produced talus. This debris, combined with debris from Taylor Glacier, supplies material to maintain the rock glacier today. The rock glacier is made up of an erosional zone at the top and a depositional zone near the bottom. The erosional zone exhibits typical boulder-protected debris tails close to Taylor Glacier. In the depositional zone below this, rocks of all sizes are perched precariously. The volume of the rock glacier is estimated at 1,091,500 m^3, and it is moving downvalley at an approximate speed of 12–16 cm yr^{-1} (White 1980). This speed is comparable to others of the Front Range. White found that the front of Taylor Rock Glacier moves twice as fast as the rest of the area, and he attributed this to an increase in bedrock slope near the front. The rock glacier also shows greater movement in the center compared with movement on the sides, a characteristic of slow-flowing rock glaciers. Other rock glaciers that have been studied in Switzerland and in Alaska have exhibited measured speeds of between 25 and 250 cm yr^{-1}. In comparison with these speeds, the rock glaciers of the Front Range appear to be moving very slowly.

Overview of the Climate

The continental climate of Colorado causes warm summers, cool to cold winters, low humidity and plenty of sunshine. Local climate is additionally affected by elevation, aspect, and position of the mountain ranges (Mutel and Emerick 1984). The orographic barrier created by the mountains generates its own weather such that sudden and extreme changes in temperature and precipitation occur at any time of the year. In the words of Mutel and Emerick (1984), only a few climatic generalizations are true.

Precipitation

The crest of the southern Rocky Mountains forms a continental-scale meteorological boundary, which influences storm trajectories and precipitation patterns. The Southern Rocky Mountains are characterized by prevailing westerly winds. Even when winds are not from due west, they have a westerly component at locations along the Front Range (Hansen et al. 1978). The seasonal and geographic distinction of storm direction has been examined by several researchers (Marr 1967; Barry 1973; Hjermstad 1980). The major sources of precipitation at locations west of the mountain crest are synoptic-scale frontal disturbances bearing Pacific moisture. These storms lose their moisture in decreasing amounts from west to east on successive orographic barriers in their path. The easternmost barrier is the Front Range of the Rocky Mountains. This type of storm is most prevalent during the winter months, when precipitation is snow. Loch Vale Watershed, located just below and east of the Front Range mountain crest, is influenced by these winter storms.

Springtime moisture originates commonly from the southwest to southeast, and the heaviest snowfall east of the Continental Divide is associated with cyclonic flow of moisture from the Gulf of Mexico. These spring storms lose their moisture along the high plains and eastern foothills, with decreasing amounts at higher elevations. Although synoptic-scale flow is still responsible for the transport of moisture into the region in summer, precipitation events most commonly result from local convective activity, as opposed to larger weather patterns. The east-facing mountain front is heated in the mornings by solar radiation, generating upslope convective flow (Toth and Johnson 1983). Summer air masses in the region are characterized by weak thermal stability, so this solar heating commonly triggers thunderstorms and showers. Cloud formation from this buildup occurs often along the Front Range near Estes Park, adjacent to Rocky Mountain National Park, (Karr and Wooten 1976; Barry 1980). The seasonal distribution of precipitation at Estes Park and Beaver Meadows (Figure 2.4a, b) emphasizes the importance of these spring and summer storms to overall precipitation inputs at Front Range locations east of the mountain crest. Loch Vale Watershed also receives precipitation from these upslope storms (Figure 2.4c), while sites west of the crest do not (Figure 2.4d, e).

A comparison of the daily precipitation records from six monitoring sites emphasized the differences in precipitation patterns according to geographic position relative to the mountain crest (Table 2.1). Eleven years of daily precipitation data for two sites at Grand Lake (2660 m), Fraser Experimental Forest (2740 m), and Estes Park (2295 m) were compared with precipitation from Beaver Meadows and Loch Vale, the monitoring site at Loch Vale Watershed (Grand Lake and Estes Park data are courtesy of N. Doesken, Colorado Climate Center; Fraser Experimental Forest data are courtesy of C. Troendle, USFS). Only those days with valid data between September and May for all sites ($n = 892$) were retained in the analysis (because the occurrence of isolated summer thunderstorms in June, July, and August decreased the correlation values). Correlation between the western slope sites (Fraser and the two from Grand Lake) and Estes Park was weak, with a much stronger relation observed between Estes Park, Beaver Meadows, and

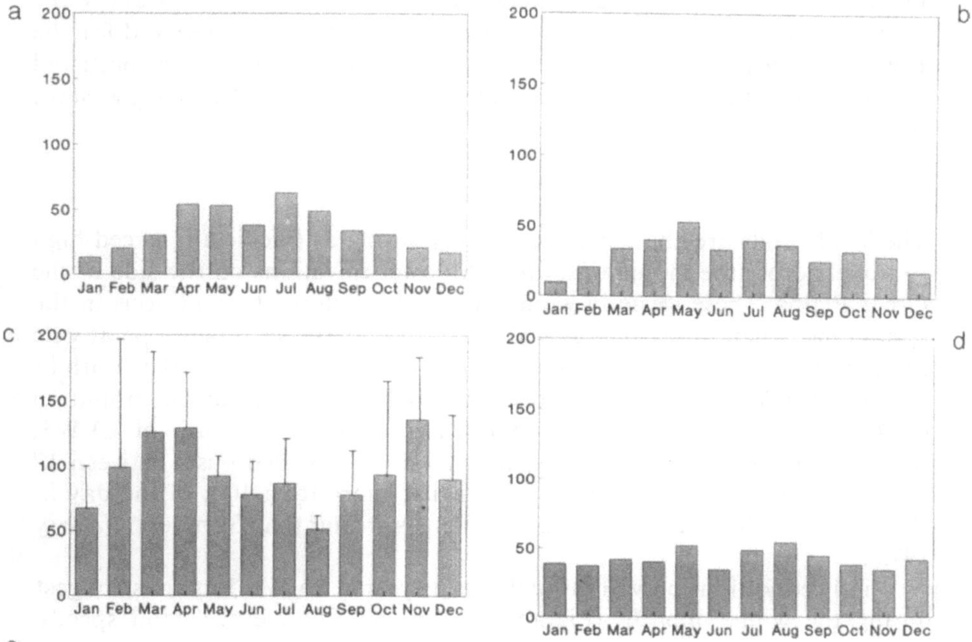

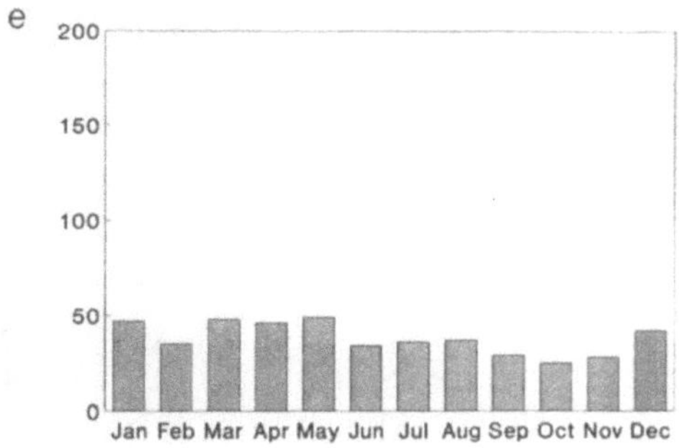

Figure 2.4. Average monthly distribution of precipitation (mm) at (A) Estes Park (1978–1988), (B) Beaver Meadows (1980–1987), (C) Loch Vale Watershed (1983–1989), (D) Grand Lake (1978–1988), and (E) Fraser Experimental Forest (1978–1988). Refer to relief map in front matter for geographic locations.

Table 2.1. Correlation Analysis of Precipitation at six Sites in North-Central Colorado[a]

Site	Correlation Matrix					
Fraser HQ	1.0000					
Grand Lake 6	0.4041	1.0000				
Grand Lake 1	0.3327	0.8479	1.0000			
Loch Vale	0.3344	0.6315	0.6157	1.0000		
Beaver Meadows	0.2084	0.4499	0.4025	0.7303	1.0000	
Estes Park	0.0850	0.2733	0.2893	0.3596	0.5815	1.0000
Variable	FR	G6	G1	LV	BM	EP

[a] Data were daily precipitation amounts excluding June, July and August, during 11-year period from 1978 to 1988. Only those days with valid observations at all six sites were used in the analysis ($n = 892$).

Loch Vale. The Loch Vale data were better correlated with the Grand Lake data than with the Estes Park data, while the reverse was observed for the Beaver Meadows site. Thus even though many of the same storms occurred at both Beaver Meadows and Loch Vale, Loch Vale reflected a strong western slope influence.

Wind

The local winds are characteristically very strong as Pacific air, forced high to travel across the mountains, drops to lower elevations on the side of the Great Plains. These westerly winds can produce chinook conditions in the springtime, when unseasonably warm and dry air rushes downslope at gale force with the strongest winds occurring near or just below the foothills (Marr 1967). Very strong winds occur throughout the year at the mountain crests. At Niwot Ridge, a 3520-m research area 40 km south of LVWS, between 20% and 50% of the days in 1965–1970 recorded gusts between 17 and 32 m sec^{-1}. Except for July and August, more than 10% of the day in 1965–1970 at Niwot Ridge had recorded gusts greater than 32 m sec^{-1} (Barry 1973).

Windy conditions prevail throughout the year at LVWS, but are strongest during the winter months (Figure 2.5). Two-minute average wind speeds,

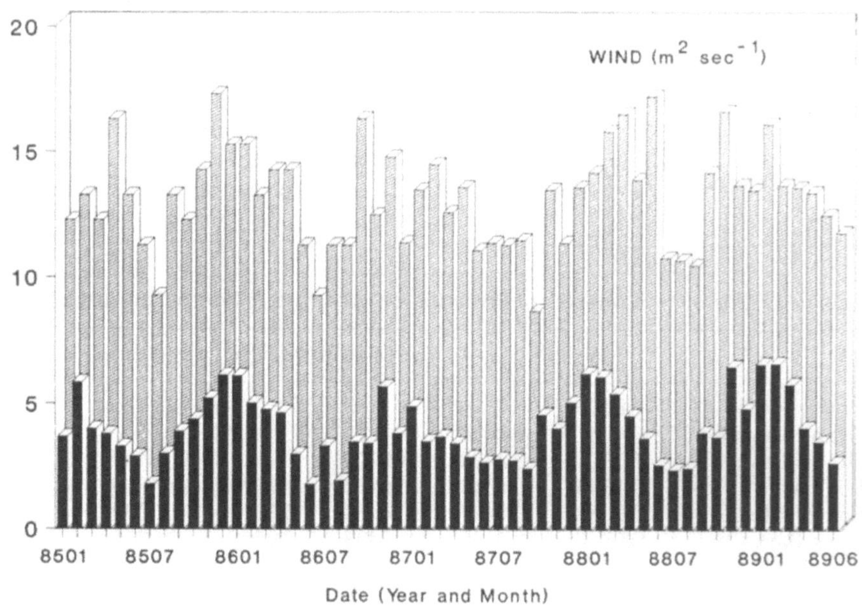

Figure 2.5. Mean (solid) and maximum (hatched) monthly wind speeds ($m^2 sec^{-1}$) at weather station in Loch Vale Watershed for 1985–1989.

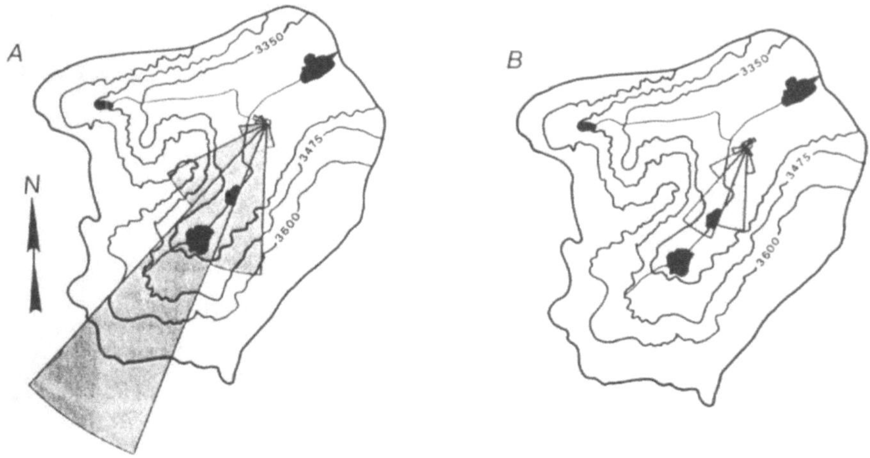

Figure 2.6. Wind roses for (A) winter (December, January, February) and (B) summer (June, July, August) superimposed on topographic map of Loch Vale Watershed.

recorded hourly, were regularly in excess of 8 m sec^{-1} (Bigelow et al. 1990), reaching 2-min average speeds of 17 m sec^{-1}. The median of daily winter wind estimates for days with measurable snowfall for 135 days between October 1987 and April 1989 was 5 m sec^{-1}, but the estimates were highly variable (Bigelow et al. 1990). Wind directions are constrained by the valley configuration to northeasterly and southwesterly directions (Figure 2.6).

Wind has probably been an important influence in ecosystem development, both regionally and at LVWS. The maintenance of treeline is controlled to some extent by wind-caused desiccation of confier needles (Wardle 1968; Ives and Webber 1980; Hadley and Smith 1986). The loss of effective precipitation as a result of redistribution of snow in alpine areas, as well as evaporation and sublimation, is directly responsible for distribution of vegetation types on the landscape (Komarkova and Webber 1980). The windy, desert-like conditions of the alpine have directly influenced soil development processes (Jenny 1941; Walthall 1985). Windblown soil, or loess, is postulated to be locally important as a soil-forming material in some areas (Burns 1980; Litaor 1987).

Solar Radiation and Temperature

Radiation from the sun is intense at the high elevations of the Front Range. The highest value recorded, 1056 W m^{-2} in June 1989, is nearly as high as the solar "constant" of 1360 W m^{-2}. the greatest influx of solar radiation occurs during May, June, and July of each year (Figure 2.7). The smallest fluxes are recorded in the winter months, with a maximum December 1989 value of 660 W m^{-2}. Steep valley walls prohibit sunlight from reaching certain

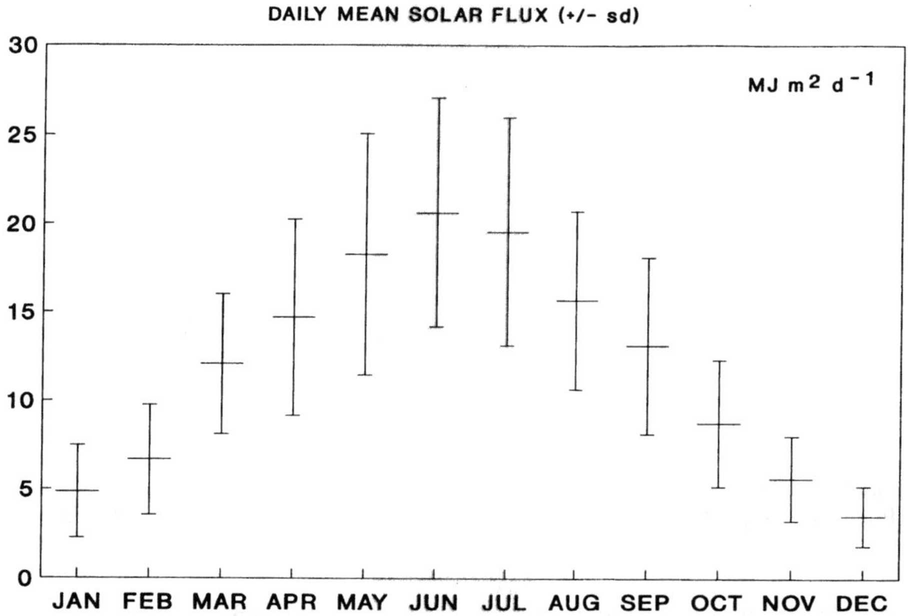

Figure 2.7. Daily mean solar flux (plus or minus 1 S.D.) for 1986–1990 (in MJ $M^{-2} d^{-1}$).

locations for part or all of the year, which has important physical and biological ramifications. The persistence of permanent snowfields and glaciers in sheltered locations in LVWS and throughout the Front Range is largely the result of shading from direct sunlight (Outcalt and MacPhail 1980). Distinct vegetational differences between north-facing and south-facing slopes in the middle elevations are caused by solar radiation. Algal blooms in The Loch occur each spring in response to direct sunlight to the lake body, which is hidden by the valley walls during the winter (Chapter 9; Spaulding 1991).

The amount of solar energy striking the ground is dependent on elevation, angle of the sun, and the degree of cloudiness. Cloud cover data are not available from LVWS, but regional estimates from Hansen et al. (1978) are that cloudy days are experienced 30%–50% of the time at elevations above 2500 m. Cloudy days were defined as days with greater than 80% cloud cover. December, January, and February are the coldest months of the year (Table 2.2); July and August are the warmest.

Vegetation

The distinct vegetation zones that occur with increasing elevation in the Rocky Mountains have been studied for more than a century. Among the ecologists who have written about this area are F. E. Clements (1910; and

Table 2.2. Absolute and mean minimum and maximum temperatures (Centigrade) in Loch Vale Watershed (1983–1990)

	Absolute minimum	Mean minimum	Mean maximum	Absolute maximum
January	−26.4	−8.8	−3.1	10.9
February	−27.0	−7.9	−1.9	12.0
March	−25.4	−5.5	1.9	18.4
April	−23.5	−5.0	6.2	18.6
May	−13.4	−0.6	11.1	20.9
June	−5.8	4.2	16.5	26.4
July	1.0	8.0	19.7	27.8
August	0.0	7.9	18.8	26.0
September	−12.6	3.6	14.1	26.1
October	−14.4	−1.9	7.1	17.1
November	−19.4	−7.2	0.7	14.8
December	−27.7	−10.9	−3.3	8.8

Goldsmith 1924). J. E. Weaver (Weaver and Clements 1938). R. F. Daubenmire (1943), D. W. Billings (1952), and J. W. Marr (1967). Descriptions of the plant associations that occur in Colorado Front Range from very detailed (Harrington 1979) to very general (Dix 1974). Marr's descriptions of the ecosystems of the Colorado Front Range are intermediate between these two extremes, and are widely followed today. He divided the vegetation systems into four distinct zones, the Lower Montane Forest, Upper Montane Forest, Subalpine Forest, and Alpine Tundra. The transition from one zone to another is gradual in some places, but more often is abrupt as moisture and temperature thresholds are encountered.

The Lower Montane Forest, or Foothills region, represents a combination of montane forest and grasslands found between 1850 and 2380 m. The zone itself forms a transition from steppe to forest. The most striking characteristic of this vegetation zone is the contrast between mountain slopes facing different directions so that exposure to direct solar radiation and consequent evapotranspiration create very different temperature and moisture regimes. North-facing slopes support ponderosa pine (*Pinus ponderosa* Dougl. ex Laws) stands at the lower elevations and mixed ponderosa pine/Douglas fir [*Pseudotsuga menzeisii* (Mirbel) Franco] stands at the higher elevations. In contrast, south-facing slopes generally support representatives of the prairie including both short grass and tall grass prairie types and shrubs such as mountain mahogany (*Cercocarpus montanus* Raf.) and chokecherry (*Prunus virginiana* L.). At higher elevations these are replaced by ponderosa pine stands.

With increasing elevation the Lower Montane Forest grades into the Upper Montane Forest, where the slopes are predominantly forested. Major tree species found in this zone include Douglas fir, ponderosa pine, lodgepole pine (*Pinus contorta* Engelm.), and aspen (*Populus tremuloides* Michx).

Lodgepole and aspen typically reforest areas disturbed by fire, logging, mining, and insect infestations.

The Subalpine Forest, extending from about 2860 m to treeline around 3400 m, contains extensive forests of Englemann spruce (*Picea englemannii* Parry) and subalpine fir (*Abies lasiocarpa* (Hook.) Nutt.). Forests of LVWS are of this vegetation type and are described in detail in Chapter 5 of this volume. Aspen and lodgepole pine are components of this system at lower elevations; limber pine (*Pinus flexilis* James) is found at the higher elevations. Fire is relatively uncommon in this zone, and moisture is abundant from both precipitation and redistribution of snow from higher elevations, so individual trees can develop very stately characteristics. Spruce in LVWS have been aged at greater than 500 years. Areas of poor drainage at these elevations have developed wet sedge meadows and willow thickets.

Subalpine Forests give way to Alpine Tundra at elevations above approximately 3400 m. Dwarfed and stunted trees, often mark the transition to the alpine. Vegetation is limited above the treeline due to the harsh environment brought about by extreme cold and high winds. The growing season is restricted to June, July, and August by temperature, and frost can occur at any time of year. Moisture is also a limiting factor in many areas, because precipitation either evaporates, sublimates, or is blown off by strong winds to lower elevations. Plant communities that are found in the alpine are characterized by low, cushiony habits, small waxy, hard, or furry leaves that retard moisture loss, and vegetative forms of reproduction. Most if not all plants are perennial. Different communities form depending on the amount of snow (moisture) and the timing of snow melt (Komarkova and Webber 1980). These range from herbaceous and woody cushion plant communities in the driest, most exposed areas to mixes of grasses and sedges such as kobresia [*Kobresia myosuroides* (D. Vill.) Fiore and Paol] in more protected zones and willows in the wettest areas. Alpine vegetation and landforms are very similar to their arctic counterparts. Permafrost has been recorded on Niwot Ridge (Ives 1980), and frost action has created polygons or patterned ground in some areas.

Land Use

The predominant land use practice today in the mountains is recreational, although mining and a mining-related logging industry altered much of the countryside in the mineralized belts beginning about 1850. Much of the land above 1800 m is now federally owned. The U.S. Forest Service practices multiple use management in national forests, with grazing, timber cutting, or development of downhill ski areas alongside, and in some cases coincident with, developed hiking, ski, and snowmobile trails and camping areas. Land use within Rocky Mountain National Park is restricted to recreation, including hiking, fishing, technical climbing, downhill skiing, and cross-country skiing. Before the park was created in 1916 much of the level land had been grazed. Park land has been allowed to revert to a natural state, in some cases with active resource management aid.

Water is one of the most valuable resources of the high mountains. Early in Colorado history many alpine and subalpine lakes had their water-holding capacity enhanced by dams. Periodic releases from these dams are used for crop irrigation or municipal supplies on the semiarid plains below. Drainage ditches and diversion tunnels exist throughout the mountains, including within Rocky Mountain National Park. These water projects divert water from the Colorado River drainage to rivers and reservoirs flowing east. The Colorado–Big Thompson River Project supports the Denver metropolitan area. As part of this project, the Alva B. Adams tunnel bores through the Front Range at least 500 m below the northwest boundary of Loch Vale Watershed. The Adams Tunnel, completed in 1910, transports Colorado River water from Shadow Mountain Reservoir along the western border of Rocky Mountain National Park, to Lake Estes and the Big Thompson River.

Land use within Loch Vale Watershed is recreational only, and there is no historical evidence indicating the valley was ever logged or grazed. A fire in 1910 affected some of the lower stands around The Loch. The Loch is a very popular destination of day hikers in Rocky Mountain National Park, being 2.5 km from the Bear Lake Road. Fewer hikers make their way to the higher lakes. The mountain peaks above Sky Pond, such as Cathedral Spires, Petit Gripon, and the Sharkstooth, are popular technical climbing areas, with as many as five parties per day during the summer time. Climbers are allowed to bivouac overnight above treeline. Sport fishing is popular in The Loch, Glass Lake, and Sky Pond.

References

Barry RG (1973) A climatological transect on the east slope of the Front Range, Colorado. Arct. Alp. Res. 5:89–110.

Barry RG (1980) Climatology overview. In: Ives JD. ed. Geoecology of the Colorado Front Range: A Study of Alpine and Subalpine Environments, pp. 259–263. Westview Press, Boulder, Colorado.

Bigelow DS, Denning AS, Baron J (1990) Differences between nipher and alter shielded Universal Belfort precipitation gages at two Colorado deposition monitoring sites. Environ. Sci. Technol. 24:758–760.

Billings WD (1952) The environmental complex in relation to plant growth and distribution. Q. Rev. Biol. 27:251–265.

Braddock WA (1969) Geology of the Empire Quadrangle: Grand, Gilpin and Clear Creek Counties. U.S. Geol. Surv. Prof. Paper 616, USGS, Denver, Colorodo.

Braddock WA, Cole JC (1979) Precambrian structural relations, metamorphic grade and intrusive rocks along the northeast flank of the Front Range in the Thompson Canyon, Poudre Canyon and Virginia Dale areas. In: Etheridge FG, ed. Guide of the Northern Front Range and Northwest Denver Basin, Colorado, pp. 105–122. Rocky Mountain section of Geological Society of America Annual Meetings, May 26–27, Colorado State University, Fort Collins, Colorado.

Burns SF (1980) Alpine Soil Distribution and Development, Indian Peaks, Colorado Front Range. Ph.D. Dissertation, University of Colorado, Boulder.

Clements FE (1910) The life history of lodgepole burn forests. USDA For. Serv. Bull. 79.

Clements FE, Goldsmith GW (1924) The Phytometer Method in Ecology. The Plant and Community as Instruments. Carnegie Institute of Washington, D.C.

Cole JC (1977) Geology of East-Central Rocky Mountain National Park and Vicinity, with Emphasis on the Emplacement of the Precambrian Silver Plume Granite in the Longs Peak–St. Vrain Batholith. Ph.D. Dissertation, University of Colorado, Boulder.

Daubenmire RF (1943) Vegetational zonation in the Rocky Mountains. Bot. Rev. 9:325–393.

Dix RL (1974) Regional ecological systems of Colorado. In: Foss PO, ed. Environment and Colorado: A Handbook, pp. 7–17. Environmental Resources Center, Colorado State University, Fort Collins.

Hadley JL, Smith WK (1986) Wind effects on needles of timberline conifers: seasonal influence on mortality. Ecology 67:12–19.

Hansen WR, Chronic J. Matelock J (1978) Climatography of the Front Range urban corridor and vicinity, Colorado. U.S.G.S. Prof. Paper No. 1019, U.S. Government Printing Office, Washington, D.C.

Harrington HD (1979) Manual of the Plants of Colorado. Sage Books, Swallow Press, Denver, Colorado.

Hjermstad LM (1980) The influence of meteorological parameters on the distribution of precipitation across central Colorado Mountains. In: Ives JD, ed. Geoecology of the Colorado Front Range: A Study of Alpine and Subalpine Environments, pp. 286–287. Westview Press, Boulder, Colorado.

Ives JD (1980) Introduction: a description of the Front Range. In: Ives JD, ed. Geoecology of the Colorado Front Range, pp. 1–8. Westview Press, Boulder, Colorado.

Ives JD, Webber PJ (1980) Overview of plant ecology. In: Ives JD, ed. Geoecology of the Colorado Front Range, pp. 326–338. Westview Press, Boulder, Colorado.

Jenny H (1941) Factors of Soil Formation: A System of Quantitative Pedology. McGraw-Hill, New York.

Karr TW, Wooten RL (1976) Summer radar echo distribution around Limon, Colorado. Mon. Weather Rev. 104:728–734.

Komarkova V, Webber PJ (1980) An alpine vegetation map of Niwot Ridge, Colorado. In: Ives JD, ed. Geoecology of the Colorado Front Range, pp. 364–392. Westview Press, Boulder, Colorado.

Litaor MI (1987) The influence of eolian dust on the genesis of alpine soils in the Front Range, Colorado. Soil Sci. Soc. Am. J. 51:142–147.

Lovering TS, Goddard EN (1959) Geology and ore deposits of the Front Range, Colorado. U.S. Geol. Surv. Prof. Paper 223, USGS, Denver, Colorado.

Madole RF (1976) Glacial geology of the Front Range, Colorado. In: Mahaney WC, ed. Quaternary Stratigraphy of North America, pp. 297–398. Dowden, Hutchinson & Ross, Stroudsburg, Pennsylvania.

Marr JW (1967) Ecosystems of the east slope of the Front Range in Colorado. University of Colorado Studies, Series in Biology No. 8, University of Colorado Press, Boulder, Colorado.

Mutel CF, Emerick JC (1984) From Grassland to Glacier. The Natural History of Colorado. Johnson Books, Boulder, Colorado.

Outcalt SI, MacPhail DD (1980) A survey of neoglaciation in the Front Range of Colorado. In: Ives JD, ed. Geoecology of the Colorado Front Range, pp. 203–208. Westview Press, Boulder, Colorado.

Richmond GM (1980) Glaciation of the east slope of Rocky Mountain National Park, Colorado. In: Ives JD, ed. Geoecology of the Colorado Front Range, pp. 24–34. Westview Press, Boulder, Colorado.
Spaulding, SA (1991) Phytoplankton Community Dynamics under Ice-Cover in The Loch, a Lake in Rocky Mountain National Park. M.S. Thesis, Colorado State University, Fort Collins.
Toth JJ, Johnson RHI (1983) An Observational Study of Summer Surface Wind Flow over Northeast Colorado. Paper No. 2, Cooperative Institute for Research in the Atmosphere, Colorado State University, Fort Collins, Colorado.
Walthall PM (1985) Acidic Deposition and the Soil Environment of Loch Vale Watershed in Rocky Mountain National Park. Ph.D. Dissertation, Colorado State University, Fort Collins.
Wardle P (1968) Engelmann spruce (*Picea engelmannii* Engel.) at its upper limits on the Front Range, Colorado. Ecology 49:483–495.
Weaver JE, Clements FE (1938) Plant Ecology, 2d Ed. McGraw-Hill, New York.
White SE (1980) Rock glacier studies in the Colorado Front Range, 1961–1968. In: Ives JD, ed. Geoecology of the Colorado Front Range, pp. 102–122. Westview Press, Boulder, Colorado.

3. Hydrologic Budget Estimates

Jill Baron and A. Scott Denning

Introduction

The Rocky Mountains serve as an important source of water in what is otherwise a continental, semiarid environment (Alford 1980). The mountains capture precipitation that supplies the headwaters of nearly every major river of the southwestern United States. An understanding of the hydrology of these souce regions is important to understanding regional water supplies. It is also vital to understanding the biogeochemical responses of alpine and subalpine ecosystems to acidic atmospheric deposition. Because the quantities and paths of water dictate biogeochemical processes, we derive the hydrologic budget in detail for Loch Vale Watershed (LVWS) in this chapter.

The hydrologic budget for any basin is described by the general equation

$$\text{INPUTS} = \text{OUTPUTS} + \text{CHANGE IN STORAGE} \qquad (1)$$

Inputs to LVWS include direct precipitation in the form of both snow and rain. An additional, indirect precipitation input is snow that blows into the catchment from outside its boundaries. The word "catchment," in fact, takes on expanded meaning in high-elevation watersheds because watershed boundaries are permeable to snow that blows into or out of the basin. Outputs from LVWS include: (a) stream discharge; (b) evaporative loss, including transpiration and sublimation; (c) snow blown out of the catchment; and (d) both deep and shallow groundwater seepage. Actual or potential storage reservoirs include four lakes, two small glaciers, a rock glacier (which stores substantial amounts of water as ice), and shallow groundwater, including soil water.

We considered the foregoing components and ranked them by their significance to the overall hydrologic budget. Additionally, we estimated the uncertainty of each term. Obviously uncertainty exists in watershed-scale studies, but there is greater confidence associated with some water terms than others. Accordingly, three values are reported for each term. The middle or "moderate" values represent measured values or those calculated directly from measured values. Because undermeasurement is one of the most common sources of error in hydrology, a maximum value for all terms is also derived. Where appropriate, minimum values are given. Confidence in water budgets increases with longer time intervals; thus, although annual measurements for the various hydrologic terms are presented, the budgets are calculated for the 5-year (1984–1988) interval. The moderate value for the hydrologic budget comes closest to balancing inputs with outputs.

Two independent estimates are compared with our hydrologic budget. The first is the hydrologic portion of the WRENSS model (Water Resources Evaluation of Non-point Silvicultural Sources; Troendle and Leaf 1980), which has been adapted for personal computer use (Bernier 1986; Charter and Swanson 1989). WRENSS is made up of nomographs developed from a large number of simulations with WATBAL, a forest hydrology model derived for regions that receive most precipitation as snow (Leaf and Brink 1973). Site-specific descriptors from LVWS are combined with WATBAL output from Fraser Experimental Forest, the Rocky Mountain regional validation site. The WRENSS model predicts water yields. We compared these to our measured flow values as a qualitative evaluation of how well we understand the hydrology of LVWS.

Chloride is treated as a conservative chemical tracer for a second estimate of the validity of the assumptions we used to develop the hydrologic budget. Successful use of chloride for reconstructing water budgets depends on having no internal mineral or biological sources or sinks for chloride within the basin, and no external source other than wet deposition. Other investigators have used this method for determining whether catchment basins are hydrologically tight (Cleaves et al. 1970, Likens et al. 1977), and for estimating the chemical inputs from dry deposition (Creasey et al. 1986). Unfortunately, chloride concentrations were near the limits of detection, weakening the confidence that we can place in them.

Methods and Calculations

Precipitation, wind speed, humidity, and temperature are measured at 3160 m (Figure 3.1) with a solar-powered Remote Area Weather Station. Precipitation at the weather station is measured continuously with a Belfort

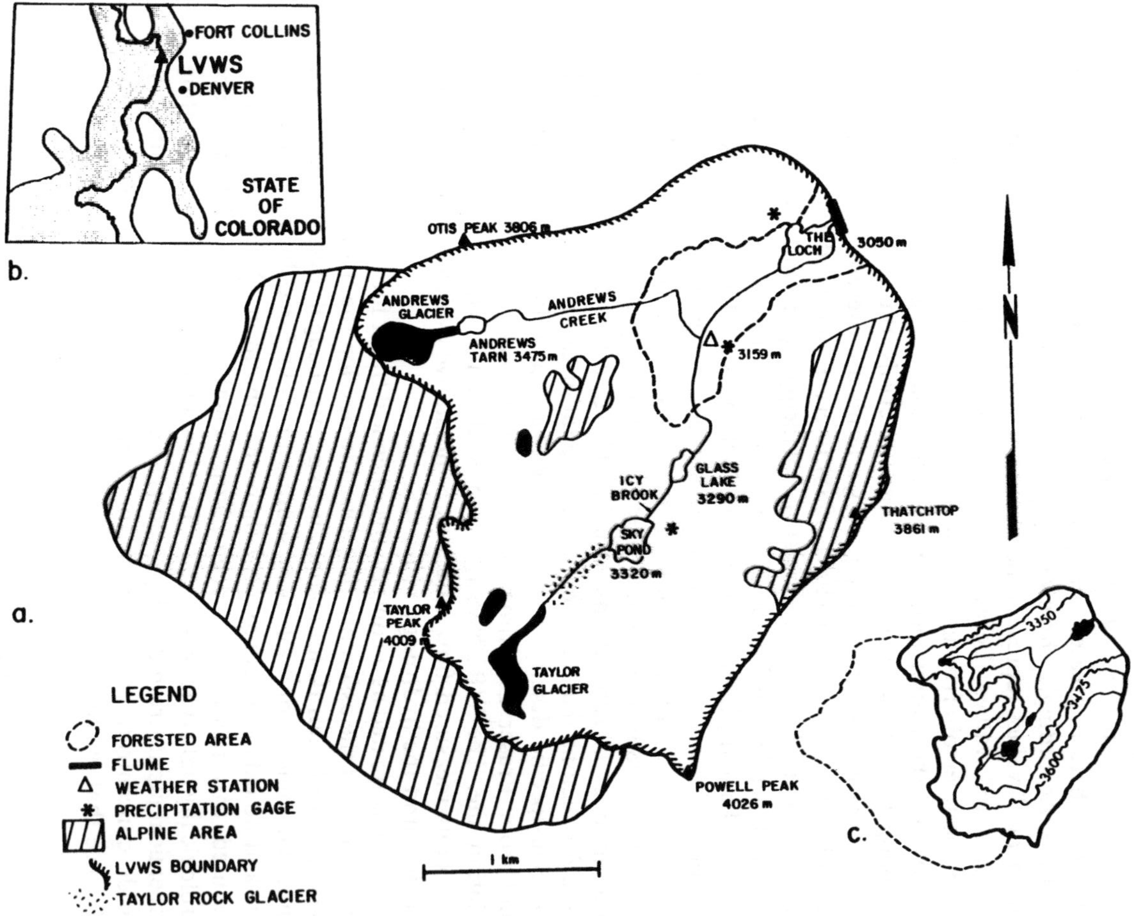

Figure 3.1. (a) Map of Loch Vale Watershed (LVWS), Rocky Mountain National Park. (b) Inset shows location of LVWS within State of Colorado. Shaded areas denote mountains. Line through state is Continental Divide. (c) Inset of LVWS shows major topographic contours (m).

Figure 3.2. Weather (in plastic bag) station at Loch Vale Watershed. Equipment, left to right: precipitation bucket, Nipher-shielded precipitation gauge, Alter-shielded precipitation gauge, temperature-humidity probe, solar panels, satellite transmitting antenna, anemometer, solar radiometer, Aerochemetrics wet deposition collector.

weighing-bucket rain gauge with Alter wind shield (Figure 3.2). Antifreeze solution is kept in the bucket throughout the year because snow is possible at any time. Recorded data are sent hourly via satellite to Colorado State University. Meteorological equipment is calibrated once a year.

Errors in estimates of precipitation inputs can include inaccurate gauge catch and the extrapolation of unrepresentative point measurements to the catchment as a whole. The uncertainty associated with measured values differ between summer and winter in our study because of seasonal differences in wind velocity and because of a summer treatment of spatial variability. Because of these differences, snow and rain are treated separately.

Rain

Precipitation from May 15 to September 30 of each year is considered to be rain, although this also includes some events of mixed rain and snow. We operate two additional Belfort rain gauges during the summer months at 3120 and 3320 m to obtain better spatial representation and to determine an elevational gradient. Neither of these additional gauges have wind shields. Minimum and maximum values for rain in the hydrologic budget (see Table 3.3 later in this chapter) represent the gauge value plus or minus 10% (Goodison et al. 1981).

Snow

All precipitation between October 1 and May 14 falls as snow. The moderate numbers in Table 3.3 are the measured values. These numbers are also used to represent the minimum values, because we do not think it possible that the Belfort overmeasures snow. We increased our measured values by 40% to obtain maximum values, after Goodison et al. (1981), who reported 60% gauge catch efficiency of snow for Belfort rain gauges with Alter shields. Snow blows past the aperture of the gauge, especially in high winds when snow is cold and dry.

Indirect Snow

We estimated the source area (400 ha) for indirect or blowing snow that contributes to LVWS water flux with aircraft observations. The maximum area from which snow could blow in was determined by planimetry of the topographic maps. Because some of this snow either stays on the slope or blows away from LVWS, we then applied the equation:

$$\text{IND}\cdot\text{SNOW} = S_{\text{ind}}(P_s/GC)(1 - \text{RET})(1 - B_{\text{other}})(1 - E_s) \qquad (2)$$

where S_{ind} = indirect source area, P_s = snow (centimeters of water) recorded at the Belfort gauge, GC = gauge catch efficiency, RET = the snow volume retained on the source area in terrain depressions or behind boulders, B_{other} = the volume of snow that blows off the source area but not into LVWS, and E_s brackets the volume of snow that sublimated from the source area according to Berg (1986). Sublimation is explained in more detail in the following section on evaporation. The minimum value assumes GC = 1.0, RET = 0.3, $B_{\text{other}} = 0.5$, and $E_s = 0.5$; the moderate value assumes GC = 1.0, RET = 0.3, $B_{\text{other}} = 0$, and $E_s = 0.4$; and the maximum value assumes GC = 0.6, RET = 0.2, $B_{\text{other}} = 0$, and $E_s = 0.3$.

Stream Discharge

We used an October 1–September 30 water year to construct annual and 5-year average hydrologic budgets. Outflow from LVWS is gauged continuously when there is enough flowing water to be measured with a Parshall flume located at the outlet of The Loch. Stilling well stage heights, measured with a float, are recorded electronically (Omnidata one-channel Datapod recorder) and calibrated to a backup Stevens Type F Strip Chart Recorder. A malfunction in 1986 caused the float mechanism to cap at a maximum of 1.1 $m^3 s^{-1}$. The narrow outlet area fills with more than 3 m of drifted snow each winter, prohibiting year-round operation of the flume. The error associated with Parshall flumes is $\pm$ 5% (Winter 1981). When the water is too low to measure in late fall and early spring, we estimate the low water discharge. This is done by multiplying half of the last measured value for any given year by the number of days until the flume becomes operational again the following spring. The minimum value of 30,000 m^3 was obtained (in 1987). The maximum value was obtained by multiplying the last measured value for the year with the greatest discharge at time of shutdown (1986) by the number of days to get a maximum value of 150,000 m^3. The error in the annual mean discharge is estimated at $\pm 10\%$, or less than 5×10^5 m^3.

Evaporation

We included all water that was returned to the atmosphere under the term evaporation; this includes evaporation, transpiration, and sublimation. Potential evaporation is calculated daily May 1–November 15 each year from a modified Penman equation (Linacre 1977). The equation used is

$$E_0 = \frac{k(T + 0.006h) + 15(T - T_d)}{1000(80 - T)} \text{ m day}^{-1} \tag{3}$$

where E_0 = evaporation, T = daily mean temperature (°C), T_d = daily mean dewpoint temperature (°C), h = elevation in meters, and k = a dimensionless constant. The k parameter has the value 11.67 for free water evaporation and 8.33 for potential evapotranspiration. Daily mean temperature and daily mean dewpoint temperature were used from days with at least 22 hours of valid data. When daily data were missing, values were substituted from the previous day.

Potential evaporation (E_0) was modified, and evaporation components of LVWS were classified by season and by an arbitrarily defined "evaporation category" as follows:

1. Summer Rocks. Rock outcrop and talus cover about 500 ha of LVWS. Carroll (1980) empirically derived a runoff coefficient of 0.85 for rock surfaces. We used Carroll's approach to estimate evaporation from rock surfaces as:

$$E_{rock} = P_r(5 \times 10^6 \text{ m}^2)(0.15) \tag{4}$$

where P_r = the annual total rain input in meters.

2. Spring Soils. Areas with well-developed soils in the flatter part of LVWS become saturated with water during the spring snowmelt. Under these conditions, actual evapotranspiration is equal to potential evapotranspiration predicted by the energy balance. We used this approximation for the 120 ha of the watershed flat enough to support standing water May 1–June 14 of each year. Springtime evapotranspiration was thus calculated as

$$E_{spring} = 1.2 \times 10^6 \text{ m}^2(E_0) \tag{5}$$

where 1.2×10^6 m^2 = area in meters and E_0 = the potential evapotranspiration given by equation (3).

3. Summer Lakes. Lake surface evaporation was calculated for 19 ha of lakes and bogs for the open water period (June 15–November 16) as

$$E_{lakes} = 1.9 \times 10^5 \text{ m}^2(E_0) \tag{6}$$

where $1.9 \times 10^5\,m^2$ = area in meters and E_0 = the free water evaporation given by equation (3).

4. Spring and Summer Forest. Water lost to evapotranspiration in the forest is expected to be somewhat less than the potential loss calculated by equation (3). We assumed the modeled value for canopy transpiration from Kaufmann (1983b). These values are derived from 4 years of daily hydrologic measurements and evapotranspiration estimates from Subunit 2, Lexen Creek watershed, Fraser Experimental Forest (Kaufmann 1983a). Subunit 2 is a northeast-facing slope at 3200 m elevation 30 km south and west of LVWS. Forest species composition and stand density in the modeled forest are nearly identical to that in LVWS (Arthur and Fahey 1990).
5. Spring and Summer Tundra. Carroll (1980) showed that evapotranspiration from tundra vegetation on Niwot Ridge, 35 km south of LVWS, is related to potential evaporation by an empirical relation. We applied this relationship to an estimated 100 ha of tundra vegetation in LVWS to obtain water loss from tundra:

$$E_{tundra} = 1.0 \times 10^6\,m^2 \times (0.212 \times E_0)(1 - 1.275E_0) \quad (7)$$

where $1.0 \times 10^6\,m^2$ = area in meters; all other terms were defined previously.

6. Winter Sublimation. Sublimation losses from the snowpack are estimated as 30%–50% of total snow water equivalent, based on the work of Berg (1986) at Niwot Ridge. Total sublimation is calculated as

$$E_{sub} = E_s(P_s + P_{ind}) \quad (8)$$

where P_s = total snow inputs, P_{ind} = indirect snow input, and E_s is the proportion of water lost to sublimation. We used $E_s = 0.3$ for the minimum estimate, $E_s = 0.4$ for the moderate, and $E_s = 0.5$ for the maximum estimate.

Blowing Snow

Field observations indicated that some snow falling on exposed areas near the eastern (downwind) margins of the catchment blows out of the watershed. We obtained the minimum value for this loss term by using recorded snow volume, an estimated 50 ha of snow loss area, and 50% water retention by the terrain. Moderate values use the mean annual snow input, 75 ha of snow loss area, and 30% water retention. The maximum estimate is obtained by using 60% gauge catch efficiency, 100 ha as the effective snow loss area, and 0% water retention.

Groundwater Seepage

Subsurface discharge from The Loch lake bottom and Icy Brook streambed is estimated by Darcy's Law

$$Q = KiA \quad (9)$$

where Q = the rate of flow, K = the permeability of the porous medium, i = the hydraulic gradient, and A = the cross-sectional area of the flow (Anonymous 1977). The depth of the glacial till was measured along 37 transects in the study area using seismic refraction techniques (Redpath 1973). These measurements indicate a mean till depth of about 2 m in the vicinity of The Loch outlet and thus a cross-sectional area of about 100 m^2. The hydraulic gradient in this area, calculated from the topographic maps, is about 10%.

Permeability of the glacial debris in the study area is unknown and probably variable even across the narrow 50-m outlet area. We used a lower limit of 0.1 m day^{-1} corresponding to a mixture of sand, silt, and clay for the minimum reported value, and an upper limit of 100 m day^{-1} for a mixture of sand and gravel for the maximum reported value (Anonymous 1977).

Lake Storage

The bathymetries of Sky Pond, Glass Lake, and The Loch were mapped with a sounding line along a series of transects across the lake (Chapter 8; McKnight et al. 1986). Andrews Tarn was not mapped. Late summer lake volume was found to be 121,700 m^3 for Sky Pond, 25,700 m^3 for Glass Lake, and 61,100 m^3 for The Loch.

Snow and Ice Storage

We estimated water content of Taylor and Andrews Glaciers by interpolating the ground surface area under the glaciers and deriving glacial volume. The volume was multiplied by a density of 0.8 for the mixture of old snow and glacial ice. An earlier study provided an estimate of the volume of interstitial ice in Taylor Rock Glacier (White 1980).

We did not measure accumulation or ablation rates for the glaciers in the study area. Using previous mass-balance studies of glaciers in the Colorado Front Range, Alford (1980) derived an empirical relationship between the three-dimensional orientation of a cirque and glacier mass balance. These values were used to calculate net ablation and accumulation and thus net storage or loss. We did not determine the amount of accumulation or ablation from Taylor Rock Glacier, and assumed it stored a constant volume of water between 1983 and 1988.

Shallow Groundwater Storage

This term includes soil water as well as that stored in porous glacial till. Till plus soil cover about 100 ha of LVWS. Mean depth of this material was determined along seismic transects to be approximately 1 m. We assumed 20% porosity for this material to derive with an estimated volume of soil and till pore space.

Deep Groundwater Storage

Seismic refraction techniques (Redpath 1973) were used to explore the possibility of large faults and fractures in the bedrock of LVWS. These were combined with literature references (Cole 1977; Mast 1989) to estimate the importance of this water loss term. Because there are no large faults or fractures within LVWS, deep groundwater storage was assumed to be insignificant.

Chloride

Chloride in precipitation is analyzed as part of the National Atmospheric Deposition Program (Peden 1986). The analytical uncertainty for chloride is near zero (Lockard 1987), but between 10 to 20% of precipitation samples are below the limits of detection (NADP/NTN Data Base 1989). The volume-weighted mean values used for the chloride budgets reflect 0.5 times these detection limit values. Between 50 and 60 samples of stream water are collected each year as part of the LVWS project. Chloride in stream water is analyzed at the U.S. Geological Survey Central Analytical Laboratory (Skougstad et al. 1979). Uncertainty of chloride concentrations in stream water at concentrations near the limits of detection ($0.01\,mg\,L^{-1}$) is $\pm 20\%$ (Shockey, personal communication).

Results

Inputs

At Grand Lake, 10 km west of LVWS, 1984 and 1986 were years of very high precipitation (Colorado Climate 1984–1988). 1985 and 1988 were similar to the long-term average, and 1987 was dry. The precipitation record in LVWS reflects these patterns (Table 3.1). February, March, April, and November had the highest total precipitation inputs, while August had the lowest (Figure 3.3). Much of the direct precipitation that contributes to a water year (61%–76% of water year total) falls as snow in November to May. Weekly precipitation records (Figure 3.4) show weeks of greatest precipitation amounts occur most often in the spring (April) and the fall (October).

The amount of rain (precipitation May 15–September 30) ranged from 27.0 to 44.9 cm for the 5 years, with a mean of 34.1 cm (Table 3.1). Given uniform distribution of this precipitation over the 660-ha catchment, $2.25 \times 10^6\,m^3$ of water enters Loch Vale as rain in the average summer. No elevational difference in precipitation amount was evident from a comparison of rain gauges located at 3120 m and 3320 m (paired t test results: $p = .586$, $n = 23$). Rain catch at these gauges differed significantly from the Alter-shielded gauge at 3160 m at the weather station ($p = .0001$, $n = 46$ for comparison of 3120-m gauge with weather station; $p = .095$, $n = 22$ for

Table 3.1. Annual Measured Precipitation and Flow Values for Loch Vale Watershed[a]

Year	Rain (cm)	Snow (cm)	Total Precipitation (cm)	Percent of Long-Term Average: Estes Park	Percent of Long-Term Average: Grand Lake	Total Precipitation ($*10^6 m^3$)	Total Flow ($*10^6 m^3$)
1984	44.9	70.4	115.3	131	150	7.61	5.81
1985	29.3	86.4	115.7	95	91	7.63	4.28
1986	35.8	92.4	128.2	108	134	8.46	6.06
1987	33.6	62.6	96.2	106	74	6.35	4.70
1988	27.0	84.5	111.5	91	95	7.36	4.67
Average	34.1	79.3	113.4			7.48	5.10

[a] Values for Estes Park to east and Grand Lake to west of LVWS (see Fig. 2.1 for location) are percentages of 1961–1980 annual average precipitation.

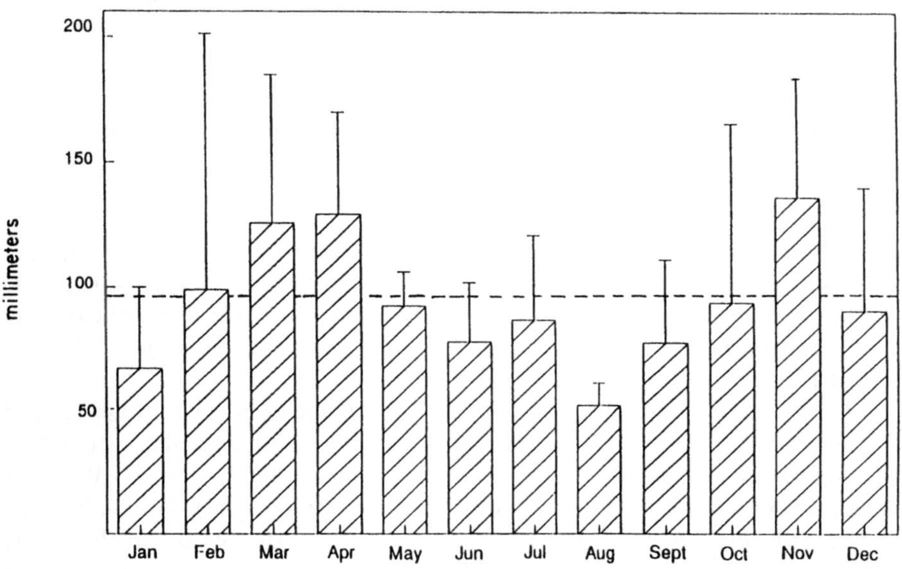

Figure 3.3. Average monthly precipitation (mm) 1983–1989, Loch Vale Watershed. Error bars denote 1 S.D. Five-year average monthly precipitation (93.8 mm), dashed line.

comparison of 3320-m gauge with weather station). The same storms are responsible for precipitation at all three stations in the summer months, and we think the difference in catch was due to the presence of the wind shield at the middle site (Goodison et al. 1981).

Water equivalent depth of snow fall (October 1–May 14) during the study ranged from 62.6 to 92.4 cm, with a mean of 79.3 cm or $5.23 \times 10^6 m^3$ of

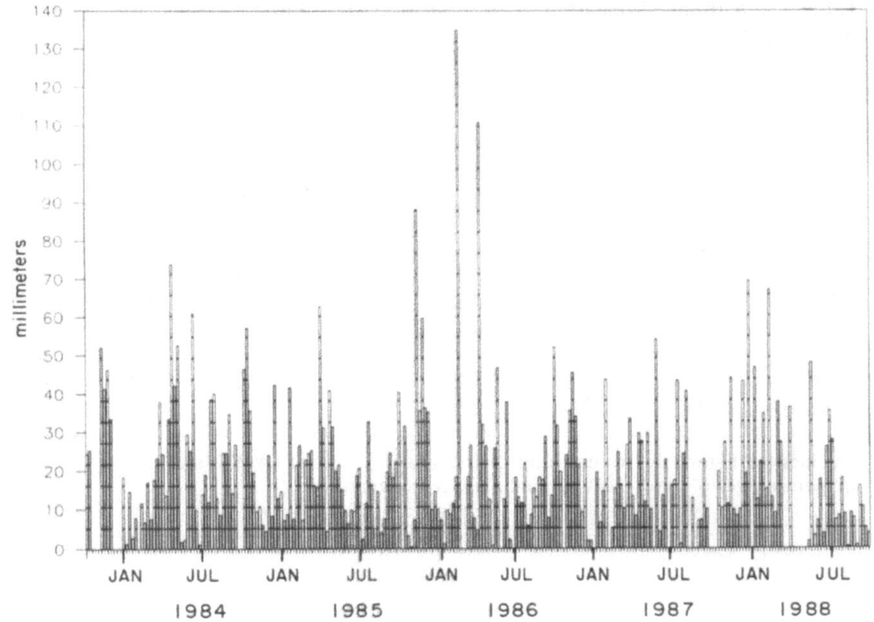

Figure 3.4. Weekly precipitation (mm) for 1984–1988 water years for Loch Vale Watershed.

water input (see Table 3.1). A two-winter comparison study of the Alter-shielded Belfort gauge with a colocated Nipher-shielded Belfort gauge did not reveal a significant difference in the amount of snow caught by one gauge compared to the other (Bigelow et al. 1990). A maximum value of $8.72 \times 10^6\,m^3$ was obtained assuming 60% catch efficiency.

Much of the snow maintaining the Taylor and Andrews Glaciers and Taylor Rock Glacier blows into LVWS from outside the watershed boundaries. If all snow that fell on these slopes blew into LVWS, the contribution would average $4.76 \times 10^6\,m^3$ of water per year. This value was considered unreasonably high because not all snow deposited on this slope contributes to LVWS. We therefore calculated the contribution from blowing snow to be 0.83, 2.00, and $4.44 \times 10^6\,m^3$ for minimum, moderate and maximum values, respectively.

Outputs

Watershed discharge is well constrained to the gauged outlet of The Loch. Flow rates are largely determined by the melting and freezing of snow in the watershed and reach a peak of about $1.5\,m^3\,sec^{-1}$ in June of each year (Figure 3.5). Base flow in late August and September is maintained by shallow groundwater drainage and glacial melt. During winter, lake stage at The

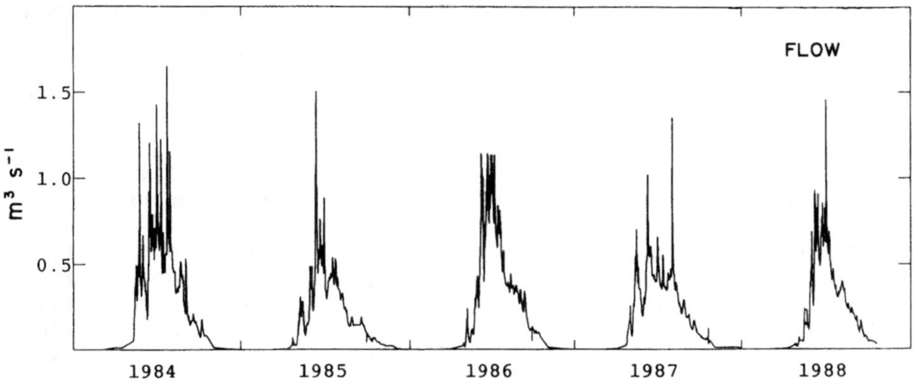

Figure 3.5. Recorded stream flow ($m^3 sec^{-1}$) at outlet of The Loch for water years 1984–1988. Malfunction of stilling well equipment caused "capping" observed at peak flows in 1986 (see text for discussion).

Loch drops below the minimum measurement level of the flume at The Loch outlet.

Recorded discharge volumes for LVWS varied over the 5 years of data collection from a low value of $4.28 \times 10^6 m^3$ in 1985 to a high of $6.06 \times 10^6 m^3$ in 1986, with an annual mean of $5.10 \times 10^6 m^3$ (see Table 3.1). The 1985 value is puzzling; precipitation for that year was not unusually low, nor were recorded temperatures lower than average. However there was less flow than 1987, the year of the lowest total annual precipitation.

Accurate values for the three different types of evaporation (evaporation, transpiration, and sublimation) are difficult to obtain (Winter 1981). Further, the estimated errors associated with different methods of measuring or computing evaporation are determined by comparing one method with another and not by comparing a calculation with a measured value (Winter 1981). Various investigators in the southern Rockies have estimated evaporation and sublimation from high elevation sites to vary from as high as 50% (Kaufmann 1983b; Berg 1986) to essentially zero (Barry 1980; Turk and Spahr 1989). When summer evaporation and transpiration compartments (Table 3.2) are summed, this calculation suggests that roughly 11% of the total moderate estimated water inputs evaporate or transpire annually. Estimated sublimation was an additional significant loss term for LVWS water; sublimitation values were 1.40, 1.87, and $3.88 \times 10^6 m^3$ for minimum, moderate, and maximum estimates, respectively (Table 3.2). Using moderate estimates, all the evaporation terms combine to account for 39% of all precipitation inputs (Table 3.3). A comparison of chloride concentrations of inputs and outputs provides an independent estimate of evaporation. Chloride in precipitation averages $0.08 mg L^{-1}$ for snow and $0.1 mg L^{-1}$ for rain compared with the 5-year volume-weighted mean (VWM) Cl concentra-

Table 3.2. Evaporation Terms (millions of cubic meters) Calculated for LVWS for Each Year 1984–1988[a]

	Evaporation			Evapotranspiration		Sublimation			Total Evaporation
Year	Rocks	Soils	Lakes	Forest	Tundra	Minimum	Moderate	Maximum	
1984	0.34	0.29	0.23	0.21	0.27	1.54	2.33	5.18	3.67
1985	0.22	0.29	0.19	0.21	0.23	1.89	2.86	6.36	4.00
1986	0.27	0.27	0.18	0.21	0.22	2.02	3.06	6.80	4.21
1987	0.25	0.25	0.19	0.21	0.23	1.37	2.07	4.61	3.20
1988	0.20	0.25	0.19	0.21	0.22	1.85	2.80	6.23	3.87

[a] Total values use the moderate sublimation term. Periods for which the values were calculated are as follows: rocks, 5/15–9/30; soils, 5/1–6/14; lakes, 6/15–11/16; forest and tundra, 5/1–11/16; sublimation, 11/17–4/30.

Table 3.3. Five-Year-Average Hydrologic Budget Estimates for LVWS[a]

Budget Terms	Stored Volume	Minimum	Moderate	Maximum	Range of Confidence
Inputs					
Rain	—	2.08	2.25	2.48	Most
Snow	—	5.23	5.23	8.72	Less
Indirect	—	0.83	2.00	4.44	Least
Total		8.14	9.48	15.64	
Outputs					
Flow	—	4.60	5.10	5.61	Most
Evaporation	—	2.90	3.79	7.00	Less
Blown snow	—	0.20	0.42	1.32	Least
Seepage*	—	0.0004	0.10	0.37	Most
Total		7.70	9.41	14.3	
Storage					
Lakes*	0.21 for 3 lakes only	0.00	0.00	0.00	Most
Glacial snow* and ice	3.30	0.03	0.03	0.03	Least
Shallow* groundwater	0.20	0.00	0.00	0.00	Less
Deep* groundwater	—	0.00	0.00	0.00	Less

[a] Values in million cubic meters of water.
* Asterisks denote water terms that are probably not significant to hydrologic budget.

tion in stream water of 0.12 mg L^{-1}. This suggests roughly 15%–30% evaporation of incoming precipitation. If chloride behaves conservatively within the watershed, with no internal sources or sinks, as other authors have suggested (Cleaves et al. 1970; Likens et al. 1977; Creasey et al. 1986), then the evaporation values calculated by us may be overestimated.

We estimated the loss of water from the catchment due to wind transport of snow using the same reasoning as for indirect inputs. The minimum and moderate values are an order of magnitude less than any loss term discussed so far, but the maximum value (assuming gross undermeasurement by the precipitation gauge), $1.3 \times 10^6\ m^2$, is quite high, suggesting water loss from blowing snow could be significant (Table 3.3).

Theoretically, seepage loss could occur through bedrock faults and fractures, till near the lower watershed boundary (including under or through the streambed) and The Loch lake bottom. Although there is no surface flow through the flume during some Decembers and Januarys, there is surface flow in Icy Brook 300 m downstream. This means water flows under or through the streambed. Additionally, a wet meadow directly below The Loch has small, flowing rivulets during the summer. The Loch is the only possible source for this water, so water must be seeping out. Calculations produced a maximum estimate of $1000\ m^3\ day^{-1}$, or $3.7 \times 10^5\ m^3\ yr^{-1}$. The minimum value of permeability produced a seepage rate of only $1\ m^3\ day^{-1}$ or $365\ m^3\ yr^{-1}$ (see Table 3.3). We concluded that seepage is an insignificant loss term for water in the overall budget.

Watershed Storage

Sky Pond, Glass Lake, and The Loch have a combined volume of $2.1 \times 10^5\ m^3$ (McKnight et al. 1986). Lake stages have been observed to fluctuate 1 m between high water during snowmelt and base flow (September–October). This seasonal fluctuation does not affect the long-term storage term, which is constrained by lake basin water-holding capacity. Interannual variability in the amount of water stored in the lakes was probably less than 10%, or $2.5 \times 10^4\ m^3$, which we judged to be insignificant.

Almost 2% of total land area of LVWS is permanently covered by snow or ice. Taylor Glacier is approximately 3.8 ha in area, with an estimated average depth of 10 m. Andrews Glacier is larger, with an area of about 7.5 ha and estimated average depth of 30 m. This produces a water storage value of $2.1 \times 10^6\ m^3$ for the two glaciers. White (1980) estimated the volume of interstitial ice in Taylor Rock Glacier to be $1.1 \times 10^6\ m^3$. Combined, about $3.3 \times 10^6\ m^3$ of water is stored in these frozen reservoirs. A net ablation rate of $1.5\ m\ yr^{-1}$ for Taylor Glacier and a net accumulation of $0.3\ m\ yr^{-1}$ for Andrews Glacier were calculated from the relations developed by Alford (1980). This converts to an estimated net loss of stored water of $3.3 \times 10^4\ m^3$ from the glaciers. The amount of accumulation or ablation from Taylor Rock

Glacier is unknown, but we considered it unlikely that the amount of water stored there varies by more than 10% (10^5 m^3) in a given year.

The volume of pore space in soils and glacial till was estimated to be 2×10^5 m^3. By late September, much of this groundwater has been lost to evaporation and drainage to streams. Soil water storage at the end of the summer may vary from year to year depending on summer rainfall and temperature patterns. We estimated that the annual variability of stored water in surficial deposits of LVWS does not exceed 20% of the total pore space, or 4×10^4 m^3.

The crystalline bedrock underlying the watershed is not porous, so groundwater storage is confined to faults and fractures. We consider this reservoir to represent an insignificant amount of stored water in LVWS.

Hydrologic Budgets

The hydrologic budget terms in Table 3.3 were ranked according to the confidence placed in each. Confidence (which ranged from most to least) was placed according to the assumptions and uncertainties associated with each term. Seepage and storage terms are at least an order of magnitude smaller than the other terms (see Table 3.3). Because of this we did not consider them to be significant to the overall hydrologic budget. We therefore calculated hydrologic budgets for the basin based on the equation

$$P_r + P_s + P_{ind} = Q + E_0 + P_b \qquad (10)$$

where Q = discharge and P_b = blown snow. The output volumes compare most closely with input volumes when the moderate budget amounts are used (Table 3.3). More than half the water inputs are direct snow precipitation; in the moderate scenario, rain and indirect inputs account about equally for the remainder. Up to 40% of the inputs are lost to evaporation. Measured precipitation (see Table 3.1) is in excess of measured flow by 24%–44%, so a sizeable evaporative water loss term seems reasonable.

Our greatest source of uncertainty is the accuracy of snow measurements, and this is reflected in the variability between minimum, moderate, and maximum water budgets (Figure 3.6). Rain does not vary greatly among the different estimates, but snow and indirect snow combined can be underestimated by as much as 5.9×10^6 m^3. However, when we assume undermeasurement of snow, the maximum input estimates do not compare well with the maximum estimates for output. Accounting for all sources of undermeasurement, maximum outputs come to 12.74×10^6 m^2, almost 3.00×10^6 m^3 less than maximum inputs.

The minimum estimates for inputs are greater by almost 1.00×10^6 m^3 than the minimum estimated outputs, suggesting that the minimum scenario undervalues the different loss terms, especially evaporation. Within the

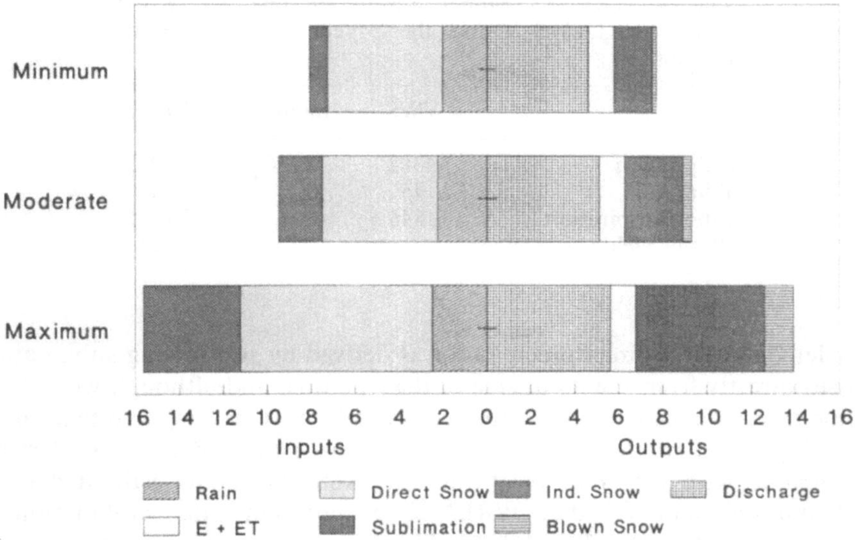

Figure 3.6. Comparison of minimum, moderate, and maximum water budget estimates (5-year average) for Loch Vale Watershed.

evaporation total, the sublimation term fluctuates most widely depending on assumptions.

WRENSS

Hydrologic budget estimates were compared against output from WRENSS. Input requirements included descriptive information on basin geography, climate, vegetation, and size of forest openings. Each watershed was divided into homogeneous units corresponding to aspect, vegetation type, and silvicultural treatment. The model computed the amount of moisture evaporated, transpired, or sublimated; the remainder became stream flow, or water yield.

We divided LVWS into 12 landscape units plus an additional "indirect" unit representing the source area for indirect snow. Although the model does not recognize alpine vegetation, it can be run successfully by defining alpine areas as clearcuts (Swanson, personal communication; Bernier 1986). WRENSS model runs were made with average monthly precipitation (average of 1984–1988) (Table 3.4). The source area for indirect snow was given precipitation only during winter months. The internal algorithm for sublimation (0.15 mm of water lost per each meter sec^{-1} wind speed) adjusted these inputs downward.

WRENSS-generated values are in close agreement with our hydrologic budget estimates. Direct precipitation (rain plus snow) is a model input. The

Table 3.4. WRENSS Model Results (in millions of cubic meters) for LVWS Using the 5-Year Average Monthly Precipitation

	LVWS	Indirect	Total
Precipitation	7.4382	1.256	8.6942
Sublimation	2.145	0	2.145
Evapotranspiration	2.1846	0	2.1846
Water yield			4.3646

model value for indirect precipitation is derived by subtracting sublimation losses directly from the input side of the equation, and although we did the same thing, we also adjusted indirect inputs by loss factors for retention on the slope. Our moderate indirect inputs were estimated at $2.00 \times 10^6\,m^3$. Compared with $1.26 \times 10^6\,m^3$ from WRENSS, the model estimated more sublimation than we did. WRENSS evaporation plus sublimation is somewhat higher at $4.33 \times 10^6\,m^3$ than our computed moderate value of $3.79 \times 10^6\,m^3$. The WRENSS water yield (flow), $4.36 \times 10^6\,m^3$, is also slightly lower than our value for minimum, $4.60 \times 10^6\,m^3$.

Chloride

As an alternative method for estimating hydrologic flux, chloride was employed as a conservative tracer for water. If there are no internal or external sources or sinks for chloride, measured chloride outputs from LVWS should represent the inputs, from which the amount of water can be calculated.

Internal sources and sinks of chloride can include parent material, vegetation or aquatic organisms, and clay adsorption or desorption. Biotite is the most likely chloride-bearing mineral within LVWS. The average chloride concentration in biotite is less than 0.01%, so there is negligible contribution from weathering (Mast 1989). Even in aggrading forests the incorporation of chloride into biomass is low (Likens et al. 1977). Vegetation in LVWS is not growing rapidly (Arthur and Fahey 1990), so the loss or contribution of chloride from vegetation is negligible and probably well within our limits of uncertainty. Chloride behavior in natural waters is generally conservative, in that it neither adsorbs nor leaches from available substrate (Bencala et al. 1990). The percentage of clay minerals available for adsorption in the watershed is small (Walthall 1985), and it is assumed that chloride is not lost to this potential sink.

Dry deposition is a potential confounding source for chloride because it would not be measured in precipitation. Fortunately, there is no substantial upwind source for chloride in the region (Charlson and Rodhe 1982; Turk

Table 3.5. Chloride Budgets for LVWS[a]

Budget Terms	Units	1984	1985	1986	1987	1988	5-Year Average
Inputs							
Rain VWM	$mg\,L^{-1}$	0.10	0.12	0.10	0.10	0.09	0.10
Snow VWM	$mg\,L^{-1}$	0.07	0.14	0.08	0.06	0.06	0.08
Input min	kg	707.29	1172.51	817.22	465.29	576.51	747.76
Input mod	kg	779.22	1364.14	946.26	534.59	667.78	858.40
Input max	kg	1165.07	2220.00	1489.25	794.50	1053.73	1344.51
Outputs							
Stream VWM	$mg\,L^{-1}$	0.11	0.12	0.11	0.12	0.13	0.12
Output min	kg	673.06	527.28	663.48	578.28	626.48	613.72
Output mod	kg	687.19	560.29	684.06	587.75	641.13	632.08
Output max	kg	745.85	697.34	769.47	627.08	701.93	708.33
Out/In min		0.95	0.45	0.81	1.24	1.09	0.82
Out/In mod		0.88	0.41	0.72	1.10	0.96	0.74
Out/In max		0.64	0.31	0.52	0.79	0.67	0.53

[a] Minimum, moderate, and maximum values were obtained by multiplying volume-weighted mean (VWM) concentration of Cl ($mg\,L^{-1}$) with ranges of uncertainty for each water term described in text.

and Spahr 1989). Dry deposition of local materials did not include chloride (Mast 1989) and differences between throughfall and wet deposition measurements in LVWS were insignificant (Arthur, personal communication), so dry deposition was not considered an important chloride source.

Chloride flux (5-year average, 614–708 kg Cl) was calculated as the total chloride output from LVWS via stream flow and blowing snow (Table 3.5). Variation in the numbers derives from different estimates for blowing snow. Calculated chloride inputs showed a broad range of uncertainty (five-year average, 748–1345 kg Cl). If our assumptions are correct, outputs should equal inputs, but, instead they are lower than even the minimum estimate of chloride inputs. The five year average ratio of chloride outputs to inputs ranges between 0.82 and 0.53 (depending on uncertainties in various water terms). If 1985, the peculiar year in which outputs were far less than inputs, is not considered, outputs come much closer to balancing inputs. The chloride (im)balance suggests that one or more of the following may be true: a) inputs, particularly from indirect sources, have been overestimated; b) outputs have been underestimated; c) discharge in 1985 was undermeasured; or d) in 1985 there was an increase in glacier-stored water that went unnoticed by us.

Conclusions

Loch Vale Watershed inputs averaged $9.48 \times 10^6\ m^3$ of water for 1984–1988. Twenty four percent fell as rain, 55% fell as snow, and another 21% blew in from outside the drainage area. Outputs during this time interval averaged

9.26×10^6 m^3. Stream flow accounts for most of this loss, at 55%. Up to 40% was lost through evaporation, of which half was sublimation. An additional 6% leaves LVWS through groundwater seepage or blowing snow.

It is clear that snow dominates the hydrologic cycle in this mountain basin. Blowing snow maintains glaciers at a lower elevation than annual temperatures suggest. Snowmelt drives the annual stream hydrograph. Peak runoff during snowmelt is about 1.5 $m^3\,sec^{-1}$ in June, compared with base flow of about 0.05 $m^3\,sec^{-1}$ in September.

The moderate hydrologic budget estimate that we have presented comes close to balancing inputs with outputs. It agrees with WRENSS model runs. It does not agree with the chloride budgets because of one unusual water year, 1985, in which measured inputs were much larger than measured outputs. Continued monitoring in future years will be used to verify and refine our estimates.

References

Alford DL (1980) The orientation gradient: regional variations of accumulation and ablation in alpine basins. In: Ives JD, ed. Geoecology of the Colorado Front Range: A Study of Alpine and Subalpine Environments, pp. 214–223. Westview Press, Boulder, Colorado.

Anonymous (1977) Groundwater Manual: A Water Resources Technical Publication. USDI Bureau of Reclamation, Denver, Colorado.

Arthur MA, Fahey TJ (1990) Mass and nutrient content of decaying boles in an Englemann spruce/subalpine fir forest, Rocky Mountain National Park, Colorado. Can. J. For. Res. 20:730–737.

Barry RG (1980) Climatology overview. In: Ives JD, ed. Geoecology of the Colorado Front Range: A Study of Alpine and Subalpine Environments, pp. 259–263. Westview Press, Boulder, Colorado.

Berg NH (1986) Blowing snow at a Colorado alpine site: measurements and implications. Arct. Alp. Res. 18:147–161.

Bernier PY (1986) A programmed procedure for evaluating the effect of forest management on water yield. Forest Management Note No. 37, Northern Forestry Centre, Edmonton, Alberta, Canada.

Bigelow DS, Denning AS, Baron J (1990) Differences between Nipher and Alter shielded Universal Belfort precipitation gages at two Colorado deposition monitoring sites. Environ. Sci. Technol. 24:758–760.

Carroll T (1980) An estimate of watershed efficiency for a Colorado alpine basin. In: Ives JD, ed. Geoecology of the Colorado Front Range: A Study of Alpine and Subalpine Environments, pp. 239–240. Westview Press, boulder, Colorado.

Charter K, Swanson R (1989) WRENSS Hydrologic Procedures for Rain and Snow Dominated Regions. Personal Computer Simulation Model. Northern Forestry Centre, Edmonton, Alberta, Canada.

Cleaves ET, Godfrey AE, Bricker OP (1970) Geochemical balance of a small watershed and its geomorphic implications. Geol. Soc. Am. Bull. 81:3015–3032.

Cole JC (1977) Geology of East-Central Rocky Mountain National Park and Vicinity, with Emphasis on the Emplacement of the Precambrian Silver Plume Granite in the Longs Peak-St. Vrain Batholith. Ph.D. Dissertation, University of Colorado, Boulder.

Colorado Climate (1984–1988) Vols. 7–11 (September issues). Colorado Climate Center, Department of Atmospheric Science, Colorado State University, Fort Collins, Colorado.

Creasey J, Edwards AC, Reid JM, MacLeod DA, Cresser MS (1986) The use of catchment studies for assessing chemical weathering rates in two contrasting upland areas in northeast Scotland. In: Colman SM, Dethier DP, eds. Rates of Chemical Weathering of Rocks and Minerals, pp. 468–502. Academic Press, New York.

Goodison BE, Ferguson HL, McKay GA (1981) Measurement and data analysis. In: Gray DM, Hale DH, eds. Handbook of Snow: Principles, Processes, Management and Use, pp. 191–274, Pergamon Press, Willowdale, Ontario.

Kaufmann MR (1983a) A canopy model (RM-CWU) for determining transpiration of subalpine forests. I. Model development. Can. J. For. Res. 14:218–226.

Kaufmann MR (1983b) A canopy model (RM-CWU) for determining transpiration of subalpine forests. II. Consumptive water use in two watersheds. Can. J. For. Res. 14:227–232.

Leaf CF, Brink, GE (1973) Hydrologic Simulation Model of a Colorado Subalpine Forest. USDA Forest Sevice Research Paper RM-107, Rocky Mountain Forest and Range Experiment Station, Fort Collins, Colorado.

Likens GE, Bormann FH, Pierce RS, Eaton JS, Johnson NM (1977) Biogeochemistry of a Forested Ecosystem. Springer-Verlag, New York.

Linacre ET (1977) A simple formula for estimating evaporation rates in various climates, using temperature data alone. Agric. Meteorol. 18:409–424.

Mast MA (1989) A Laboratory and Field Study of Chemical Weathering with Special Reference to Acid Deposition. Ph.D. dissertation, University of Wyoming, Laramie.

McKnight D, Brenner M, Smith R, Baron J (1986) Seasonal Changes in Phytoplankton Populations and Related Chemical and Physical Characteristics in Lakes in Loch Vale, Rocky Mountain National Park, Colorado. Water-Resources Investigations Report 86-4101, U.S. Geological Survey, Denver, Colorado.

NADP/NTN Data Base (1989) National Atmospheric Deposition Program. Tape of weekly data. National Atmospheric Deposition Program (IR-7)/National Trends Network July 1978–January 1988. [Magnetic tape, 9 track, 1600 cpi, ASCII]. NADP/NTN Coordination Office, Natural Resource Ecology Laboratory, Colorado State University, Fort Collins.

Outcalt SI, MacPhail DD (1980) A survey of neoglaciation in the Front Range of Colorado. In: Ives JD, ed. Geoecology of the Colorado Front Range: A Study of Alpine and Subalpine Environments, pp. 203–208. Westview Press, Boulder, Colorado.

Redpath BB (1973) Seismic Refraction Exploration for Engineering Site Investigations. Technical Report E-73-4, U.S. Army Engineer Waterways Experiment Station, Livermore, California.

Troendle CA, Leaf CF (1980) Hydrology. In: An Approach to Water Resources Evaluation of Non-Point Silvicultural Sources. EPA 60018-80-012, Environmental Research Laboratory, U.S. Environmental Protection Agency, Athens, Georgia.

Turk JT, Spahr NE (1989) Chemistry of rocky mountain lakes. In: Adriano DC, Havas M, eds. Acid Precipitation, Vol. 1: Case Studies, pp. 181–208. Advances in Environmental Science, Springer-Verlag, New York.

Winter TC (1981) Uncertainties in estimating the water balance of lakes. Water Resour. Bull. 17:82–115.

White SE (1980) Rock glacier studies in the Colorado Front Range, 1961 to 1968. In: Ives JD, ed. Geoecology of the Colorado Front Range: A Study of Alpine and Subalpine Environments, pp. 102–122. Westview Press, Boulder Colorado.

4. Deposition

Jill Baron, A. Scott Denning, and Paul McLaughlin

We have found that the sources of moisture and the sources of atmospheric pollution are spatially and temporally uncoupled for the Loch Vale Watershed (LVWS) and that this has resulted in the maintenance of fairly clean precipitation to the site over time, although summertime acidity does occur. Central to the question of acid deposition effects, we found no evidence that either current or historic precipitation contains enough strong acid anions to deplete ecosystem acid-neutralizing capacity. Wet and dry deposition chemistry are characterized for Loch Vale Watershed in the first two sections of this chapter. Sources of precipitation were presented in Chapter 2 (this volume). The combination of precipitation and emissions that lead to acidic wet deposition to Rocky Mountain National Park are discussed in the section on sources of precipitation acidity. Paleolimnological reconstructions of deposition trends in metals and acidity since approximately 1850 are presented in a fourth section.

Where appropriate, we compare deposition with other high-elevation sites, especially Beaver Meadows, another deposition monitoring site within Rocky Mountain National Park (Figure 4.1). The Beaver Meadows monitoring site is located less than 15 km north of LVWS. Wet deposition collected at Beaver Meadows differs from LVWS deposition in its amount, timing, and the stronger association of acidity with NO_3 than SO_4. In spite of these differences, there are many similarities, and some of the results presented here were developed using Beaver Meadows data because of its longer period

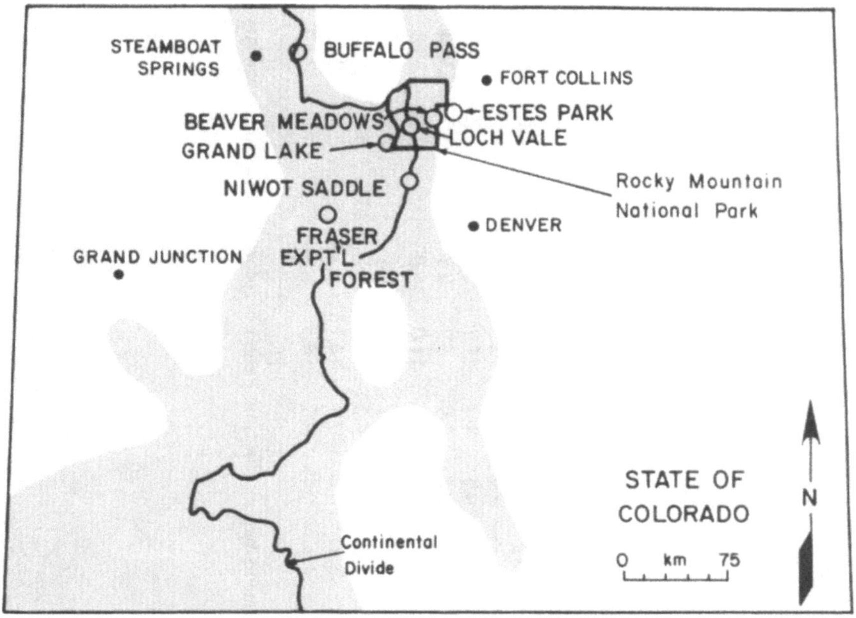

Figure 4.1. Map of northcentral Colorado show locations of Loch Vale and Beaver Meadows wet deposition monitoring sites within Rocky Mountain National Park and other precipitation collection sites used for this study.

of operation. Wet deposition has been collected since September 1983 at Loch Vale and at Beaver Meadows since June 1980 (NADP/NTN Data Base 1990).

The Composition of Wet Deposition

Precipitation at the LVWS monitoring site, Loch Vale, is a dilute solution of nitrate and sulfate salts. Nitrate and SO_4 are the dominant anions in precipitation at both Loch Vale and Beaver Meadows, and Ca, H, and NH_4 are the dominant cations (Table 4.1). Sodium and Cl are present in small amounts. Values presented throughout this chapter are volume-weighted means (VWM) in microequivalents per liter ($\mu eq\, L^{-1}$). The VWM is calculated as

$$VWM = \sum(C_i P_i)/\sum P_i \qquad (1)$$

where C_i is the concentration of the ith sample, and P_i is the precipitation volume of that sample.

Precipitation in Rocky Mountain National Part is two to three times less acidic than eastern high-elevation deposition monitoring sites (Table 4.1;

Table 4.1. Volume-Weighted Mean Precipitation concentrations ($\mu eq\, L^{-1}$) for High-Elevation Sites in the United States[a]

Site	Elevation (m)	Start-Up Date	n	Field H	Lab H	Ca	Mg	K	Na	NH_4	NO_3	SO_4	Cl	Ppt (cm)
Whiteface Mountain, NY	622	84 07 03	203	42.0	38.2	4.1	1.3	0.3	1.7	11.3	20.8	38.4	2.2	111.0
Clingman's Peak, NC	1987	85 11 26	120	31.7	24.5	2.1	0.9	0.2	2.9	6.9	9.7	27.6	3.0	136.4
Beaver Meadows, CO	2487	80 05 29	327	15.3	8.3	12.2	3.1	1.3	3.7	11.0	16.0	18.3	3.1	37.9
Loch Vale, CO	3159	83 08 16	214	12.3	7.0	8.5	2.3	0.5	3.3	6.0	10.9	13.5	2.4	98.7
Niwot Saddle, CO	3520	84 06 05	127	17.3	6.7	8.7	2.2	0.8	4.5	6.4	12.2	20.8	3.2	133.2
Buffalo Pass, CO	3221	84 02 07	163	18.8	10.0	7.0	1.8	0.4	2.7	4.1	9.1	15.7	2.4	85.3
Yosemite, CA	1408	81 12 08	144	12.0	3.7	2.3	1.3	0.4	3.7	6.5	6.8	6.2	4.1	127.3

[a] Location of Colorado sites can be found on Figure 4.1. Start-up date was when the instruments began collection. Period of record for this table is January following the start-up date through January 1990.

NADP/NTN Data Base 1990). Annual VWM field H concentrations of two high-elevation eastern U.S. sites, Whiteface Mountain, New York (42.0 μeq H L^{-1}) and Clingman's Peak, North Carolina (31.7 μeq H L^{-1}) are much higher than at any of the Colorado sites and Yosemite National Park in the Sierra Nevada Mountains of California (12.0–18.8 μeq H L^{-1}). Most of the acidity appears to be caused by SO_4 at the eastern sites, where SO_4 concentrations are two to three times more concentrated than NO_3 in precipitation. At the western sites the concentrations of strong acid anions are more evenly distributed between SO_4 and NO_3.

The base cations are more concentrated in precipitation at the Colorado sites than at the eastern locations and Yosemite. This may reflect the location and climatic conditions associated with the dry western North American interior (Charlson and Rodhe 1982). Similarly, Cl concentrations are greatest at Yosemite, which probably has the greatest marine influence of these comparison sites.

Annual volume-weighted mean (1984–1989) field H concentrations at Loch Vale were between 11 and 15 μeq L^{-1}. Acidic deposition, however, is a regular, seasonal occurrence at both Beaver Meadows and Loch Vale (Table 4.2). A time series of concentrations in precipitation shows this seasonality (Figure 4.2). Summertime field pH values averaged 4.7 (19.6 μeq H L^{-1}) for Loch Vale and 4.6 (25.6 μeq H L^{-1}) for Beaver Meadows. After accounting for differences in laboratory and field pH from sampling procedures and equipment, Bigelow et al. (1989) found that there is a systematic bias of -5 μeq H L^{-1} for both sites between field and laboratory samples when field concentrations are subtracted from laboratory concentrations. After thorough analysis of internal and external quality assurance procedures, these authors attributed this bias to seasonal and geographic differences in the abundance of soil-derived materials (Ca, Mg, and NH_4) in the sample buckets. These cations, and their accompanying neutral salt anions, neutralize some acidity in the sample buckets during the interval between the field measurements in Colorado and the laboratory measurements made in Illinois. In the statistical treatments that follow, laboratory H concentrations are used to address the relationship between H, NO_3, and SO_4 to maintain consistency in the data set.

In spite of physical proximity, solute composition at the Beaver Meadows and Loch Vale stations is quite different when examined closely. All the major ions are more concentrated at Beaver Meadows between June and August, with the strong acid anions and NH_4 almost twice their winter concentrations (see Table 4.2). Summer acidity at Beaver Meadows is caused by greater increases in deposition of NO_3 and SO_4 over base cations. At Loch Vale, NO_3 and SO_4 are slightly more concentrated in the summer than in the winter, but cations are not noticeably different from their winter values.

Summer precipitation at most monitoring sites, including those in Rocky Mountain National Park, is more acidic than winter precipitation

Table 4.2. Annual and Seasonal Volume-Weighted Mean Precipitation Concentrations ($\mu eq\ L^{-1}$) for Two NADP/NTN Sites in Rocky Mountain National Park[a]

Site	Year	Field H	Lab H	Ca	Mg	K	Na	NH_4	NO_3	SO_4	Cl	Ppt (cm)
Loch Vale	1984	12.4	7.3	8.1	3.1	0.5	3.1	6.8	10.9	14.5	2.6	111.1
	1985	14.3	9.0	12.1	3.6	1.0	3.5	6.2	13.0	17.4	3.4	109.8
	1986	—	—	—	—	—	—	—	—	—	—	106.3
	1987	12.3	6.9	5.0	1.3	0.4	3.0	5.6	9.7	10.0	1.9	96.4
	1988	11.9	7.8	5.7	1.2	0.3	2.4	2.7	8.7	11.0	1.7	78.0
	1989	13.2	4.6	8.9	1.9	0.4	3.9	6.9	10.8	12.4	2.2	90.5
	Sept–May	10.7	5.9	8.4	2.3	0.5	3.4	5.3	9.8	12.8	2.3	69.1
	Jun–Aug	19.6	12.8	8.6	2.2	0.9	2.2	8.8	15.3	17.8	3.0	29.6
Beaver Meadows	1981	29.7	10.3	18.9	6.7	1.2	5.1	18.6	23.5	32.4	4.3	33.4
	1982	15.3	10.8	10.6	2.9	1.2	2.4	7.5	15.0	19.0	2.6	36.8
	1983	12.7	8.6	11.8	3.1	1.0	5.0	8.9	15.0	18.0	3.1	45.3
	1984	12.9	7.9	13.4	4.4	2.0	4.9	13.5	17.6	20.0	3.4	43.1
	1985	17.5	10.2	10.5	3.2	1.0	2.6	9.7	15.2	19.0	3.5	41.7
	1986	13.5	8.5	8.2	1.7	2.8	2.6	12.8	15.0	16.1	3.5	44.0
	1987	—	—	—	—	—	—	—	—	—	—	—
	1988	16.3	6.5	12.4	2.3	0.7	4.3	5.8	15.1	15.6	2.6	25.4
	1989	11.2	5.5	13.1	2.3	0.7	3.8	14.5	18.0	16.6	3.2	33.2
	Sept–May	13.0	6.4	10.3	2.7	1.2	3.7	9.0	13.1	15.5	3.1	26.5
	Jun–Aug	25.6	14.3	15.2	4.1	1.7	3.5	17.8	25.3	28.2	3.7	11.4

[a] Data are missing for Loch Vale in 1986 and Beaver Meadows in 1987.

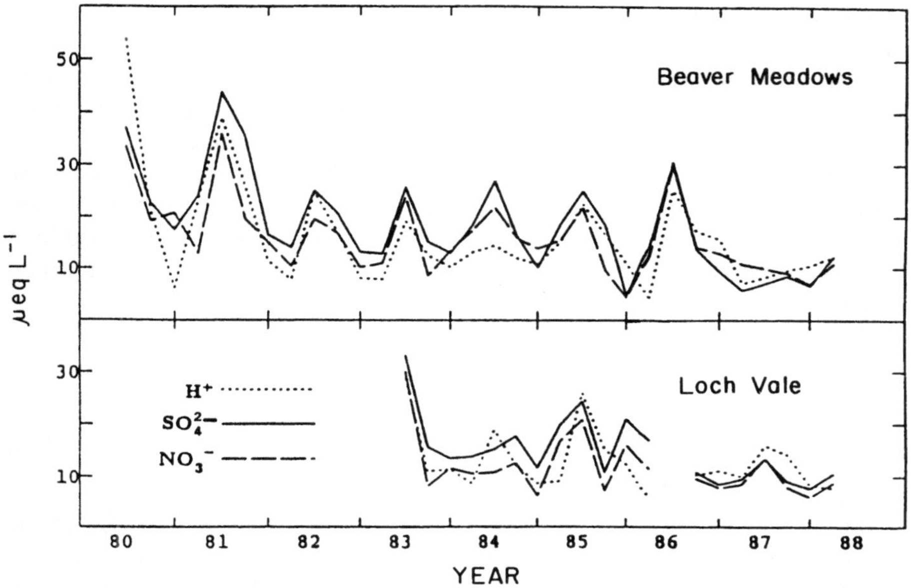

Figure 4.2. Seasonal volume-weighted mean concentrations (μeq L^{-1}) of H, NO_3, and SO_4 in wet deposition for Beaver Meadows and Loch Vale. Values reported are for field H.

(NADP/NTN Data Base 1990), because atmospheric reactions producing NO_3 and SO_4 from their gaseous precursors are more vigorous in the summer as the result of warmer temperatures and longer daily photoperiods. Two additional processes enhance the seasonal acid deposition pattern for the two Rocky Mountain sites: the development of winter inversion layers and seasonal shifts in weather patterns. The development of winter inversion layers can cause winter air masses to have much shallower mixing depths, preventing polluted air parcels close to the earth's surface from becoming incorporated into upper air masses. Upper-level flow is generally much faster in winter than summer, and this reduces the residence time of air parcels over sources of pollutant emissions.

Seasonal shifts in weather patterns bring different air masses with their different pollutant sources to the southern Rocky Mountains. Because summer weather patterns pass over more urban or industrial areas than winter weather patterns, we hypothesize that this enhances seasonal acidity. This is also indicated by different seasonal ratios of Na to Cl. Sodium to Cl ratios have been used to distinguish precipitation influenced by coal-fired power plants from that influenced by natural marine aerosols or soil dust (Wagner and Steele 1989). Ratios at or greater than the sea salt equivalent (0.86) are attributed to natural sources; lower ratios indicate industrial influences. Summer (June, July, and August) ratios for both Loch Vale (0.77) and Beaver Meadows (0.72) fall below the sea salt ratio. Ratios for the rest of the year are 1.16 and 1.35 for the two sites, respectively.

A stratification technique developed by Verry and Harris (1988) was used to determine whether precipitation acidity could be apportioned between NO_3 and SO_4. In the full, unstratified data set most ions are significantly correlated with each other. Verry and Harris (1988) attributed this to the presence of SO_4 and NO_3 from soil dust, marine aerosols, or other sources, which obscures the relationship of H to these ions. Stratification to exclude *high-salt* samples, those that are influenced by such salts, yields a much clearer relationship. *Low salt* is defined as

$$H > (Ca + Mg + Na + K + NH_4)(\mu eq\, L^{-1}) \tag{2a}$$

or

$$(Ca + Mg + Na + K + NH_4) < 50\ (\mu eq\, L^{-1}) \tag{2b}$$

(Verry and Harris 1988).

Simple linear regressions (BMDP 1985) of H versus NO_3 or SO_4 for the stratified data (Table 4.3) reveal different associations at the two sites. By the foregoing definition, 67% of the weekly samples from Beaver Meadows and 80% from Loch Vale fall in the low-salt stratum. During the summer months (June, July, and August) when pH is lowest, acidity in the low-salt samples at Beaver Meadows is associated with NO_3, which accounts for 42% of the H variance; sulfate accounts for 29% of the H variance. For the remainder of the year, the correlation between acidity and both strong acid anions in the Beaver Meadows low-salt data is weaker ($r^2 = .2$). At Loch Vale, 21% of the summer H variance in the low-salt samples is attributable to NO_3, with no significant relationship to SO_4 ($p > .01$). The difference between summer and the rest of the year at Loch Vale is not pronounced.

Table 4.3. Simple Linear Regressions for H versus NO_3 or SO_4 in Low-Salt (Verry and Harris 1988) Wet Deposition from Beaver Meadows and Loch Vale[a]

	Beaver Meadows			Loch Vale		
	n	Slope	r^2	n	Slope	r^2
All NO_3	201	0.57	0.34	162	0.47	0.29
All SO_4	201	0.47	0.29	162	0.36	0.21
Summer NO_3	45	0.55	0.42	35	0.41	0.21
Summer SO_4	45	0.46	0.29	35	0.31	0.16*
Winter NO_3	156	0.48	0.20	127	0.41	0.23
Winter SO_4	156	0.39	0.19	127	0.31	0.17

[a] Excluding samples with notes, trace volumes, or invalid lab pH, NO_3, or SO_4). Prob(F) was < .00005 for all regressions except Summer SO_4 at LV (indicated by *), when p(F) = .0179. Summer is defined as June, July, and August, winter, September through May. Data from Beaver Meadows, 5/80–12/87; data from Loch Vale, 8/83–12/87.

Table 4.4. Partial Regression Results for H versus NO_3 and SO_4 in Low-Salt (Verry and Harris 1988) Wet Deposition from Beaver Meadows and Loch Vale[a]

	N	βNO_3	βSO_4	r^2	prob(F)
Beaver Meadows					
All data	201	0.56	0.39	0.31	0.0000
Summer	45	0.60	0.08	0.47	0.0000
Winter	156	0.49	0.30	0.19	0.0000
Loch Vale					
All data	162	0.16	0.29	0.10	0.0002
Summer	35	0.01	0.33	0.04	0.4879
Winter	127	0.11	0.24	0.06	0.0196

[a] Excluding samples with notes, trace volumes, or invalid lab pH, NO_3, or SO_4) by season. Summer is defined as June, July, and August; winter, September through May. Data from Beaver Meadows, 5/80–12/87; data from Loch Vale, 8/83–12.87.

Partial regression coefficients give another measure of the relative influences of SO_4 and NO_3 on precipitation acidity (Table 4.4). The model applied is

$$H = \beta_0 + \beta_1(NO_3) + \beta_2(SO_4) \quad (3)$$

where the coefficients β_0, β_1, and β_2 were fitted to the measured concentrations in the low-salt data. The coefficient β_1, provides a measure of the increase in H for a unit increase in NO_3 concentration, holding SO_4 constant. Similarly, β_2 indicates the expected rise in H for an increase in SO_4 concentration by 1 $\mu eq\,L^{-1}$ with no change in NO_3. This analysis also reveals differences between the two sites. Beaver Meadows summer acidity responds much more strongly to changes in NO_3 concentration than it does to changes in SO_4 concentration ($\beta_1 = 0.60$, $\beta_2 = 0.08$). The NO_3 coefficient remains higher than the SO_4 coefficient throughout the rest of the year at Beaver Meadows as well, although the differences are not as dramatic. In contrast, partial regressions for the low-salt data at Loch Vale did not yield significant relationships ($p > .01$) except when the seasonal data are combined. In this combined data set, H concentrations respond nearly twice as much to changes in SO_4 ($\beta_2 = 0.29$) than to changes in NO_3 ($\beta_1 = 0.16$), although neither coefficient is as high as those determined for the Beaver Meadows data.

Dry Deposition

In some areas of North America, dry deposition of SO_2 or fine-particle SO_4 can deliver as much sulfur to ecosystems as SO_4 that falls in precipitation (Galloway et al. 1984). Nitric acid (Heubert and Robert 1985) and ammonia gas (NAPAP 1989) can also be enhanced in overall loading by dry deposition processes. Dry deposition is not believed to be an important source of materials from the atmosphere at the highest elevations in the Sierra Nevada Mountains (Williams and Melack 1991), and we have not found it to be important at Loch Vale Watershed. At midelevation sites, such as Beaver Meadows, dry deposition may assume greater importance in total deposition. However, annual (1982–1985) average fine-particle SO_4 concentrations from Beaver Meadows were 0.9 $\mu g\,m^{-3}$ (Cahill et al. 1987), far lower than fine-particle SO_4 values (4.4–9.0 $\mu g\,m^{-3}$) measured east of the Mississippi River (Cahill et al. 1987). The importance of dry SO_4 deposition is therefore limited by the availability of particles to be deposited. We present evidence here, based on analyses of the mechanisms by which materials can be deposited and on mass balance of two conservative ion species, to support our conclusion that dry deposition of distant-source materials is not important to the biogeochemistry of LVWS.

Dry deposition is thought to occur via three separate processes: gravitational settling of large particles, turbulent exchange of small particles and trace gases to surfaces, and interception and impaction of cloud and fog droplets to surfaces (NAPAP 1989). Gravitational settling is strongly influenced by the size, mass, and shape of the particle and additionally by the viscosity of the air. Sources of particulates include agricultural, industrial, and urban activity, almost all of which are confined to the lower elevations in Colorado. These activities are altitudinally disjunct from subalpine LVWS, and we expect the influence of gravitational settling to decrease with increasing altitude. Preliminary results from an elevational study in the southeastern United States showed the concentrations of major ions decrease with elevation in wet deposition (NAPAP 1989). Similar results were found for Colorado, where annual and seasonal concentrations of all major ions except H in both wet and bulk deposition were negatively correlated with elevation (Lewis et al. 1984; Klein 1989).

Clay- and silt-sized sediments collected from snow surfaces were found, without exception, to have mineralogy similar to that of local soils derived from bedrock materials (Mast 1989). Ti/Zr ratios of the sediments from snow samples, soil samples $< 2\,\mu m$ in size, and local bedrock were also found to be similar, suggesting the same origin for all samples (Mast 1989). Several investigators suggest eolian-derived material has historically, and even recently, been important to the genesis of alpine soils (Birkeland 1974; Litaor 1987) of the southern Rockies, while others maintain the ultimate source of the eolian material is local and that it does not travel very far (Thorn and Darmody 1980). The studies from LVWS suggest particles deposited are of local origin.

Turbulent exchange is influenced by mechanical friction caused by obstacles at the ground suface. The mountains themselves are large obstacles to air movement. Within the mountains, large expanses of open tundra or boulder fields do not offer much impediment to air movement, while the presence of a forest canopy does. The presence of forests appears to be important to turbulent exchange as a means of dry atmospheric deposition, although this is not yet clearly understood (NAPAP 1989). The forest cover within LVWS is only 6% of the total land area, so the effects are not as great as in a watershed that is completely forested. A comparison of bulk precipitation to canopy throughfall from May to October 1986 and 1987 (Table 4.5) for 35 storm events showed significant enrichment of the base cations, total alkalinity, and dissolved organic carbon (DOC) in throughfall samples (Arthur 1990). Some constituents, such as K and Mg, were undoubtedly leached from the foliage (Tukey 1970), although others might represent washing of dry deposited materials. Note that SO_4 and NO_3 were not significantly different between bulk and throughfall, and H concentrations were significantly lower in throughfall, presumably because of a large increase in total alkalinity. The greater concentrations in base cations and alkalinity are suggestive of soil dust, and this is not inconsistent with results from the analyses of snow surface sediments. The interpretation of throughfall data for estimating dry deposition requires large numbers of precipitation events to capture spatial and temporal variance (Lovett and Lindberg 1984), so conclusions regarding the importance of turbulent diffusion are premature. The data do not, however, contradict the hypothesis that dry deposition is confined to locally derived soil materials.

Fog and impaction by cloud droplets is important to total deposition only in areas that are immersed in clouds or fog frequently or for long periods (NAPAP 1989). Whiteface Mountain, New York (1483 m) and Clingman's Peak, North Carolina (1950 m) were immersed in clouds an average of 37%

Table 4.5. Volume-Weighted Mean Chemistry of Throughfall (TF) and bulk precipitation (BP), May–October 1986 and 1987 in Loch Vale Watershed[a]

Element	TF (μeq L^{-1})	BP (μeq L^{-1})
[H]	13.0a (pH = 4.88)	15.2b (pH = 4.81)
Ca	67.9a	22.5b
Mg	18.2a	4.7b
Na	10.9a	3.8b
K	42.0a	3.5b
NH_4	10.8a	11.5a
Cl	21.1a	18.2a
NO_3	24.9a	18.6a
SO_4	26.6a	17.7a
PO_4	1.8a	0.4a
Total alkalinity	8.2a	1.3b
DOC	14.0a	2.6b

[a] Values with different letters are different at $p = .05$. Includes total of 35 TF and BP events. DOC, dissolved organic carbon. From Arthur 1990.

and 29% of the time in 1986–1988 (percent of hours in cloud), respectively. The percentage of total days experiencing some cloud during this time period was 77% for Whiteface and 76% for Clingman's Peak (NAPAP 1989). The occurrence of fog in the Southern Rocky Mountains is 20–40 days per year (5%–11%), and these episodes are concentrated in valleys (Baldwin 1973). This is substantially less than the percentage of time observed at the eastern sites and implies a less important role for cloud impaction and fogs to the total dry deposition of chemical species to LVWS.

Sources of Precipitation Acidity

Precipitation acidity is characterized by an excess of the strong acid anions SO_4 and NO_3 over available base cations and NH_4. Emissions of sulfur dioxide (SO_2) and nitrogen oxides (NO_x) are widely accepted as the major sources of precipitation acidity (Gibson 1986; NAPAP 1989). Industrial processes such as the burning of coal and smelting of metal ores emit both SO_2, the precursor of SO_4, and NO_x, the precursors of NO_3. Automobile emissions, on the other hand, contribute NO_x but only negligible amounts of SO_2 (Knudson 1986). Lightning is an important source of NO_x on a global scale (Logan 1983), but its contribution to local or regional concentrations of NO_3 is unknown. Nitrate in Antarctic snow has been attributed to lightning from temperate latitudes (LeGrand and Delmas 1986). This requires long-range transport and does not support or refute the idea of local increases in NO_3 from lightning.

Once emitted into the atmosphere, SO_2 or SO_4 can be transported great distances because of their relatively long residence time (Gillani 1984). Nitric oxides are readily transformed via gas-phase reactions into nitric acids in the atmosphere. Nitric acid is very soluble, making it susceptible to precipitation scavenging when there is moisture in the atmosphere (Sperber and Hameed 1986). Summer thunderstorms, then, might be a very effective mechanism for regularly removing HNO_3 from the atmosphere. In the absence of precipitation, NO, NO_2, HNO_3, and nitrate aerosols can be transported long distances. A transport distance of several hundred kilometers between source and receptor was measured in central Ontario (Anlauf et al. 1986). Derwent and Nodop (1986) derived mean residence times of 1.2 days and mean transport distances of 800 km for nitrogen species in northwestern Europe from model output. Given the aridity of the western United States, it is possible for nitrogen species to be carried great distances.

The western United States has few sources of SO_2 and NO_x, which has made it easier for investigators to identify possible urban, mining-related, or coal-fired industrial pollutant sources in the West than in other, more populated, parts of the world. Source areas that we and others have identified for acid precursors include the Denver metropolitan area (Lewis et al. 1986; Colorado Department of Health 1987; Parrish et al. 1990), the southern

California metropolitan area (Bresch et al. 1987; Borys et al. 1987), the El Paso, Texas/Cicudad Juarez, Mexico metropolitan area (Texas Air Control Board 1990), metal (mostly copper) refineries of the Southwest and Mexico (Oppenheimer et al. 1985; Epstein and Oppenheimer 1986; Borys et al. 1987), coal-fired electric generating plants through the Inter-Mountain West (Nochumson 1983; Lewis et al. 1984; Borys et al. 1987), oil and gas refineries of southwestern Wyoming and western Texas (Bresch et al. 1987; Texas Air Control Board 1990; Figure 4.3).

We explored the derivation of air masses bringing two types of polluted precipitation to Rocky Mountain National Park by coupling output from

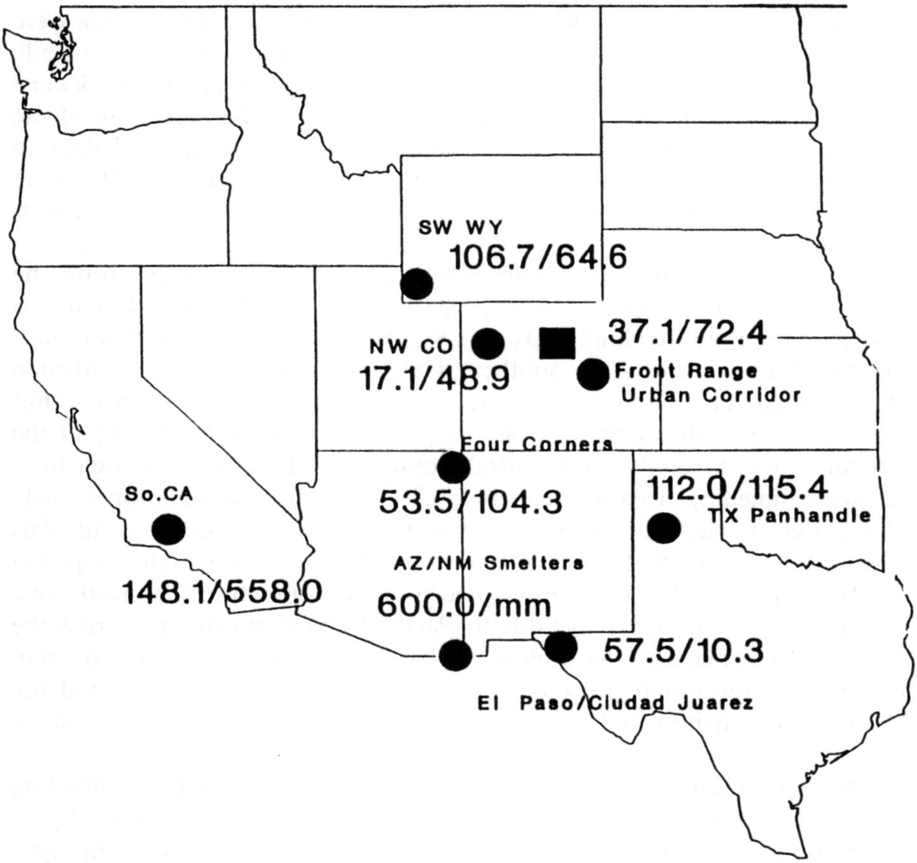

Figure 4.3. Map of SO_2 and NO_x emission sources (solid dots) located along back trajectories of high SO_4 plus NO_3 or high-acidity/low-salt stratified precipitation events. Values shown as SO_2/NO_x in thousand tons per year. NO_x emissions data were unavailable from smelter regions of AZ/NM and are noted as mm. References for emissions figures are in text. Solid square shows location of Rocky Mountain National Park in Colorado.

a computer model, the ARL-ATAD model (Heffter 1981) with a residence time analysis of air parcels passing though each 1° latitude by 1° longitude geographic grid square (Ashbaugh et al. 1985). The two precipitation types were (1) high NO_3 plus SO_4 rain events, defined as those precipitation events where NO_3 plus SO_4 concentrations exceeded the geometric mean concentration by two or more standard deviations (Ashbaugh et al. 1985), which turned out to be NO_3 plus SO_4 greater than $80\,\mu eq\,L^{-1}$; and (2) acidic, low-salt rain events (Verry and Harris 1988). High acidity was defined according to Verry and Harris (1988) as $13\,\mu eq\,L^{-1}$ (pH 4.9) or greater. The ARL-ATAD model calculates back trajectories of air parcels from wind observations obtained from the National Weather Service radiosonde network. Input to the model consisted of the date and time of the onset of precipitation at the Beaver Meadows monitoring site (insufficient data were available from Loch Vale). Wind vectors were calculated from the radiosonde data, and the vector for the given time at the site was extrapolated backward for 3 hours. The air parcel arriving at the station at the beginning of the precipitation event was thus followed backward for as long as 5 days in 3-hour time steps. Only those weeks of June 1980–December 1987 during which all precipitation fell in a single event were used in this analysis (McLaughlin 1988).

Residence time analysis (Ashbaugh et al. 1985) involved counting the 3-hour trajectory endpoints at each grid point in the model domain. A residence time plot for high NO_3 plus SO_4 precipitation events emphasized air parcel trajectories from southern California and the Gulf of Mexico (Figure 4.4a). The isopleth lines stretched east slightly but did not range beyond the eastern border of Colorado. Because all parcels arriving at the site must pass through north-central Colorado, a "bull's eye" of high total residence time appeared in the vicinity of Rocky Mountain National Park. These tightly packed contours also suggested that stagnant conditions contribute to high NO_3 plus SO_4 events. The other two highs in parcel residence time on this map represented sources of moisture. The peak over northern Baja California was due to Pacific tropical storms that cross the Gulf of California, and the peak over New Mexico represented the overlap of summer monsoonal trajectories from both the Gulf of Mexico and the Pacific which then moved north along the Front Range to support upslope storms.

Spurious features of the residence time maps were removed by dividing the number of high-concentration endpoints in a given 1° latitude by 1° longitude grid square by the total number of endpoints in the same grid square for the same time period. This yielded plots of the conditional probability that an air parcel trajectory passing through that specific grid square contained a high ionic concentration. Such a plot for high SO_4 plus NO_3 events (Figure 4.4b) showed the Los Angeles basin, the Four Corners area, western Texas, and southwestern Wyoming to be likely source areas for these events.

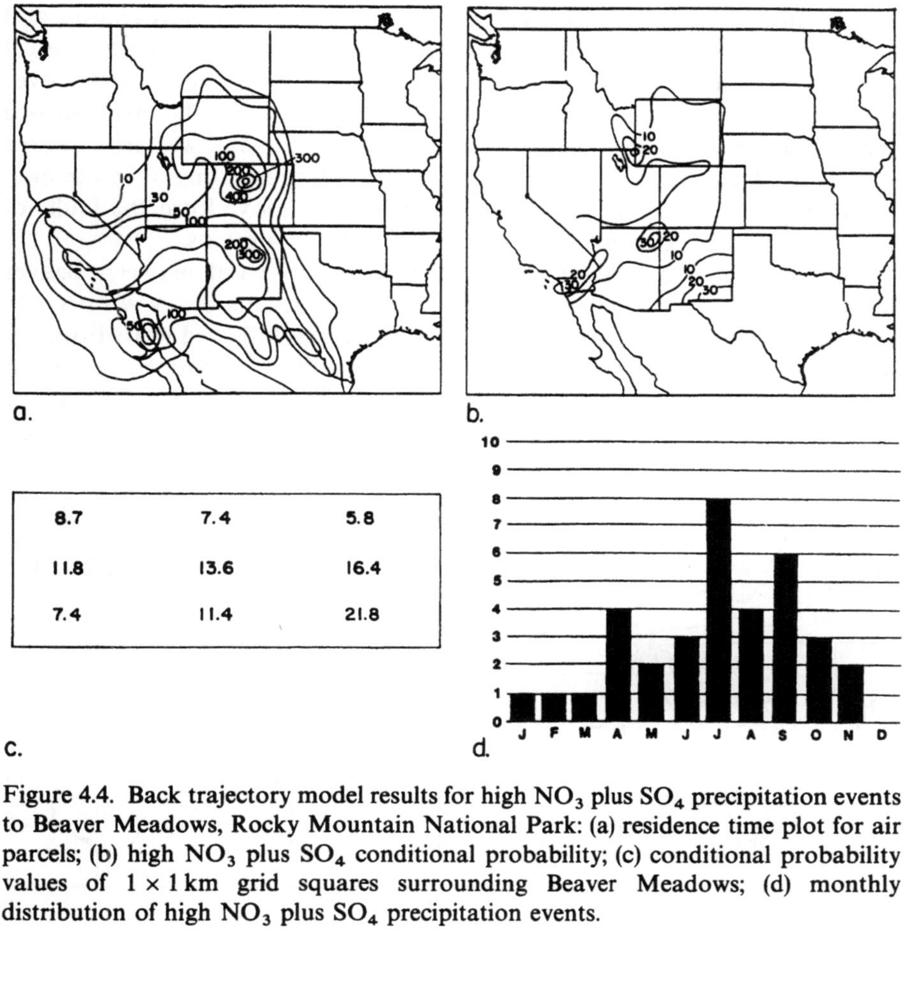

Figure 4.4. Back trajectory model results for high NO_3 plus SO_4 precipitation events to Beaver Meadows, Rocky Mountain National Park: (a) residence time plot for air parcels; (b) high NO_3 plus SO_4 conditional probability; (c) conditional probability values of 1×1 km grid squares surrounding Beaver Meadows; (d) monthly distribution of high NO_3 plus SO_4 precipitation events.

The ARL-ATAD model has poor resolution of local- to mesoscale wind patterns. Because of the large distances between the synoptic radiosonde network used by the model, the wind field must be interpolated over hundreds of kilometers. Nevertheless, model output does suggest that the Front Range urban corridor may contribute to acidic deposition at the study site. A numerical array of the nine grid squares including and immediately adjacent to Beaver Meadows (Figure 4.4c) indicates that trajectories arriving from the east and southeast are twice as likely to bring high NO_3 plus SO_4 precipitation to Rocky Mountain National Park as those from other directions. High NO_3 plus SO_4 events occurred most frequently in the spring and summer months (Figure 4.4d).

Only nine acidic, low-salt events (Verry and Harris 1988) occurred during the period analyzed with the ARL-ATAD model, but the residence time plot (Figure 4.5a) suggested a strong dependence on local air movement for these precipitation events. Arizona, the southern Texas–Mexican border area, and the Texas Panhandle are also implicated as source areas for the air parcel trajectories. The conditional probability contour plot shows peaks over northern and southwestern Arizona (Figure 4.5b), and the grid square conditional probability values also suggested a southwesterly source (Figure 4.5c). Convectively driven upslope circulation is strongest in summer, but this is also when monsoonal precipitation can move from the Gulfs of California and Mexico into the Southern Rocky Mountains.

Because of the differences in wet deposition observed between Beaver Meadows and Loch Vale, extrapolation of model results of LVWS need to be made with caution. However, the back trajectories followed were distinctly regional. During spring and summer, when precipitation patterns are most similar between the sites, it may be that the same pollution sources impinge on both sites. June, July, and August were the months of greatest acidity

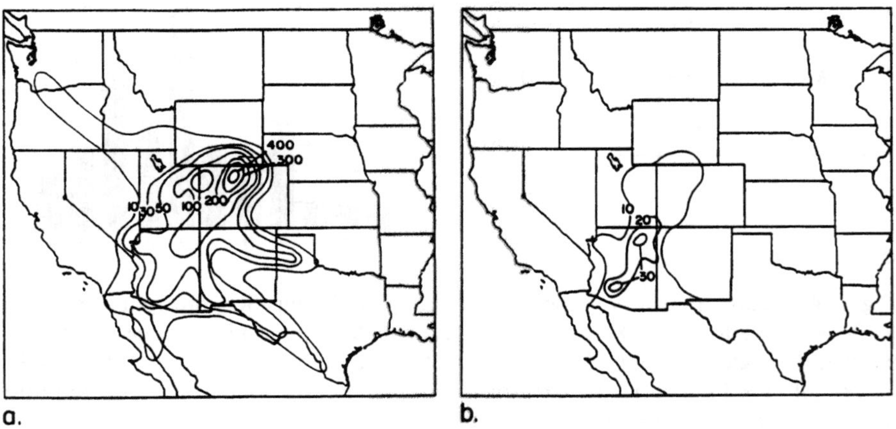

a. b.

11.4	18.6	3.3
16.8	14.2	7.1
21.8	1.5	0.0

c.

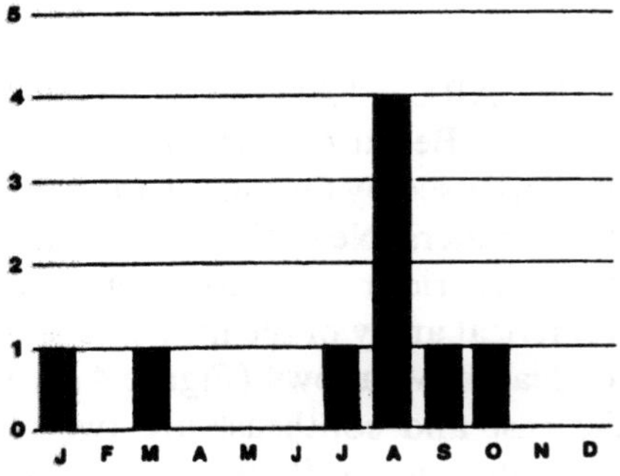

d.

Figure 4.5. Back trajectory model results for acidic, low-salt stratified precipitation events to Beaver Meadows, Rocky Mountain National Park: (a) residence time plot for air parcels; (b) high acidity, low-salt stratified conditional probability; (c) conditional probability values of 1 × 1 km grid squares surrounding Beaver Meadows; (d) monthly distribution of high acidity, low-salt precipitation events.

and greatest concentrations of NO_3 plus SO_4 in precipitation at both sites (Table 4.2, Figure 4.5d).

Emissions from the southern California metropolitan area were conservatively estimated at 148,100 tons SO_2 and 558,000 tons NO_x (values are half the statewide emissions estimates for 1985 from Kohout et al. 1987). Even though this is the largest regional source of NO_x and the second largest for SO_2, a trace element signature analysis by Borys et al. (1987) suggested this contribution is only 6% of the SO_4 aerosol.

Nonferrous metal smelters in Arizona, New Mexico, Utah, and Nevada have been found to generate enough SO_2 to account for 41% of the variability in seasonally adjusted monthly SO_4 concentrations in precipitation throughout the southern Rocky Mountains (Oppenheimer et al. 1985; Epstein and Oppenheimer 1986). In 1985, SO_2 emissions from Arizona and New Mexico alone were approximately 600,000 tons (Blanchard 1991). Smelter emissions from Arizona have declined since 1988. This regional source accounted for about 27% of the industrially derived SO_4 deposited in precipitation during 1984–1985 at a site 100 km northwest of Beaver Meadows, according to trace element signature analyses (Borys et al. 1987).

Emissions from the city of El Paso, 57,520 and 10,280 tons SO_2 and NO_x per year, respectively, do not appear high when compared with emissions from other source areas. Emissions from Ciudad Juarez, Mexico, over the border from El Paso, have not been measured. Juarez is a slightly larger city than El Paso, so doubling the emissions from this area to account for the combined emissions from Ciudad Juarez/El Paso area does not seem unreasonable. South and east of Ciudad Juarez is Monterey, an industrial city of unknown emissions that may also contribute pollutants to air masses from this region.

Emissions from the Four Corners region before 1985 were 117,900 tons SO_2 per year and 104,300 tons NO_x per year. Since 1985, SO_2 emissions have declined to 53,500 tons per year wile NO_x emissions have remained the same (C. V. Mathai, Arizona Public Service Co., and R. Williams, New Mexico Public Service Co., personal communications). Emissions from the Green River Basin of southwest Wyoming were 64,600 tons SO_2 and 31,700 tons NO_x for 1985 from one power plant (C. A. Collins, Wyoming Department of Environmental Quality, personal communication). Two power plants in northwestern Colorado emit 17,100 tons SO_2 and 48,900 tons NO_x per year (Colorado Department of Health 1987). The fine-particulate signature analyses conducted by Borys et al. (1987) suggested that most of the SO_4 aerosol for this region is generated from the combustion of coal.

West Texas is an area of oil and gas refining and storage factories (Texas Air Control Board 1990). Studies of fine particulate sulfate (Bresch et al. 1987) support our finding that this region may be an important source for SO_4 and NO_3. Current emissions from the Panhandle region are 111,980 tons SO_2 per year and 115,453 tons NO_x per year (Texas Air Control Board

1990). Emissions from the Odessa region of Texas are 151,223 tons SO_2 per year and 112,089 tons NO_x per year. Natural gas refineries in southwest Wyoming contributed 42,100 tons SO_2 and 32,900 tons NO_x for 1985 (C. A. Collins, Wyoming Department of Environmental Quality, personal communication).

Emissions from the Front Range urban corridor include more than 37,000 tons SO_2 yr^{-1} and almost 72,400 tons yr^{-1} NO_x from both stationary and mobile sources (Lewis et al. 1986; Colorado Department of Health 1987). And although prevailing winds are westerly, there is strong evidence that the highest concentrations of NO_x and HNO_3 in the local atmosphere occurs during easterly upslope conditions (Parrish et al. 1990). Parrish et al. (1990) suggested that convectively driven mountain-valley winds are an important mechanism by which urban pollutants are transported up into the mountains.

We investigated the influence of upslope winds on precipitation solute composition at Loch Vale by comparing the westerly wind component

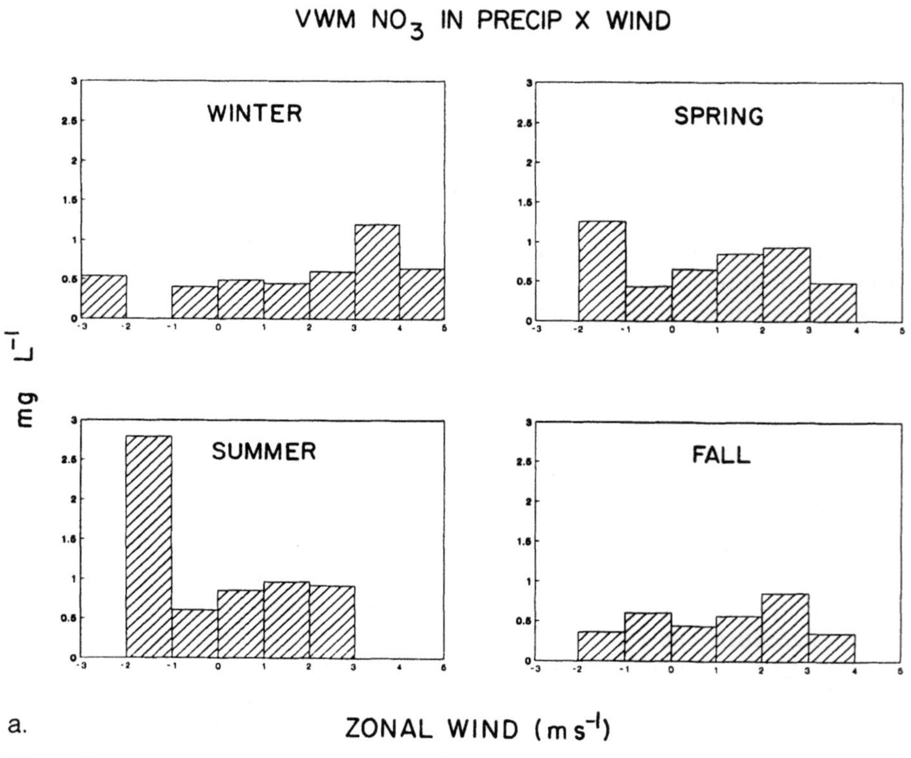

Figure 4.6. Comparison of precipitation and acid anion concentrations with zonal winds (μ) at Loch Vale: (a) Volume-weighted mean (VWM) NO_3 concentration versus mean zonal wind component; (b) VWM SO_4 concentration versus mean zonal wind component; and (c) distribution of precipitation by zonal wind speed.

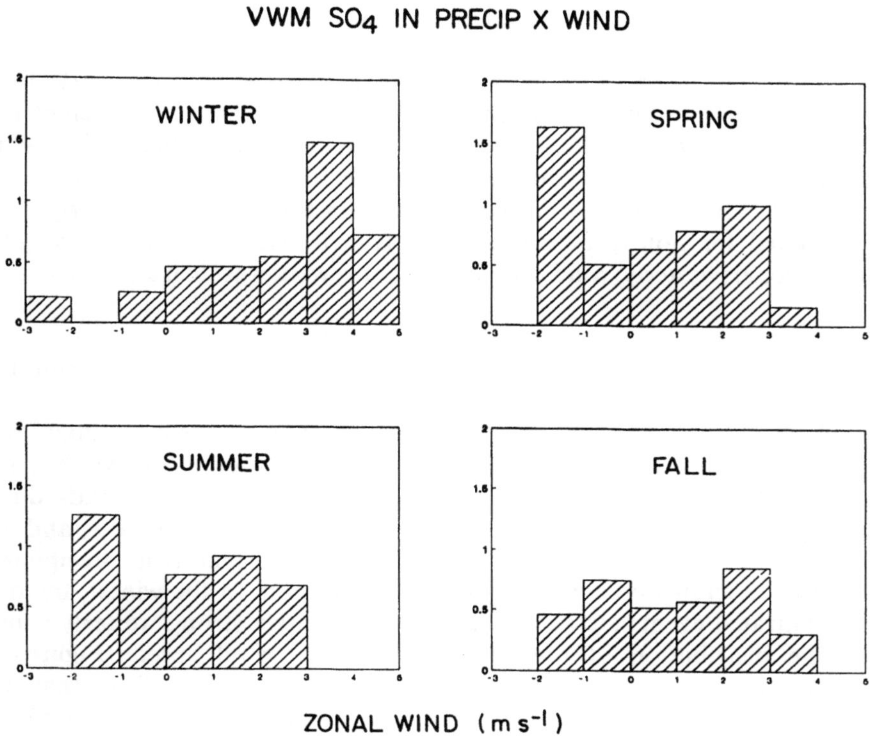
VWM SO_4 IN PRECIP X WIND
WINTER
SPRING
SUMMER
FALL
mg L^{-1}
0
0.5
1
1.5
2
-3
-2
-1
0
1
2
3
4
5
ZONAL WIND (m s^{-1})

b.

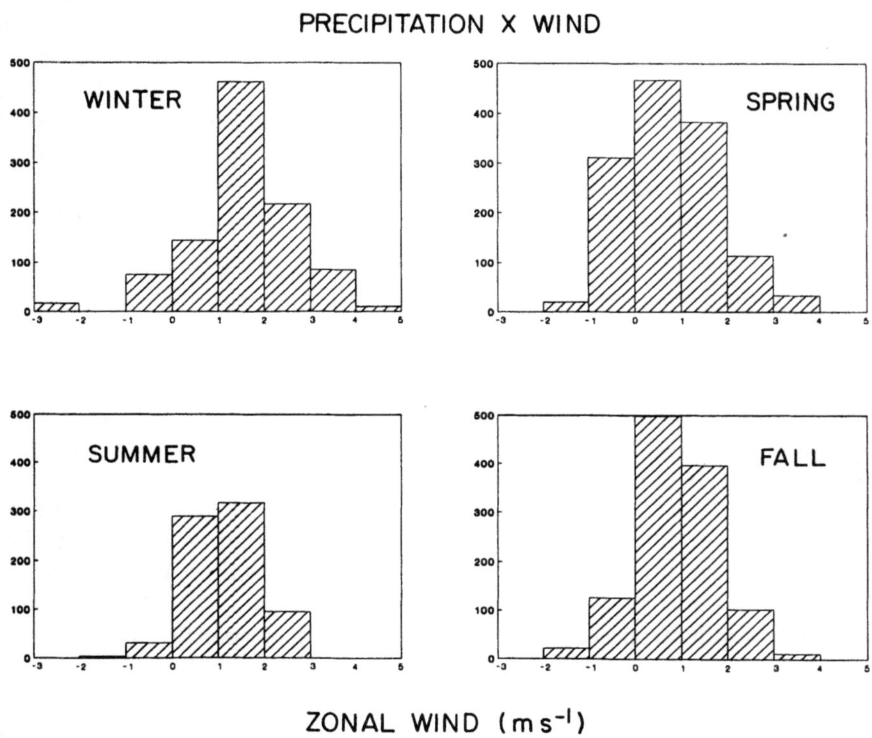
PRECIPITATION X WIND
WINTER
SPRING
SUMMER
FALL
mm
0
100
200
300
400
500
-3
-2
-1
0
1
2
3
4
5
ZONAL WIND (m s^{-1})

c.

measured during precipitation events with the sample chemistry for 190 weeks of NADP/NTN data. Although surface winds may be quite different from winds aloft, terrain geometry at LVWS generally constrains the airflow into downvalley or upvalley directions (see the wind roses in Chapter 2, this volume). Orthogonal wind components (westerly and southerly) were computed for hourly observations at LVWS. The precipitation-weighted mean zonal (westerly) wind component for the week was then calculated as

$$u = \sum(u_i p_i)/\sum(p_i) \tag{4}$$

where u_i and p_i are the hourly zonal wind component and precipitation, respectively.

Plots of the VWM NO_3 and SO_4 concentration versus mean zonal wind component reveal strong seasonal differences (Figure 4.6a, b). Negative values of zonal wind (u) in these plots (indicating easterly winds during precipitation) were associated with higher concentrations of NO_3 and SO_4 in spring and summer. Concentrations of the acid anions in precipitation during fall show no dependency on wind direction. In the winter, the most concentrated events are associated with moderately strong westerly winds.

The distribution of precipitation by zonal wind speed (Figure 4.6c) reveals that easterly events are comparatively rare at LVWS, so that total deposition of SO_4 and NO_3 is greatest during precipitation associated with weak westerly winds. Convective rainfall in summer is most likely to be accompanied by strong downdrafts (Hansen et al. 1978) that will be measured as downvalley (westerly) events in LVWS, even though these storms are triggered by upslope lifting along the entire mountain front. That the few easterly events in spring and summer are measurably more concentrated in these acid anions than westerly events strongly suggests a source to the east. Although the nearest obvious source is the Front Range urban corridor, contributions from the more distant sources discussed here cannot be ruled out.

Historical Reconstructions of Deposition

Before 1980 there had been no systematic collection of precipitation for chemical analysis for the southern Rocky Mountains. Thus, we used paleolimnological techniques to reconstruct trends in deposition over time from the mid-1800s to the present (Baron et al. 1986). Lake sediments were cored and analyzed for metals and diatom community assemblages. Sequential layers from the sediment cores were radiometrically dated to reconstruct patterns of metal deposition that could be attributed to an increase in industrial emissions. Diatoms have very specific pH preferences, which have been documented, so that knowledge of diatom community changes over time can be used to interpret lake chemical composition over time (Charles and Norton 1986). Diatoms were assigned a pH preference according to Husted (1939) within one of the five following pH occurrence categories: (1) acidobiontic, optimum distribution below pH 5.5; (2) acidophilic, widest distribution at less than pH 7.0; (3) indifferent, distributed equally above and below pH 7.0; (4) alkaliphilic, widest distribution at greater than pH 7.0; and (5) alkalibiontic, occurs only at greater than pH 7.0. Indices based ratios of percentages of diatoms within each of the pH preference categories have been developed to infer the pH of the water from which the diatom community assemblages were taken (index α, Nygaard 1956; index B, Renberg and Hellberg 1982).

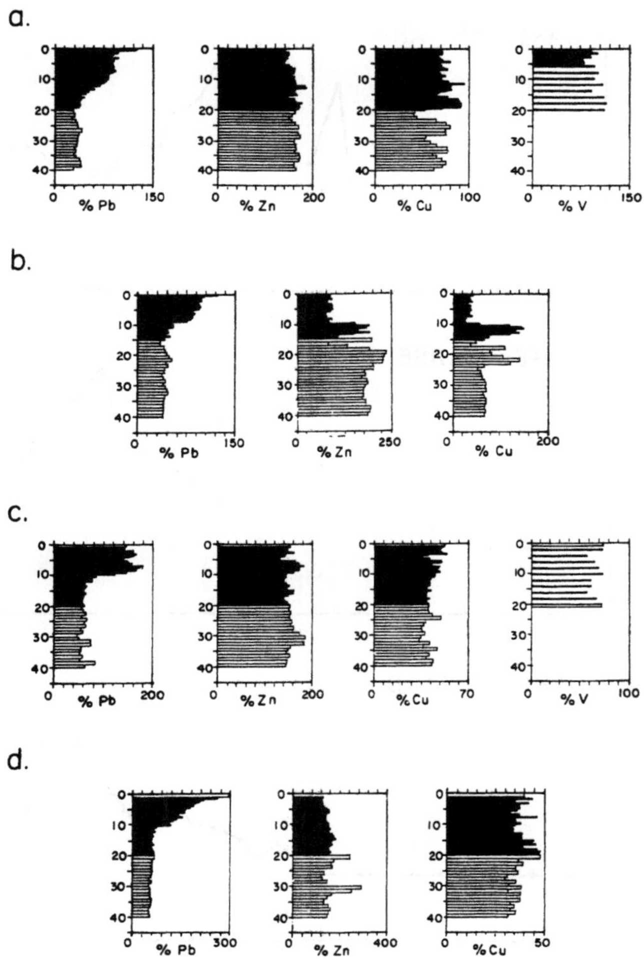

Figure 4.7. Selected sediment chemistry profiles for four lakes in Rocky Mountain National Park: (a) Emerald Lake, (b) Lake Haiyaha, (c) Lake Louise, and (d) Lake Husted. Values are percentage ignition for each metal versus depth below lake sediment/water interface. (From Baron et al. 1986; reprinted with permision.)

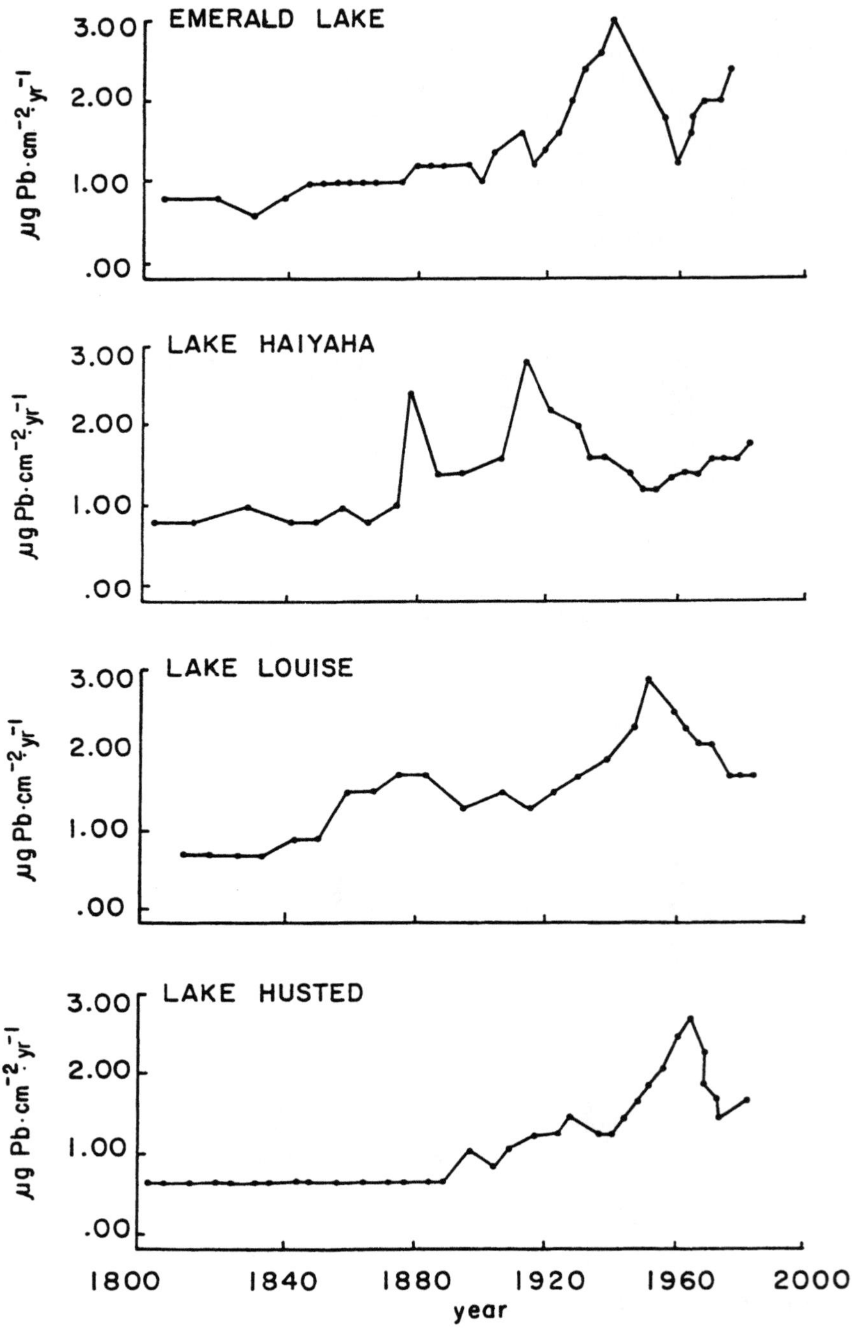

Figure 4.8. Atmospheric deposition rate of Pb over time from four lakes in Rocky Mountain National Park. (From Baron et al. 1986; reprinted with permission.)

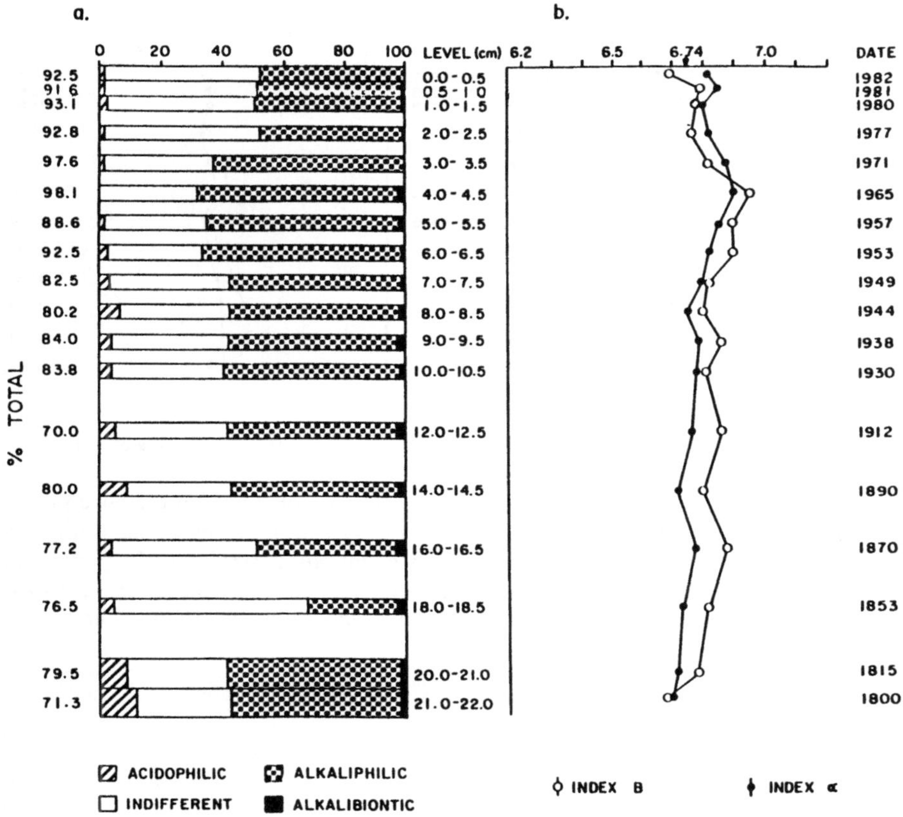

Figure 4.9. (a) Diatom stratigraphy showing distribution of pH preference groups downcore in sediments from Emerald Lake, Rocky Mountain National Park. Left column indicates percentage of each diatom count categorized by pH preference. (b) Inferred pH values downcore (over time) using both Index B and Index α for Emerald Lake. Modern surface water pH ($n = 4$) indicated with arrow. (From Baron et al. 1986; reprinted with permission.)

The metal analyses from Rocky Mountain lakes do not indicate these lakes were receiving aerosols attributable to combustion of coal or oil (Figure 4.7a–d). The only notable metal signal at all is an increase in the atmospheric deposition of Pb beginning between 1850 and 1890 (Figure 4.8). Early increases in Pb deposition can be related to lead mining and processing in Colorado, while later inputs are probably related to increases in the numbers of automobiles burning leaded gasoline both regionally and nationally.

The lake chemistry reconstructions from diatom assemblages also do not indicate any trend of increasing acidity over time. Three of the four study lakes showed increasing numbers of alkaliphilous taxa from about 1800 to the present (Figures 4.9a, 4.10a, 4.11a, 4.12a). The diatom stratigraphy from

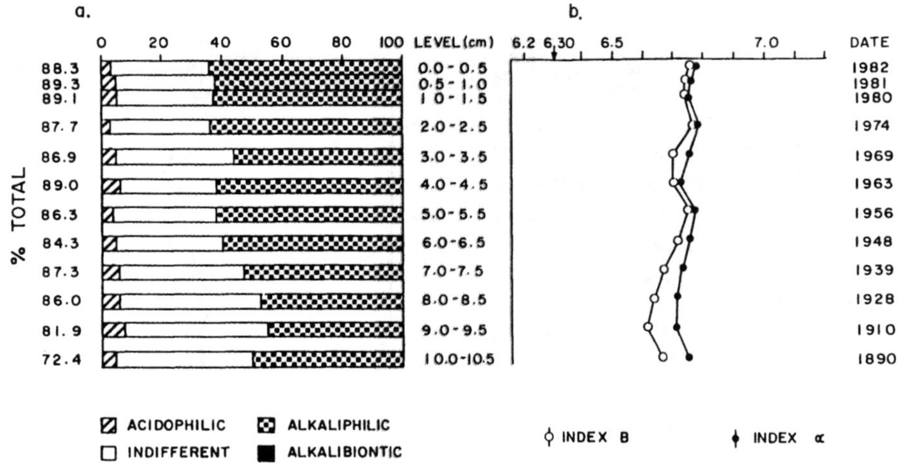

Figure 4.10. (a) Diatom stratigraphy showing distribution of pH preference groups downcore in sediments from Lake Haiyaha, Rocky Mountain National Park. Left column indicates percentage of each diatom count categorized by pH preference. (b) Inferred pH values downcore (over time) using both Index B and Index α for Lake Haiyaha. Modern surface water pH ($n = 4$) indicated with arrow. (From Baron et al. 1986; reprinted with permission.)

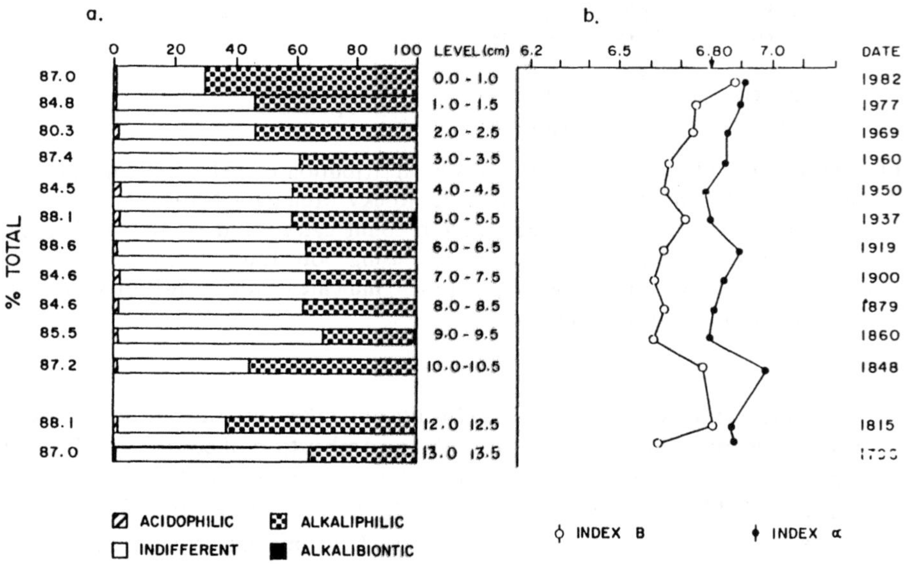

Figure 4.11. (a) Diatom stratigraphy showing distribution of pH preference groups downcore in sediments from Lake Louise, Rocky Mountain National Park. Left column indicates percentage of each diatom count categorized by pH preference. (b) Inferred pH values downcore (over time) using both Index B and Index α for Lake Louise. Modern surface water pH ($n = 4$) indicated with arrow. (From Baron et al. 1986; reprinted with permission.)

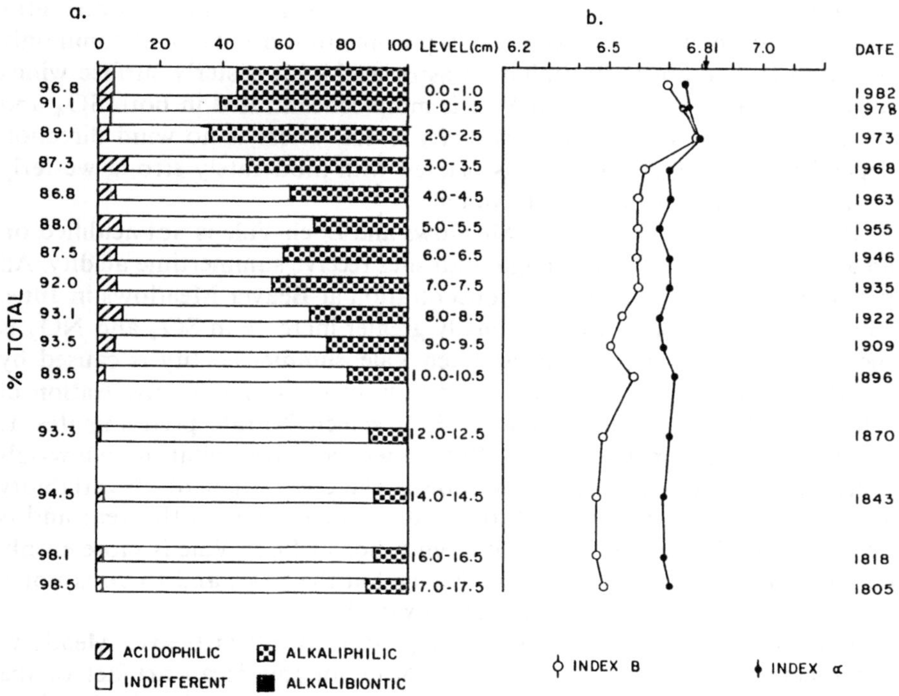

Figure 4.12. (a) Diatom stratigraphy showing distribution of pH preference groups downcore in sediments from Lake Husted, Rocky Mountain National Park. Left column indicates percentage of each diatom count categorized by pH preference. (b) Inferred pH values downcore (over time) using both Index B and Index α for Lake Husted. Modern surface water pH ($n = 4$) indicated with arrow. (From Baron et al. 1986; reprinted with permission.)

all four lakes is dominated by taxa assigned to indifferent or alkaliphilous pH preference categories. The inferred pH profiles indicate lake water pH values ranged from about 6.4 to 7.0 during the past 100 years, with slight trends, if any, toward increasing pH in three of the four lakes (Figures 4.9b, 4.10b, 4.11b, 4.12b).

It was concluded from this work that industrial emissions were not substantial enough before 1982 to cause metals other than Pb to be transported to high elevations or to cause any acidification of lakes (Baron et al. 1986).

Summary

Precipitation at Loch Vale reflects Pacific moisture sources in the winter months, and upslope moisture from the southeast, southwest, and local convective storms in the spring and summer. Because Beaver Meadows falls in the rain shadow of the Rockies, upslope storms are a major contributor to precipitation at the site. As a result, precipitation occurs most commonly in spring and summer. Precipitation associated with easterly surface winds in spring and summer at Loch Vale is more concentrated in both SO_4 and NO_3 than that associated with westerly winds. There is no wind direction dependence in fall, and storms associated with moderately strong westerly winds are most concentrated in winter.

Precipitation at both Beaver Meadows and Loch Vale is not acidified on an average annual basis, although both sites receive summertime acidity. All precipitation solutes increase in concentration at Beaver Meadows in June, July, and August, but a proportionally greater increase in SO_4 and NO_3 is the cause of acidic deposition. At Loch Vale, summer acidity is caused by increasing amounts of SO_4, NO_3, and H in precipitation; concentration of other ions does not increase during the summer. Nitrate predominates as the strong acid anion associated with summer acidic precipitation in low-salt precipitation weeks at Beaver Meadows. The correlation of H variability with SO_4 and NO_3 is equally distributed during the rest of the year and is not as strong. The variation in H concentration at Loch Vale is more evenly distributed beween NO_3 and SO_4 throughout the year, and there is not a strong correlation between either anion with H.

Most of the high NO_3 plus SO_4 precipitation events at Beaver Meadows occur between March and September. Some of this is an artifact of the winter precipitation minimum. Air parcel trajectories associated with these precipitation events are most commonly from the southeast and southwest, with occasional inputs from the northwest. High SO_2 and NO_x emissions to the southeast and southwest can be transported from the West Texas oil refineries, from nonferrous metal smelters in New Mexico, Arizona, and northern Mexico, from the Los Angeles metropolitan area, and from the Four Corners area coal-fired electricity plants. All air masses that result in upslope precipitation additionally pass through and over the Front Range urban corridor. This urban area contributes substantially to the loading of air pollutants to the Front Range Rocky Mountains, particularly during the summer. Coal-fired electric generating plants in northwestern Colorado, and mining and energy-related activity in the Green River Basin are additional sources of emissions from the north and west.

It is important to note the heterogeneity in both precipitation and precipitation chemistry between two sites located close to each other. Regional extrapolations from local observations of atmospheric deposition should be made only with extensive knowledge of the many factors that influence Rocky Mountain climates.

References

Anlauf KG, Bottenheim JW, Brice KA, Weibe HA (1986) A comparison of summer and winter measurements of atmospheric nitrogen and sulphur compounds. Water Air Soil Pollut. 30:153–160.

Arthur MA (1990) The effects of vegetation on watershed biogeochemistry at Loch Vale Watershed, Rocky Mountain National Park, Colorado. Ph.D. Dissertation, Cornell University, Ithaca, NY.

Ashbaugh LL, Malm WC, Sadeh WZ (1985) A residence time probability analysis of sulfur concentrations at Grand Canyon National Park. Atmos. Environ. 19:1263–1270.

Baldwin JL (1973) Climates of the United States. National Oceanic Atmospheric Administration, Washington, D.C.

Baron J, Norton SA, Beeson DR, Herrmann R (1986) Sediment diatom and metal stratigraphy from Rocky Mountain lakes with special reference to atmospheric deposition. Can. J. Fish. Aquat. Sci. 43:1350–1362.

Bigelow DS, Sisterson DL, Schroder LJ (1989) An interpretation of differences between field and laboratory pH values reported by the National Atmospheric Deposition Program/National Trends Network monitoring program. Environ. Sci. Technol. 23:881–887.

Birkeland PW (1974) Pedology, weathering and geomorphological research. Oxford University Press, New York.

Blanchard CA (1991) Emissions that affect the arid west. In: Mangis D, Baron J, Stolte KA, eds. Acid Rain and Air Quality in Desert Parks. Air Quality Division Report NPS/NRAQD/NRTR-91/01, National Park Service, Washington, D.C.

BMDP (1985) Statistical Software. University of California, Berkeley, California.

Borys RD, Lowenthal DH, Rahn KA (1987) Contributions of smelters and other sources to pollution sulfate at a mountaintop site in Northwestern Colorado. In: Pielke RA, ed. Acid Deposition in Colorado—A Potential or Current Problem; Local versus Long-Distance Transport Into the State, pp. 167–174. Cooperative Institute for Research in the Atmosphere, Colorado State University, Fort Collins, Colorado.

Bresch JF, Reiter ER, Klitsch MA, Iyer HK, Malm WC, Gebhart K (1987) Origins of sulfur-laden air at national parks in the continental U.S. In: Bhardwaja PS, ed. APCA International Specialty Conference: Visibility Protection: Research and Policy Aspects, pp. 695–708, Jackson, Wyoming.

Cahill TA, Eldred RA, Feeney PJ (1987) Particulate monitoring and data analysis for the National Park Service, 1982–1985. Air Quality Group, University of California, Davis.

Charles DF, Norton SA (1986) Paleolimnological evidence for trends in atmospheric deposition of acids and metals. In: Gibson JH, ed. Acid Deposition: Long-Term Trends, pp. 335–434. National Academy Press, Washington, D.C.

Charlson RJ, Rodhe H (1982) Factors controlling the acidity of natural rainwater. Nature (London) 295:683–685.

Colorado Department of Health (1987) Emissions Inventory/Point Source Subsystem for the State of Colorado. Colorado Department of Health, Denver, Colorado.

Derwent RG, Nodop K (1986) Long-range transport and deposition of acidic nitrogen species in north-west Europe. Nature (London) 324:356–358.

Epstein CB, Oppenheimer M (1986) Empirical relation between sulphur dioxide emissions and acid deposition derived from monthly data. Nature (London) 323:245–247.

Galloway JN, Whelpdale DM, Wolff GT (1984) The flux of S and N eastward from North America. Atmos. Environ. 18:2595–2607.

Gibson JH (ed.) (1986) Acid Deposition: Long-Term Trends. National Academy Press. Washington, D.C.

Gillani NV (1984) Transport processes. In: Acidic Deposition Phenomenon and Its Effects. Critical Assessment Review Papers, Vol. 1. Atmospheric Science. EPA-600/8-83-016A, U.S. Environmental Protection Agency, Washington, D.C.

Hansen WR, Chronic J, Matelock J (1978) Climatography of the Front Range Urban

Corridor and Vicinity. Colorado Geological Survey Professional Paper No. 1019, U.S. Govt. Printing Office, Washington, D.C.

Heffter JL (1981) Air Resources Laboratories Atmospheric Transport and Dispersion Model. Technical memorandum ERL ARL-81, National Oceanic and Atmospheric Administration Washington, D.C.

Heubert BJ, Robert CH (1985) The dry deposition of nitric acid to grass. J. Geophys. Res. 90:2085–2090.

Hustedt F (1939) Systematische und ökologische Untersuchungen uber die Diatomeen-Flora von Java, Bali, und Sumatra nach dem Material der Deutschen Limnologischen Sunda-Expedition III. Die ökologischen Factorin und ihr Einfluss auf dieDiatomeenflora. Arch. Hydrobiol. Suppl. 16: 274–394.

Klein EJ (1989) The Variation in Wet Precipitation Chemistry with Elevation in Colorado. M.S. Thesis, Colorado State University, Fort Collins.

Knudson DA (1986) Estimated Monthly Emissions of Sulfur Dioxide and Oxides of Nitrogen for the 48 Contiguous States, 1975–1984. Vols. 1 and 2. ANL/EES-TM 318, Argonne National Laboratory, Argonne, Illinois.

Kohout EJ, Knudson DA, Saricks CL, Miller DJ (1987) Estimated Monthly Emissions of Sulfur Dioxide and Oxides of Nitrogen for the 48 Contiguous States, 1975–1984. Vol. 2: Sectoral Emissions by Month for States. ANL/EES-TM-318, Argonne National Laboratory, Argonne, Illinois.

LeGrand MR, Delmas RJ (1986) Relative contributions of tropospheric and stratospheric sources to nitrate in Antarctic snow. Tellus 3873:236–249.

Lewis CW, Baumgardner RE, Stevens RK (1986) Receptor modeling study of Denver winter haze. Environ. Sci. Technol. 20:1126–1136.

Lewis WM, Jr., Grant MC, Saunders JF (1984) Chemical patterns of bulk atmospheric deposition in the state of Colorado. Water Resour. Res. 20:1691–1704.

Litaor MI (1987) The influence of eolian dust on the genesis of alpine soils in the Front Range, Colorado. Soil Sci. Soc. Am. J. 51:142–147.

Logan J (1983) Nitrogen oxides in the troposphere: global and regional budgets. J. Geophys. Res. 88:10785–10808.

Lovett GM, Lindberg SE (1984) Dry deposition and canopy exchange in a mixed oak forest as determined by analysis of throughfall. J. Appl. Ecol. 21:1013–1027.

Mast MA (1989) A Laboratory and Field Study of Chemical Weathering with Special Reference to Acid Deposition. Ph.D. Dissertation, University of Wyoming, Laramie.

McLaughlin P (1988) The Effects of Storm Trajectory on Precipitation Chemistry in Rocky Mountain National Park. M.S. thesis, Colorado State University, Fort Collins.

NADP/NTN Data Base (1990) National Atmospheric Deposition Program. Tape of weekly data. National Atmospheric Deposition Program (IR-7)/National Trends Network. July 1978–January 1989. [Magnetic tape, 9 track, 1600 cpi, ASCII.] NADP/NTN Coordination Office, Natural Resource Ecology Laboratory, Colorado State University, Fort Collins, Colorado.

NAPAP (1989) 1987 Annual Report to the President and Congress. National Acid Precipitation Assessment Program, Washington, D.C.

Nochumson DH (1983) Regional air quality in the Four Corners area. APCA J. 33:670–677.

Nygaard G (1956) Ancient and recent flora of diatoms and Chrysophyceae in Lake Gribso. In: Berg K, Peterson IC, eds. Folia Limnol. Scand. 8:1–273.

Oppenheimer M, Epstein CB, Yuhnke RE (1985) Acid deposition, smelter emissions, and the linearity issue in the western United States. Science 229:859–862.

Parrish DD, Hahn CH, Fahey DW, Williams EJ, Bollinger MJ, Hubler G, Buhr MP, Murphy PC, Trainer M, Hsie EY, Liu SC, Fehsenfeld FC (1990) Systematic variations in the concentration of NO_x (NO plus NO_2) at Niwot Ridge, Colorado. J. geophys. Res. 95(D2):1817–1836.

Renberg I, Hellberg T (1982) The pH history of lakes in southwestern Sweden, as calculated from the subfossil diatom flora of the sediments. Ambio 11:30–33.
Sperber KR, Hameed S (1986) Rate of precipitation scavenging of nitrates in Central Long Island. J. Geophys. Res. 91(D11):11833–11839.
Texas Air Control Board (1990) County Emissions Data Base. Control Strategy Division, Texas.
Thorn CE, Darmody RG (1980) Contemporary eolian sediments in the alpine zone, Colorado Front Range. Phys. Geogr. 1/2:162–171.
Tukey HB (1970) The leaching of substances from plants. Annu. Rev. Plant Phys. 21:305–324.
Verry ES, Harris AR (1988) A description of low- and high-acid precipitation. Water Resour. Res. 24:481–492.
Wagner GH, Steele KF (1989) Na^+/Cl^- ratios in rain across the USA, 1982–1986. Tellus 41B:444–451.
Williams MW, Melack JM (1991) Precipitation chemistry and ionic loading to an alpine basin, Sierra Nevada. Water Resour. Res. 27:1563–1588.

5. Vegetation

Mary A. Arthur

Although only 18% of Loch Vale Watershed (LVWS) is vegetated, forest and meadows may be important sources or sinks for nutrient as well as nonnutrient ions. For example, forest growth can generate H ions in excess of incoming precipitation (Andersson et al., 1980; Sollins et al., 1980). Recently disturbed vegetation may exhibit increased rates of nitrification and consequent acidification of stream water (Bormann and Likens 1979), while areas with anaerobic soils (e.g., wet meadows) may serve as temporary sinks for acidity. In this chapter, the different types of vegetation in LVWS are described and the role of each type in the biogeochemistry of the watershed is considered. Particular attention is focused on the dynamics of the forest at LVWS, including biomass accumulation, annual production of above- and belowground components, detrital dynamics, nutrient uptake, and the role of disturbance.

The vegetation is typical of the subalpine-alpine vegetation found throughout the Rocky Mountains, from central Alberta and British Columbia to Arizona and especially in the central Rockies (Daubenmire 1978; Peet 1988). The subalpine Engelmann spruce–subalpine fir [*Picea engelmannii* (Parry)–*Abies lasiocarpa* (Hook.) Nutt.] forest occurs just below timberline (at approximately 3350 m in Rocky Mountain National Park; Wardle 1968), and grades into lodgepole pine [*Pinus contorta* var. *latifolia* (Engelm.)] stands below about 3000 m. These forests are not generally considered commercially important because of their inaccessibility and the low value of fir lumber;

together with associated wet sedge meadows they are an important component of National Forest and National Park lands in the Rockies, providing habitat for large and small mammals and recreation and scenic views for people. Above timberline, alpine tundra vegetation takes the form of low, mat-like plants that are able to withstand the harsh, desiccating conditions that prevent tree survival.

General Description of Vegetation Types

Alpine Tundra

Alpine tundra vegetation occupies about 11% of the watershed area. "Alpine tundra" refers to a high-mountain ecosystem in which the plant cover consists of low herbaceous, dwarf-shrub or lichen vegetation, and where timberline is broadly correlated with summer temperatures (Wardle 1968; Billings 1974; Daubenmire 1978). The transition from forest to tundra plants, the alpine timberline, is often marked by a dwarfing of the trees of the subalpine zone (krummholz). Although the position of timberline correlates with summer temperatures, the explanations given for the occurrence of a tree limit are varied and numerous. The most widely accepted explanation is the inability of shoots to withstand the effects of dry winter winds, resulting in desiccation. Engelmann spruce shoots growing above the snowpack are exposed to winds and exhibit winter desiccation and mortality (Hadley and Smith 1983, 1986). At LVWS, krummholz vegetation is restricted to two small areas along the two first-order streams, above 3300 m. In other areas of the watershed, tree growth apparently is limited instead by a combination of frequent snow avalanches and minimal soil development; in these areas, alpine vegetation may extend well below the usual timberline.

The alpine environment is cold and windy, with frequent freeze-thaw cycles at the soil surface. Microtopography affords protection to alpine plants, resulting in patchy vegetation distribution. Soil moisture is closely tied to local temperature gradients and microtopography (Billings 1974), and snowpack depth and distribution are very important in the alpine environment. Plant temperature may be raised significantly above that of ambient air because of leaf shape, size, color, pubescence, arrangement, and density (Billings 1974). Plant adaptations to this harsh environment include compact cushion growth forms, exhibited by the moss campion (*Silene acaulis*); the mat growth form, less compact but equally low-growing and including plants like alpine clover (*Trifolium dasyphyllum*); succulence, which provides protection against desiccation as in the yellow stonecrop (*sedum lanceolatum*); pubescence, exhibited in numerous species, that serves both to reduce the intensity of radiation and to reduce water loss; and vegetative reproduction (alpine bistort, *Polygonum viviparum*). In addition, Engelmann spruce may form islands of "cushion krummholz," with adventitious rooting from lower branches (Wardle 1968).

Numerous researchers have investigated the ecology of the alpine tundra in the central Rockies, and interested readers should turn for a more extensive discussion to primary sources, including Komarkova (1979), Billings (1974), Willard (1979), and Marr (1961).

Subalpine Meadows

Subalpine meadows occupy only 1% of the total watershed area at LVWS. They are typically interspersed with forest in areas of low topographic relief along streambanks and bases of slopes. The most likely causal factors for the absence of trees in such areas are excess soil moisture, high snow accumulation, and paludification (Peet 1988).

The two types of meadows at LVWS are distinguished by the dominant vegetation: wet sedge meadows with low species diversity, dominated by sedge (*Carex utriculata*) and arctic rush (*Juncus arcticus*); and somewhat drier, riparian meadows with much higher species diversity, including *Carex scopulorum, Caltha leptosepala* (marsh-marigold), *Trollius laxus* (globeflower), *Mertensia* sp. (bluebells), and *Sedum rhodanthum* (rose crown). Wet sedge meadows at LVWS are exemplified by one large meadow that appears to be a filled-in lake. In this meadow water remains at or above the surface throughout the snow-free period, and trees encroach only at the edges. The riparian meadows are more extensive and occur primarily in association with streams. The boundary between forest and meadow vegetation in riparian areas is gradual, and islands of trees may occur where topography rises slightly amidst otherwise well-developed meadows. These meadow soils remain wet throughout the snow-free period, but pooled water at the surface gradually dissipates as snowmelt ends. Thick organic soils (see Chapter 7, this volume) and proximity to streams may result in a disproportionate influence of meadow soils on surface water chemistry.

Forest

The Engelmann spruce-subalpine fir forests of the central Rockies occur in the subalpine zone. Above 3100 m, and down to 2500 m in cool valleys protected from frequent fires, these two species form the climax vegetation (Peet 1981). Six community types of Engelmann spruce-subalpine fir were identified by Peet (1981): montane, bog, wet, mesic, xeric, and subalpine. The stands at LVWS are representative of mesic, xeric, and, to a minor extent, bog sites. The mesic sites occur in cool, sheltered, well-drained areas, and the canopy is exclusively dominated by spruce and fir. Xeric sites occur on open slopes and occasionally include individuals of limber pine [*Pinus flexilis* (James)]. Tree diameters and basal area are lower on xeric than mesic sites, as expected (Peet 1981). Spruce-fir islands of the bog type occur as a mosaic associated with meadows situated in areas of low topographic relief.

Forest development is extremely heterogeneous as the result of variability in topography, soil development, and the extent and frequency of disturbance.

Table 5.1. Selected Stand Characteristics of the Forested Zone[a]

Species	Mean Density (stems ha^{-1})	Mean Basal Area ($m^2 ha^{-1}$)	Mean Basal Area (range)	Canopy height (m)	Maximum dbh (cm)
Engelmann spruce	290	21.1	(8.1–53.3)	17–25	68
Subalpine fir	610	17.9	(8.3–26.9)	15–20	60
Limber pine	8	1.2	(0–13.3)	n.d.	47
Total	908	40.2	(16.4–72.9)		

[a] Density and mean basal area values are for all stems > 10 cm dbh, based on 20 plots, 0.04 ha each; n.d., not done. (From Arthur and Fahey 1990; reprinted with permission.)

Old-growth Engelmann spruce-subalpine fir forest, with the oldest and largest trees in the watershed, is located along the central drainages in areas with relatively deep soils, low relief, and less probability of large-scale disturbance. On rocky side-slopes, soils are generally thin, bedrock and talus are often exposed, and the forest is poorly developed. Individual trees are smaller on these slopes, and stand basal area is lower. The spruce-fir forests of Rocky Mountain National Park were never logged (Peet 1981) and in that sense may be considered old growth; however, this does not imply large stature.

Mean basal area of the forest was $40\,m^2\,ha^{-1}$ (range, 16.4–72.9), with slightly more basal area of spruce than fir (Table 5.1), and a very few limber pines confined to rocky outcroppings. The height of the canopy ranged from 17 to 25 m for spruce and from 15 to 20 m for fir. Maximum tree diameters were 68 and 69 cm for spruce and fir, respectively. Maximum age of the dominant trees was greater than 500 years for spruce, and 250 years for fir (Arthur 1990). There were many more individuals of fir than spruce, although in a declining ratio with increasing size, indicating greater recruitment of fir (Figure 5.1). Greater numbers of fir seedlings and saplings are apparently balanced by greater mortality, with ratios of fir to spruce of 2:1 for standing dead boles and 7:1 for decay class I dead boles, (the least decayed lying boles) respectively (Table 5.2). The most common pattern of spruce and fir coexistence in the Rocky Mountains consists of high spruce basal area, associated with many large individuals and high numbers of fir seedlings and saplings (Peet 1988). The forest at LVWS differs from this in that spruce basal area was only slightly higher than fir, although there were twice as many fir as spruce stems greater than 10 cm dbh (see Table 5.1). Veblen (1986) attributed the patterns of Engelmann spruce and subalpine fir coexistence to the greater longevity of spruce and greater abundance of fir and suggested that spruce-fir forests in the Colorado Front Range are near equilibrium and forest structure will remain relatively constant. Aplet et al. (1988) suggested that, although equilibrium biomass levels may be maintained in Engelmann spruce-subalpine fir forests over long periods of time (several hundred years), there is a shift in dominance from fir to spruce with time. These interpretations

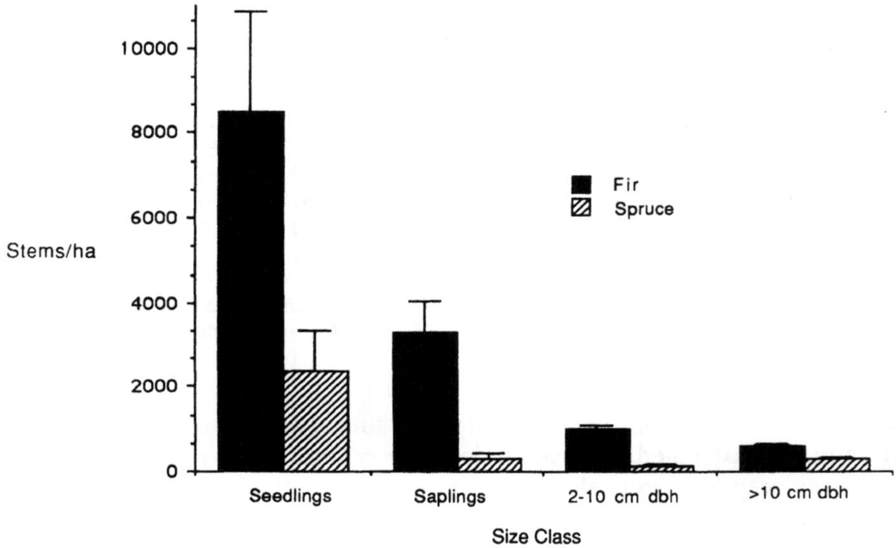

Figure 5.1. Stems per hectare of subalpine fir and Engelmann spruce in Loch Vale Watershed by size class: seedlings (stems < 50 cm height) saplings (> 50 cm height and < 2 cm dbh), 2–10 cm dbh, and > 10 cm dbh. Error bars are standard error.

Table 5.2. Live Stems and Dead Boles (stems ha^{-1}) of Spruce and Fir, and Ratio of Fir to Spruce, in Loch Vale Watershed[a]

	Stems ha^{-1}		Fir/Spruce
	Spruce	Fir	Ratio
Stems > 10 cm dbh	290 (47)	610 (59)	2.1
Stems 2–10 cm dbh	148 (31)	1000 (100)	6.8
Saplings	310 (110)	3300 (720)	10.6
Seedlings	2375 (970)	8500 (2350)	3.4
Standing dead	48 (11)	100 (22)	2.1
Decay Class I	8.8 (3.7)	60 (17)	6.8

[a] All values are based on 20 plots, 0.04 ha each. Saplings were defined as stems > 50 cm in height and < 2 cm dbh. Seedlings were defined as stems < 50 cm in height. Standard errors are in parentheses. (From Arthur and Fahey 1990; reprinted with permission.)

differ in their assessments of the relative ecological roles of spruce and fir, but both authors concur that mixed stands of spruce and fir are successional and that one or the other species will eventually dominate. The relative dominance of each species would reflect the successional status of a stand in terms of both stem density and basal area. The forest at LVWS may be considered a mosaic of stands with varying positions along a postdisturbance continuum.

Understory vegetation in this forest is poorly developed, consisting of fir regeneration occasionally complemented by spruce and by extensive coverage of *Vaccinium myrtillus* (myrtleleaf blueberry). Such species-poor understory vegetation is common in subalpine Engelmann spruce-subalpine fir forests across the Rockies (Peet 1988). Common herbaceous species in the forest were *Arnica cordifolia* (heartleaf arnica), *Erigeron peregrinus* (subalpine daisy), and *Polemonium delicatum* (subalpine Jacobs ladder).

Disturbance

Rocky Mountain forests have been described as "disturbance forests," nearly all of which are in "some stage of recovery from prior disturbance" (Peet 1988), and the Engelmann spruce-subalpine fir forests of LVWS are no exception. It is useful to think of the forest at LVWS as a "disturbance forest," both in a very long-term sense (on the order of hundreds to thousands of years) and in the shorter term, decades to centuries. Neoglacial advances of cirque glaciers have occurred three times since the Pinedale glaciation, leaving glacial deposits between 3300 and 3750 m. Forest vegetation at LVWS may have been present only since the end of the first and most extensive neoglacial advance, the Triple Lakes advance, 3000 to 5000 years B.P. (Madole 1976), or even more recently following later advances. Viewed from this perspective, the entire forest could be considered a "disturbance forest," having developed relatively recently in an environment that was probably cooler than the present environment. Viewed from the short-term perspective, disturbance (both small- and large-scale) is an ongoing factor in forest development and structure.

In areas of well-developed forest along the central drainages of the watershed, the predominant types of disturbance are single treefalls resulting from lightning strikes and wind. These areas, with the oldest and most mature trees on the watershed, may be particularly susceptible to blowdowns on a scale larger than single treefalls, because the old-aged trees are susceptible to breakage. Peet (1981) considered large blowdowns a common form of disturbance in spruce-fir forests that may result in the release of the large number of suppressed fir seedlings and saplings in the understory, thereby giving fir a head start in stand regeneration.

Snow avalanches are probably the most extensive and destructive type of disturbance along the valley walls at LVWS. An avalanche completely felled one study plot in the spring of 1986. That same year, another stand (approximately 0.1 ha) was also felled. The stands in these areas were 80 and 250 years old, respectively, indicating high variability in the return time of this type of disturbance. Stand regeneration after an avalanche disturbance should follow a pattern similar to that following a blowdown because the smaller trees are unbroken in an avalanche and the soil is relatively undisturbed (Johnson 1987).

Fire is an infrequent but potentially extensive type of disturbance to the subalpine forest. Despite the limited evidence of fire in this watershed, extensive and devastating fires have been documented in this forest type. A fire return time of 150–400 years appears common in subalpine forests of the Rockies (Heinselman 1981; Peet 1981; Romme and Knight 1981). Fire has occurred only in one small area (< 2 ha) of LVWS, and it destroyed all trees in the affected area about 80 years ago; the new stand is a dense, even-aged forest of both spruce and fir.

Insect pests, such as the Engelmann spruce beetle (*Dendroctonus engelmannii* Hopk.), are capable of destroying vast areas of spruce-fir forest. In the 1940s nearly all of the Engelmann spruce and most of the subalpine fir greater than 10 cm dbh were killed over thousands of hectares in the White River Plateau in northwestern Colorado (Miller 1970). Although the forest at LVWS may be protected to some degree by topographic barriers separating it from other forested areas, it is probably not entirely immune from such disturbance.

Biogeochemistry and Vegetation Processes

Although only 6% of the watershed supports woody plants, forest plays a crucial role in the biogeochemistry of the watershed. Little is known about nutrient cycling in subalpine meadows and alpine tundra, but certainly these areas play a role in watershed biogeochemistry as well.

Alpine Tundra

Annual aboveground biomass production on Niwot Ridge, Colorado, ranged from 100 to 300 $g\,m^{-2}\,yr^{-1}$ (Webber and May 1977). Production rates in tundra areas of LVWS are probably much lower, due to shallower soils and steeper slopes (Walthall 1985). Much greater production and accumulation in belowground than aboveground components is typical of tundra ecosystems, and on Niwot Ridge aboveground-to-belowground biomass ratios average 1:17 over seven types of sites, ranging from 1:3 to 1:25 (Webber and May 1977). Researchers have attributed low primary production in tundra communities to low levels of available nutrients, especially N and P (Haag 1974; Chapin 1980). Despite large accumulations of dead organic matter and nutrients, this pool is generally unavailable to plants, resulting in strong internal plant controls on nutrient recycling (Bunnell et al. 1975). The low availability of soil nutrient pool is the result of low soil temperatures, which limit organic decomposition and microbial transformations of nutrients, especially N (Haag 1974).

Interaction between alpine soils and surface waters at LVWS occurs primarily during snowmelt, when most of the annual precipitation (approximately 70%) flows through the system during a relatively short period (see Chapter 3, this volume). Alpine soils are often frozen during this time, so their influence on the biogeochemistry of the watershed is limited. A comparison of the chemistry of waters flowing from alpine and subalpine soils on Niwot Ridge (Litaor 1988) revealed much lower (two-to threefold) concentrations of dissolved organic carbon and total alkalinity in alpine soil solutions than in solutions from subalpine forest soils, suggesting a more limited contribution to surface water chemistry from alpine soils than from forested soils.

Subalpine Meadows

Meadow soils are generally located along streambanks, and much of the water flowing from forest soils to streams must flow through meadow soils. The interaction of snowmelt water with meadow soils may be minimized during peak snowmelt, when much of this water flows over saturated soils. However, because meadows are frequently situated in areas of low topography, meltwater in some meadow areas may have ample opportunity to interact with soils before flowing into stream channels.

Samples of solutions were collected from meadow soils, but because they equilibrated with the atmosphere during collection, some oxidation of reduced compounds probably occurred and possibly changed the chemistry, especially for SO_4, pH, and alkalinity (Arthur 1990). However, it is apparent that the meadow soils altered the solutions because they have much lower concentrations of K, Mg, and Ca, as well as major anions and dissolved organic carbon, than the forest soils. Total alkalinity of soil solutions is twice as high in meadows as in the forest despite the acidifying effects of sample oxidation, indicating that meadow soils may provide an important source of acid-neutralizing capacity (ANC). Generation of ANC in meadow soils probably results from reduction of SO_4 in these anaerobic soils. Although annual aboveground biomass production in the meadows was only half as high as in the forest (200 compared to 440 $g\,m^{-2}\,yr^{-1}$) on an area basis, annual root uptake of nutrients in the meadows was similar to that in the forest. Thus, the meadows may be important in regulating surface water chemistry through the activity of vegetation and biogeochemical transformations.

Forest

Biomass Accumulation

Total forest biomass (living biomass, forest floor, and dead bole biomass) averaged 29,000 $g\,m^{-2}$ in twenty 0.04-ha plots, with average stand density of 900 stems ha^{-1} (stems > 10 cm dbh; Arthur 1990). Living biomass accounts for about 52%, whereas detritus accounts for the remaining 48% of forest biomass (Table 5.3; Arthur and Fahey 1990). Compared to mean values from 13 temperate coniferous and 3 boreal coniferous sites examined in the International Biosphere Program (Cole and Rapp 1981), total organic matter accumulation at LVWS is less than half the average for temperate coniferous forests (61,800 $g\,m^{-2}$), and only slightly higher than the average for boreal

Table 5.3. Percent of Biomass and Nutrients in Living and Detrital Biomass at Loch Vale Watershed[a]

Component	Biomass	N	P	Ca	K	Mg
Aboveground	43	26	32	38	42	28
Belowground	9	11	8	7	14	7
Total living	52	37	40	45	56	35
Forest floor	24	56	55	43	18	48
Dead wood	24	7	5	12	26	17
Total detrital	48	63	60	55	44	65

[a] (From Arthur 1990.)

forests (22,600 g m^{-2}). The proportion of aboveground tree biomass to total ecosystem biomass at LVWS (0.43) is low relative to other temperate coniferous forests (0.50), but higher than that in boreal coniferous forests (0.23; Cole and Rapp 1981), suggesting that rates of biomass production and decomposition at LVWS are lower than in most temperate coniferous forests (Arthur 1990).

Although the biomass of dead boles at LVWS (7,000 g m^{-2}) is low compared to other coniferous forests (range, 1,000–51,100 g m^{-2}; Harmon et al. 1986), dead bole biomass composes a comparatively high percentage of total aboveground biomass: Harmon et al. (1986) reported a range in the ratio of dead wood biomass to total aboveground biomass of 1.3% to 45.1%, compared to 34% at LVWS (Arthur and Fahey 1990). This relatively high percentage of dead bole biomass in LVWS forests may be attributed to several factors, including the general observation that high accumulations of deadwood frequently are associated with very old stands (Spies et al. 1988). Also, low rates of decomposition resulting from cold temperatures in this subalpine ecosystem would maximize deadwood accumulation as would the absence of fire in recent history. Finally, the repeated disturbances such as wind and snow avalanches that are common in this environment tend to increase the number of dead boles (Arthur and Fahey 1990).

Belowground biomass is 18% of total living biomass, and live plus dead fine roots (< 2 mm diameter) comprise 32% of total belowground biomass (Arthur 1990). In lodgepole pine stands in southeastern Wyoming, Pearson et al. (1984) reported an average percentage of belowground to total living biomass of 26%, however, only 23% of belowground biomass is fine roots at that site.

Nutrient Pools and Forest Floor Residence Time

With the exception of K, more nutrients are stored in detrital than in living biomass (see Table 5.3). Forest floor and dead wood comprise nearly equal proportions of total forest biomass, 23% and 24%, respectively, but the forest floor contains considerably larger amounts of nutrients than dead boles, again with the exception of K (Table 5.4). This observation reflects the low nutrient concentrations in bole material, and the high concentrations of forest floor, which is composed primarily of decomposing foliage, twigs, and cones (Arthur and Fahey 1990).

Forest floor residence time (forest floor accumulation divided by annual litterfall) at LVWS averages 39 years (Table 5.5), but between-plot variability is extremely high. Residence times of N, P, and Mg in the forest floor are longer than organic matter residence times, whereas those for K and Ca are slightly shorter (Arthur 1990). Immobilization of N, P, and, to a limited extent, Mg by decomposer microorganisms apparently extends the residence time of these nutrients. Vogt et al. (1986) observed that residence times of N and P in the forest floor are larger than for C in cold temperate and boreal forests because of nutrient immobilization by decomposer microorganisms.

Table 5.4. Biomass and Nutrient Pools in Loch Vale Watershed[a]

Component	Biomass	N	P	Ca	K	Mg
Total aboveground	12500	36.2	3.1	45.9	18.8	4.4
	(4140)	(10.1)	(1.1)	(15.7)	(8.3)	(1.6)
Total belowground	2680	15.3	0.8	8.2	6.0	1.1
	(960)	(7.4)	(0.4)	(4.0)	(2.6)	(0.7)
Total forest floor	6770	76.5	5.4	51.1	7.9	7.6
	(3550)	(35.4)	(2.6)	(23.4)	(4.2)	(4.2)
Total deadwood	7040	8.9	0.5	15.1	11.8	2.8
	(5960)	(8.8)	(0.6)	(25.9)	(6.3)	(2.7)
Ecosystem total	29000	136.0	9.8	120.9	45.0	16.0
	(8140)	(39)	(2.9)	(38.0)	(12.0)	(5.0)

[a] Data are in grams per meter squared ($g\,m^{-2}$). Standard deviations are in parentheses. (From Arthur 1990.)

Table 5.5. Biomass and Nutrients in Forest Floor and Annual Litterfall, and Estimated Turnover Rates[a]

Component	Biomass	N	P	Ca	K	Mg
Forest floor ($g\,m^{-2}$)	6770	76.5	5.4	51.1	7.9	7.6
Annual litterfall ($g\,m^{-2}\,yr^{-1}$)	174	0.89	0.09	2.01	0.29	0.11
Turnover time (years)	39	86	60	25	27	69

[a] (From Arthur 1990).

Net Primary Production

In general, forests at LVWS are very slow growing, but there is high spatial heterogeneity. For example, annual bole biomass accumulation in twenty 0.04-ha plots ranged from 40 to 150 $g\,m^{-2}\,yr^{-1}$ (Arthur 1990). Most of this heterogeneity is apparently related to topographic influences on soil development and disturbance, as well as to disturbance history. A considerable portion of the forested area, resembling the mesic community type, is located along the central drainage of LVWS where topographic relief is low and soils are relatively deep. The forest in this area is typical of old-growth forest: an uneven-aged stand with some very large and very old inviduals and low rates of bole biomass production (92 $g\,m^{-2}\,yr^{-1}$; Table 5.6; Figure 5.2). A small area of the forest, burned approximately 80 years ago, is an even-aged stand with the highest rate of bole biomass production in the watershed (110 $g\,m^{-2}\,yr^{-1}$; Figure 5.3). The remaining forested areas in LVWS are located on hillslopes on areas with exposed bedrock, thin soils, and frequent disturbance from avalanches and resemble the Xeric community type; these areas have low basal area and very low bole biomass production (61 $g\,m^{-2}\,yr^{-1}$; Arthur 1990).

Total net primary productivity (above- and belowground) averages 440 $g\,m^{-2}\,yr^{-1}$ (Table 5.7; Arthur 1990). Aboveground net primary production at LVWS (289 $g\,m^{-2}\,yr^{-1}$) is low relative to average values for temperate coniferous forests (835 $g\,m^{-2}\,yr^{-1}$), but higher than boreal coniferous forests (121 $g\,m^{-2}\,yr^{-1}$; Cole and Rapp 1981), and again reflects the fact that the forest at LVWS is exposed to climatic conditions intermediate between temperate and boreal because of its subalpine location.

The belowground component can be a large proportion of ecosystem biomass and production. Total mass of fine roots at LVWS is 870 $g\,m^{-2}$ (Arthur 1990), similar to the average fine root mass for both cold temperate and boreal needleleaf evergreen forests reported in an extensive review by Vogt et al. (1986). The proportion of live to dead fine root mass at LVWS is estimated at 0.22, compared to 0.24 for boreal conifer forests and approximately 0.50 for cold temperate conifer forests (Vogt et al. 1986). However, 0.22 is probably an underestimate of the proportion of live to total fine root

Table 5.6. Bole Biomass Increment and Estimated Annual Bole Mortality in Three Spruce-Fir Stand Types at Loch Vale Watershed[a]

Stand Type	NPP	Mortality
Mesic	92	50
Xeric	61	84
80-year-old stand, post fire	110	17

[a] Data are $M\,g\,m^{-2}\,ha^{-1}$. (From Arthur 1990.)

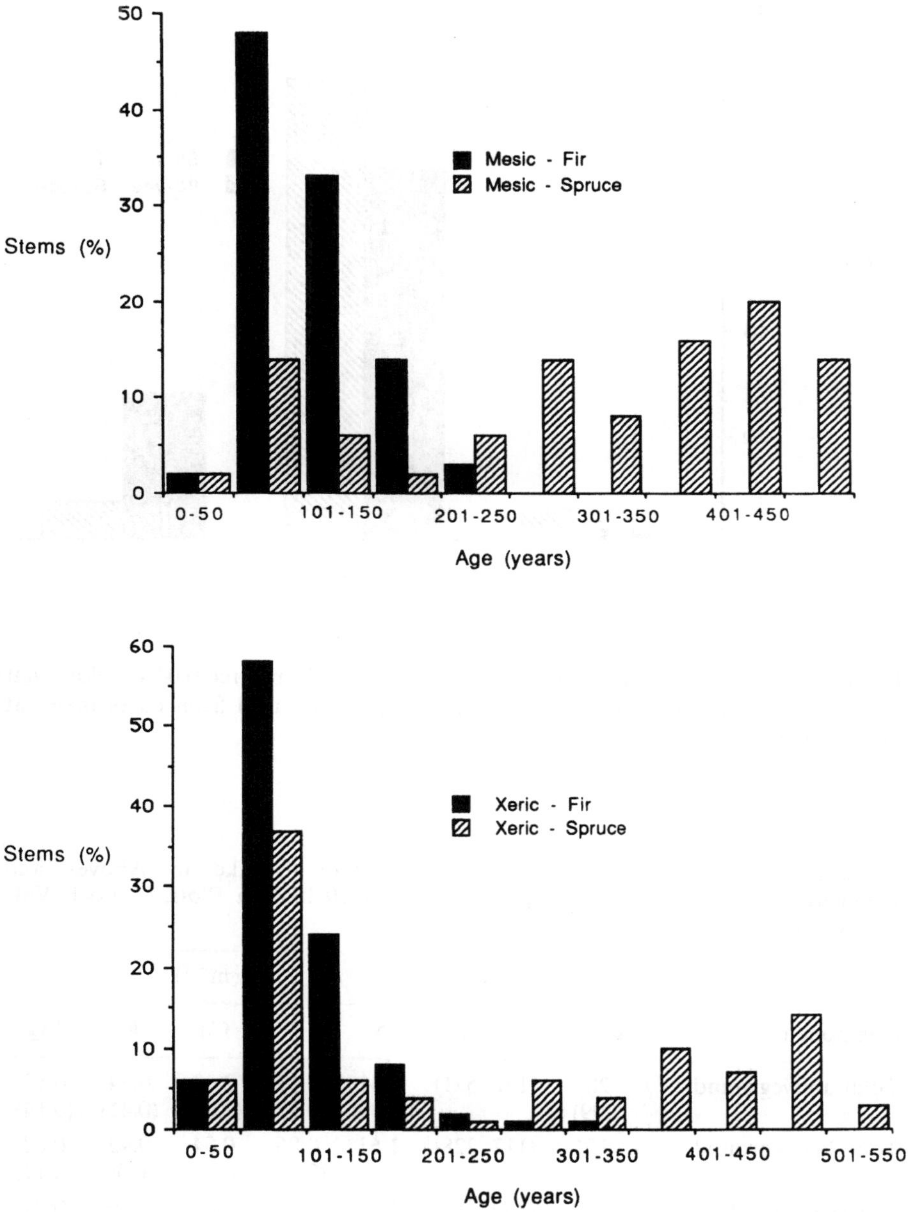

Figure 5.2. Age class distributions for stems > 10 cm dbh in 4 mesic (top) and 12 xeric (bottom) 0.04-ha plots in Loch Vale Watershed. Trees were aged from cores taken at breast height.

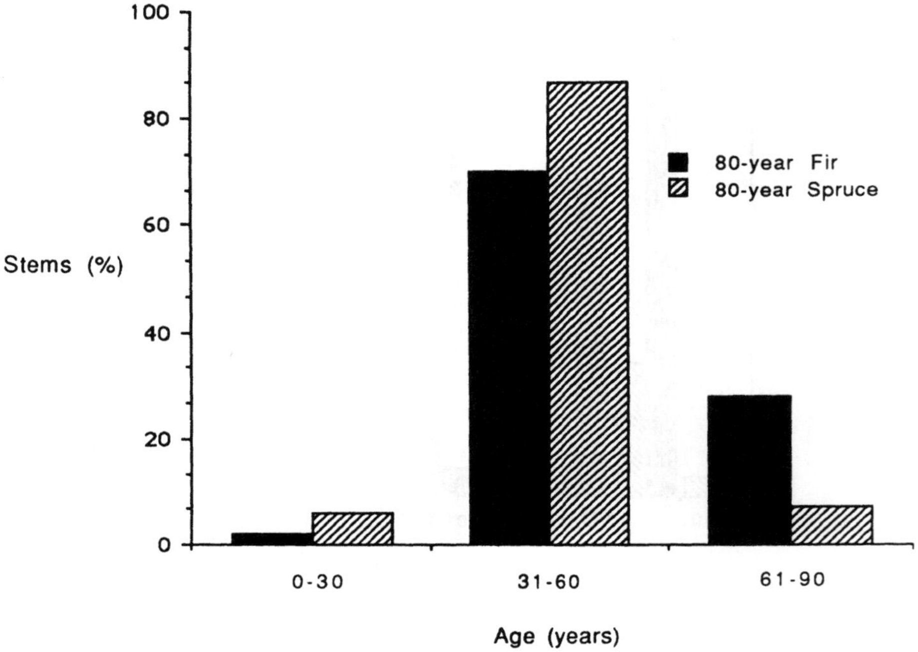

Figure 5.3. Age class distribution for stems > 10 cm dbh in three 0.04-ha plots that were disturbed approximately 80 years ago. Trees were aged from cores taken at breast height.

Table 5.7. Annual Biomass Production and Root Uptake in Above- and Belowground Biomass, and Canopy Leaching for 20 0.04-ha Plots, in Loch Vale Watershed[a]

	Biomass		$(g\,m^{-2})$				
Component	$(g\,m^{-2})$	(range)	N	P	Ca	K	Mg
Total aboveground	289 (99)	(182–521)	1.41 (1.39)	0.13 (0.12)	2.30 (2.05)	0.54 (0.41)	0.19 (0.14)
Total belowground	151 (43)	(117–228)	1.63 (1.85)	0.06 (0.07)	0.54 (0.63)	0.42 (0.39	0.08 (0.11)
Ecosystem total	440 (108)	(300–580)	3.04 (2.31)	0.19 (0.23)	2.88 (2.14)	0.96 (0.57)	0.27 (0.18)
Canopy leaching			Negative	Negative	0.04	0.45	0.04
Total root uptake			3.04 (2.31)	0.19 (0.23)	2.92 (2.89)	1.41 (0.57)	0.31 (0.28)

[a] Standard deviations are in parentheses. (From Arthur 1990).

biomass because of delays as long as 3 months before fine root separation from soil cores. Annual fine root production at LVWS (150 g m^{-2}; Arthur 1990), estimated using an N budgeting method (Aber et al. 1985; Fahey et al. 1985), is much lower than the average for cold temperate coniferous forests (615 g m^{-2}) but similar to that for boreal forests (100 g m^{-2}; Vogt et al. 1986).

Nitrogen mineralization studies combined with root core data indicated that annual fine root production was half of the total amount of live fine roots, suggesting a fine root turnover time of roughly 2 years. A fine root turnover time of slower than once per year is corroborated by the absence of any dead fine roots in root screens that remained in place 1.5 years (Arthur 1990). Because live fine root biomass probably was underestimated, 2 years should be considered a minimum turnover time for fine roots. Here again, the cool subalpine environment likely limits fine root production and decomposition rates, resulting in lower fine root production and lower ratios of live to dead fine root biomass. Fine root production represents a major proportion of net primary production (NPP) (34%) and is nearly equal to foliar production (174 g m^{-2} yr^{-1}). Together, these two components account for 74% of NPP (Arthur 1990).

Root Uptake

As a measure of how much root uptake activity occurs on an annual basis, annual root uptake (accounting for nutrient resorption before leaf fall) is compared to precipitation inputs and streamwater outputs (Table 5.8). For example, annual root uptake of Ca, Mg, K, and N are 15, 9, 54, and 10 times greater, respectively, on an areal basis than precipitation inputs. Of course many of the nutrients for root uptake are supplied by mineral weathering and recycling of nutrients among the biotic compartments, including decomposition of forest floor material and fine roots and canopy leaching.

As noted previously, there is very high spatial heterogeneity in forest growth and rates of tree mortality in LVWS (see Table 5.6), as well as in soil development and hydrology. Thus, the biogeochemical effects of nutrient

Table 5.8. Estimated Annual Nutrient Uptake (mean of 20 0.04-ha plots), Precipitation Input, and Streamwater Output of Major Nutrients in Loch Vale Watershed, Colorado[a]

	Ca	mg	K	NH_4-N	NO_3-N
Annual nutrient uptake	732	131	376	—	2255
Precipitation input	50	15	7	81	139
Streamwater output	169	47	25	7	95

[a] Data are in moles ha^{-1} yr^{-1}. (From Arthur 1990.)

uptake are expected to vary considerably among forest types within a single waershed. Areas with low biomass accumulation and high mortality are offset by areas that are accumulating biomass relatively rapidly, such as the area disturbed by fire 80 years ago. In addition, any large-scale perturbations within the forest that greatly alter biomass accumulation rates or decomposition could be expected to have an impact on the entire watershed.

Disturbance

Immediately after disturbance, interruption of root uptake and increased rates of decomposition can increase loss of nutrients from the watershed, whereas rapid regrowth may cause increased soil acidification resulting from root uptake of nutrient cations in excess of decomposition. Following a major avalanche in 1986, significant changes in soil solution chemistry and nitrification rates were observed during 1987 and 1988. Concentrations of K and Cl increased by twofold in leaching solutions. Concentrations of NO_3 and NH_4 were much higher (two to seven times) than in soil solutions from intact forest areas. In addition, N mineralization rates increased by sixfold in the disturbed forest, and nitrification by more than an order of magnitude (Arthur 1990). Immediately following large-scale natural disturbances, there is the potential for resultant changes in the cycling of nutrients within the disturbed area. A study of the effects of tree removal on biogeochemistry at nearby Fraser Experimental Forest, for example, demonstrated a pronounced increase in NO_3 in drainage water (Stottlemyer 1987). Although the effects of disturbance at LVWS may not be measurable as changes in streamwater chemistry because of intervening intact ecosystems, a disturbance affecting most or all the forested portions of the watershed would undoubtedly have a large impact.

The regrowth of vegetation gradually exerts control over the cycling of nutrients, slowing losses of nutrients to stream water and reducing the pool of N available to nitrifying bacteria. Depending on the peak rate of NPP during regrowth, nutrient uptake could greatly exceed nutrient mineralization from detritus and perhaps from soil weathering, so that the large quantities of H released are not entirely neutralized and short-term acidification occurs (Nilsson et al. 1982). Several pieces of information are required to determine whether acidification of streams and lakes could follow 20 to 40 years of biomass accumulation following a disturbance. These include (1) the rate of nutrient accumulation in living biomass during rapid regrowth as the stand becomes reestablished; (2) the rate of detrital release of nutrients; and (3) the maximum rate of mineral weathering. If nutrient accumulation in living biomass exceeds the sum of detrital nutrient release and mineral weathering, acidification of surface waters could result. Disturbance effects on hydrology would also influence watershed biogeochemistry; for example, greater snow capture in disturbed areas, combined with reduced transpiration, could result in an increase in leaching losses of nutrients. In addition, spatial heterogeneity in growth rates, site fertility, and mineral buffering confounds the effects of forest regrowth on watershed biogeochemistry following disturbance. For example, on rocky side-slopes where growth rates are low compared to the valley floor, the acidifying effects of forest regrowth following disturbance may be minimal. Thus, predictions pertaining to the effects of large-scale disturbance, and recovery from disturbance, on watershed biogeochemistry must incorporate these differences between sites within a watershed.

References

Aber JD, Melillo JM, Nadelhoffer KJ, McClaugherty CA, Pastor J (1985) Fine root turnover in forest ecosystems in relation to quantity and form of nitrogen availability: a comparison of two methods. Oecologia 66:317–321.

Andersson F, Fagerstrom T, Nilsson I (1980) Forest ecosystem responses to acid deposition—H-ion budget and N/tree growth model approaches. In: Hutchinson TC, Havas M, eds. Effects of Acid Precipitation on Terrestrial Ecosystems, pp. 319–334. Plenum, New York.

Aplet GH, Smith FW, Laven RD (1989) Stemwood biomass and production during spruce-fir stand development. J. Ecol. 77:70–77.

Arthur MA (1990) The Effects of Vegetation on Watershed Biogeochemistry at Loch Vale Watershed, Rocky Mountain National Park, Colorado. Ph.D. Dissertation, Cornell University, Ithaca, New York.

Arthur MA, Fahey TJ (1990) Mass and nutrient content of decaying boles in an Engelmann spruce/subalpine fir forest, Rocky Mountain National Park, Colorado. Can. J. For. Res. 20:730–737.

Billings WD (1974) Arctic and alpine vegetation: plant adaptations to cold summer climates. In: Ives JD, Barry RG, eds. Arctic and Alpine Environments, pp. 403–444. Methuen & Co., London.

Bormann FH, Likens GE (1979) Pattern and Process in a Forested Ecosystem. Springer-Verlag, New York.

Bunnell FL, Maclean SF, Jr., Brown J (1975) Barrow, Alaska, U.S.A. In: Rosswall T, Heal OW, eds. Structure and Function of Tundra Ecosystems. Ecol. Bull. (Stockholm) 20:73–124.

Chapin, III FS (1980) Nutrient allocation and responses to defoliation in tundra plants. Arct. Alp. Res. 12:553–563.

Cole DW, Rapp, M (1981) Elemental cycling in forest ecosystems. In: Reichle DE, ed. Dynamic Properties of Forest Ecosystems, pp. 341–409. International Biological Programme 23, Cambridge University Press, Malta.

Daubenmire R (1978) Plant Geography, pp. 66–118. Academic Press, New York.

Fahey TJ, Yavitt JB, Pearson JA, Knight DH (1985) The nitrogen cycle in lodgepole pine forests, southeastern Wyoming. Biogeochemistry 1:257–278.

Haag RW (1974) Nutrient limitations to plant production in two tundra communities. Can. J. Bot. 52:103–116.

Hadley JL, Smith WK (1983) Influence of wind exposure on needle desiccation and mortality for timberline conifers in Wyoming, U.S.A. Arct. Alp. Res. 15:127–135.

Hadley JL, Smith WK (1986) Wind effects on needles of timberline conifers: seasonal influence on mortality. Ecology 67:12–19.

Harmon ME, Franklin JF, Swanson FJ, Sollins P, Gregory SV, Lattin JD, Anderson NH, Cline SP, Aumen NG, Sedell JR, Lienkamper GW, Cromack K. Jr., Cummins JW (1986) Ecology of coarse woody debris in temperate ecosystems. Adv. Ecol. Res. 15:133–302.

Heinselman ML (1981) Fire intensity and frequency as factors in the distribution and structure of northern ecosystems. In: Mooney HA, Bonicksen TM, Christensen NL, Lotan JE, Reiners WA, eds. Fire Regimes and Ecosystem Properties, pp. 7–57. USDA Forest Service Gen. Tech. Rep. WO-26.

Johnson EA (1987) The relative importance of snow avalanche disturbance and thinning on canopy plant populations. Ecology 68:43:53.

Komarkova V (1979) Alpine Vegetation of the Indian Peaks Area, Front Range, Colorado Rocky Mountains. J. Cramer, Vaduz.

Litaor MI (1988) Soil solution chemistry in an alpine watershed, Front Range, Colorado, U.S.A. Arct. Alp. Res. 20:485–491.

Madole RF (1976) Glacial geology of the Front Range, Colorado. In: Mahaney WC, ed. Quaternary stratigraphy of North America, pp. 319–351. Dowden, Hutchinson & Ross, Stroudsburg, Pennsylvania.

Marr JW (1961) Ecosystems of the East Slope of the Front Range in Colorado. University of Colorado Stud. Ser. Biol. 8, Boulder, Colorado.

Miller PC (1970) Age distributions of spruce and fir in beetle-killed forests on the White River Plateau, Colorado. Am. Midl. Nat. 83:206–212.

Nilsson SI, Miller HG, Miller JD (1982) Forest growth as a possible cause of soil and water acidification: an examination of the concepts. Oikos 39:40–49.

Pearson JA, Fahey TJ, Knight DH (1984) Biomass and leaf area in contrasting lodgepole pine forests. Can. J. For. Res. 14:259–265.

Peet RK (1981) Forest vegetation of the Colorado Front Range. Vegetatio 45:3–75.

Peet RK (1988) Forests of the Rocky Mountains. In: Barbour MG, Billings WD, eds. North American Terrestrial Vegetation, pp. 63–102. Cambridge University Press, Cambridge, Massachusetts.

Romme WH, Knight DH (1981) Fire frequency and subalpine forest succession along a topographic gradient in Wyoming. Ecology 62:319–326.

Sollins P, Grier CC, McCorison FM, Cromack, Jr. K, Fogel R (1980) The internal element cycles of an old-growth Douglas-fir ecosystem in W. Oregon. Ecol. Monogr. 50:261–285.

Spies TA, Franklin JF, Thomas TB (1988) Coarse woody debris in Douglas-fir forests of western Oregon and Washington. Ecology 69:1689–1702.

Stottlemyer R (1987) Natural and anthropic factors as determinants of long-term streamwater chemistry. In: Management of Subalpine Forests: Building on 50 Years of Research, Proceedings of a Technical Conference. USFS Gen. Tech. Rep. RM-149.

Veblen TT (1986) Age and size structure of subalpine forest in the Colorado Front Range. Bull. Torrey Bot. Club 113:225–240.

Vogt KA, Grier CC, Vogt DJ (1986) Production, turnover, and nutrient dynamics of above- and belowground detritus of world forests. Adv. Ecol. Res. 15:303–377.

Walthall PM (1985) Acidic Deposition and the Soil Environment of Loch Vale Watershed in Rocky Mountain National Park. Ph.D. Dissertation, Colorado State University, Fort Collins.

Wardle P (1968) Engelmann spruce (*Picea engelmannii* Engel.) at its upper limits on the Front Range, Colorado. Ecology 49:483–495.

Webber PJ, May DE (1977) The magnitude and distribution of belowground plant structures in the alpine tundra of Niwot Ridge, Colorado. Arct. Alp. Res. 9:157–174.

Willard BE (1979) Plant sociology of the alpine tundra, Trail Ridge, Rocky Mountain National Park, Colorado. Colo. Sch. Mines Q. 74:1–119.

6. Geochemical Characteristics

M. Alisa Mast

Introduction

Rock outcrop is the predominant landscape feature of much of the southern Rocky Mountains, including Rocky Mountain National Park and Loch Vale Watershed (LVWS) within it. On an areal basis, 82% of LVWS is composed of rock outcrop and talus slopes derived from glacial activities or frequent debris flows and avalanches that characterize this youthful and steep terrain. The two dominant rock units found within LVWS are a biotite gneiss and the Silver Plume Granite, both Precambrian in age. The purpose of this chapter is to characterize these bedrock materials and interpret weathering in light of current water chemistry to develop arguments for weathering rates and mechanisms in LVWS.

Bedrock Mineralogy and Chemistry

Biotite Gneiss

Biotite gneiss is the dominant metamorphic unit in Rocky Mountain National Park and accounts for 80% of the bedrock in LVWS (Figure 6.1). The major minerals are quartz (41%), plagioclase (28%), microcline (9%), biotite (16%), and sillimanite (6%) (Cole 1977). Minor and accessory minerals include magnetite, ilmenite, garnet, epidote, apatite, zircon, orthpyroxene, and iron

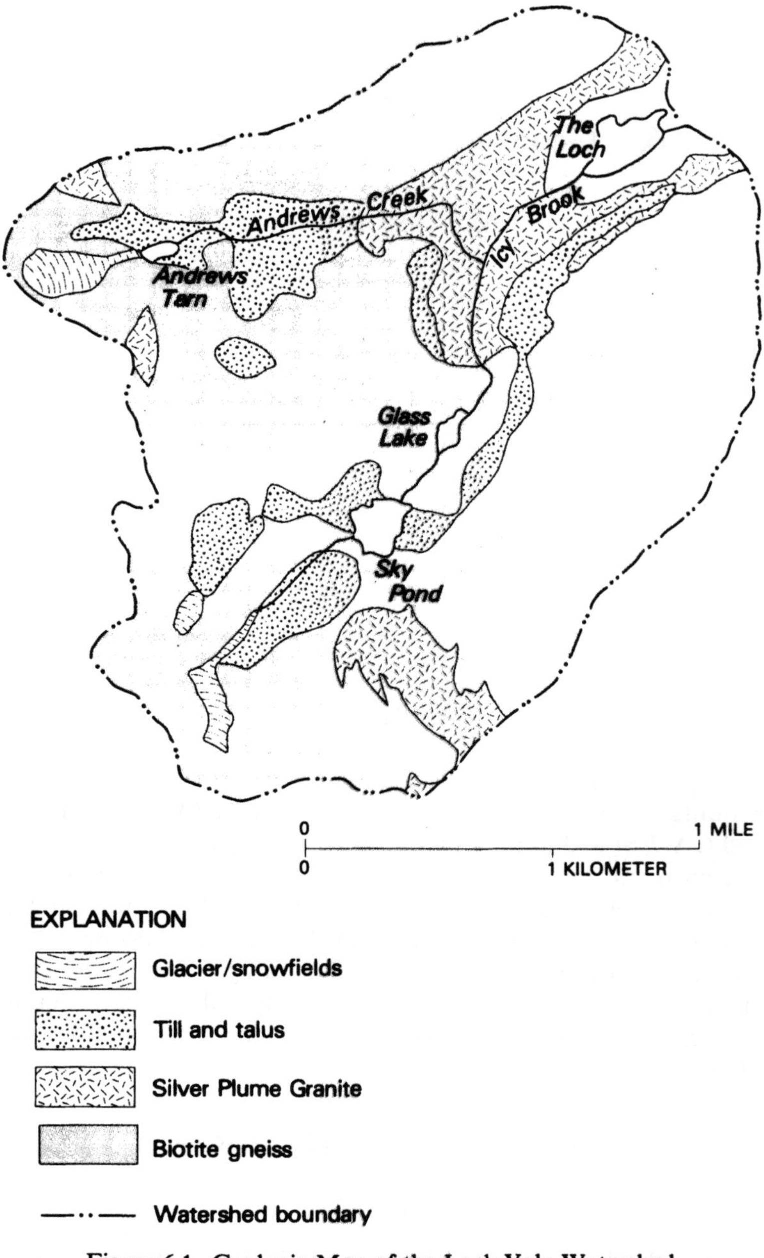

Figure 6.1. Geologic Map of the Loch Vale Watershed.

sulfides (Cole 1977). The gneiss is fine grained and displays well-developed compositional banding of alternating biotite-rich and biotite-poor layers ranging from millimeters to centimeters in thickness. Biotite is strongly segregated into biotite-rich layers and grain orientations tend to parallel the compositional layering. The biotite-rich layers often contain needles of sillimanite oriented parallel to compositional banding. Texture of the quartzo-feldspathic layers is granoblastic, displaying no preferred orientation of grains.

Microprobe results (Table 6.1) indicated plagioclase is in the middle oligoclase range (An27) and is not zoned. Microcline is perthitic and commonly displays twinning in a cross-hatched pattern. The biotite is titaniferous (to 3% TiO_2) and, in addition, some grains contain as much as 2% fluoride although chloride was not detected. Garnet and pyroxene were

Table 6.1. Chemical Composition of Minerals in LVWS Bedrock Measured by Electron Microprobe

	Plagioclase		Microcline	
Wt%	Granite	Gneiss	Granite	Gneiss
SiO_2	61.22	61.51	63.59	63.51
Al_2O_3	24.58	24.98	19.31	19.58
CaO	5.38	4.89	0.04	0.08
Na_2O	8.44	8.45	1.16	0.92
K_2O	0.19	0.33	15.01	15.28
Sum	99.86	100.27	99.17	99.43
N[a]	23	14	4	5

	Biotite		Chlorite	
Wt%	Granite	Gneiss	Granite	Gneiss
SiO_2	34.82	34.58	25.70	25.80
TiO_2	2.46	3.14	0.59	0.13
Al_2O_3	18.20	18.72	20.15	20.55
FeO	21.00	20.64	29.27	29.92
MnO	0.25	0.20	0.50	0.44
MgO	9.46	8.75	12.38	11.34
CaO	0.05	nd[b]	0.03	0.03
Na_2O	0.08	0.07	nd	nd
K_2O	9.37	9.82	nd	0.06
H_2O	4.05	4.21	11.24	11.17
F≡O	0.26	0.49	nd	nd
Sum	99.49	99.64	99.99	9.62
N	29	20	9	10

[a] *N*, number of analyses average.
[b] nd, not detected.

rarely seen in the field, and both are reported to be very minor constituents of the bedrock within Rocky Mountain National Park (Cole 1977). Opaque minerals were present in all thin sections examined. Magnetite and ilmenite are the most common, occurring as irregular grains often altered to goethite, hematite, or anatase. Examination of thin sections under reflected light revealed only three small pyrite grains in 15 samples.

Within the gneiss, there are occasional calc-silicate beds or pods. These pods are thought to have formed from calcareous layers in the original sediments and contain the minerals epidote, plagioclase (highly altered), and quartz (Cole 1977). Only three small outcrops were located in LVWS. No marble layers have been observed or have been reported to occur within this rock unit.

Silver Plume Granite

The Silver Plume Granite accounts for the remaining 20% of bedrock in LVWS and is found primarily along the valley bottom overlain by forest soils (see Figure 6.1). The Silver Plume Granite and associated pegmatites are part of the Longs Peak–St. Vrain batholith. They are fine- to medium-grained porphyritic granites with an average composition of quartz monzonite (Braddock 1969). Mineralogically and chemically, the granite and gneiss are very similar (Table 6.1 and 6.2) although mineral abundances do vary slightly. Major minerals in the granite are microcline (34%), quartz (30%), plagioclase (26%), and biotite (10%), amounts of garnet, sillimanite, zircon, and opaques (Cole 1977). Associated pegmatite veins are composed of quartz, potassium feldspar, plagioclase, and muscovite.

Postmetamorphic Alteration

Weathering and postmetamorphic alteration have affected these rocks to a minor degree (Cole 1977). In the gneiss, weathering occurs preferentially along biotite-rich layers, and weathered rocks tend to split along these layers

Table 6.2. Bulk Chemical Composition of Bedrock Samples from the Loch Vale Watershed

Rock type	SiO_2	Al_2O_3	CaO	MgO	TiO_2	Na_2O	K_2O	Fe_2O_3	MnO	H_2O
Granite	69.29	16.30	0.59	0.39	0.27	3.04	6.05	1.97	0.01	0.74
Granite	70.18	17.01	0.14	0.56	0.23	3.65	5.53	2.11	0.02	1.18
Granite	73.53	15.98	0.47	0.33	0.22	3.39	5.13	1.68	0.02	0.88
Granite	70.71	15.69	0.48	0.32	0.18	3.01	6.18	1.76	0.02	0.78
Gneiss	71.70	13.22	1.41	1.03	0.62	3.34	1.31	5.51	0.05	0.81
Gneiss	67.10	15.25	0.66	1.79	0.81	2.51	3.03	7.07	0.05	0.76
Gneiss	76.86	10.85	1.02	0.76	0.58	2.65	1.29	4.59	0.03	0.77
Gneiss	66.90	16.43	0.59	1.65	0.94	2.23	2.57	6.99	0.06	1.95

or form undercuts in outcrop. Biotite layers containing sillimanite, however, are more resistent to weathering and form resistant ridges commonly seen on the surfaces of weathered boulders.

Hydrothermal alteration occurs mainly along faults and fractures that may be associated with Laramide faulting and hydrothermal activity. The alteration assemblage includes sericite and occasionally epidote, which replace plagioclase, and chlorite, which replaces biotite. In hand samples, the altered fractures often have a bleached appearance and are light green or white in color. A dark green chlorite also occurs as a coating along fault surfaces in outcrop. Quartz is the only bedrock mineral that does not exhibit some degree of alteration. Cathodoluminescence (CL) microscopy revealed that small to moderate amounts of calcite also occur with the hydrothermal assemblage (Figure 6.2). Cathodoluminescence is visible light emitted from a specimen when it is bombarded with electrons (Kopp 1981). When seen under transmitted light the calcite is nearly impossible to identify because of its fine grain size and occurrence with other alteration products such as sericite. Calcite was also found along grain boundaries and microfractures in the granite in what otherwise appears to be mineralogically unaltered rock. The carbonate content of whole rock samples, determined by coulometric titration of CO_2 (Lee and Macalady 1989) liberated with HCl

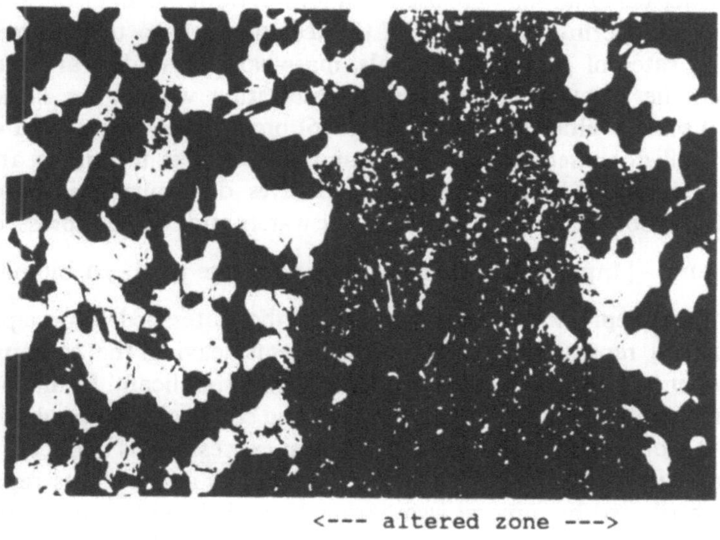

Figure 6.2. Cathodoluminescence photomicrograph showing fine-grained calcite (white areas) along hydrothermally altered zone in gneiss. Large white grains outside altered zone are plagioclase. Photomicrograph is 3.5 mm across. (From Mast MA, Drever JI, Baron J (1990) Chemical Weathering in the Loch Vale Watershed, Rocky Mountain National Park, Colorado. Water Resour. Res. 26:2971–2978, copyright by the American Geophysical Union.)

Table 6.3. Weight Percent (wt%) $CaCO_3$ in Bedrock Samples Determined by Coulometric Titration of Powdered Rock Samples (detection limit 0.001%)

Wt% CaC_3	Rock Type
0.400	Gneiss
0.022	Gneiss
0.003	Gneiss
0.010	Gneiss
0.012	Gneiss
0.006	Granite
0.025	Granite
0.018	Granite
0.005	Granite

from powdered rock samples, ranged from 0.005 to 0.400 weight percent $CaCO_3$ (Table 6.3).

Weathering Mechanisms and Rates

Mass-balance calculations were used to interpret both mechanisms and rates of mineral weathering in LVWS. Of the techniques used to determine the weathering rates of geologic materials, mass-balance studies are generally considered the most reliable means of calculating weathering rates on a watershed scale (Velbel 1985; Clayton 1986) provided the input and output data are well documented and the assumptions of this technique are fully explored. The mass-balance approach involves determining input-output budgets for dissolved constituents in surface waters and can be expressed as

$$[\text{Output}] - [\text{Input}] = [\text{Weathering} \pm \Delta \text{ exchange pool} \pm \Delta \text{ biomass}]$$

where [Input] represents elements added to the watershed in precipitation and [Output] represents dissolved constituents leaving the catchment in stream water or groundwater if subsurface flow is significant. In undisturbed catchments, annual changes in the biomass and soil exchange pool are often assumed to be negligible and the foregoing expression is reduced to the simpler case in which the difference between inputs and outputs represents solutes contributed by mineral weathering. Provided the stoichiometries for individual weathering reactions are known (or assumed), these expressions can be solved by matrix techniques to yield reaction rates for individual minerals as a function of time and surface area of the watershed (Velbel 1985).

Three assumptions are required for the application of this technique in LVWS. First, the biomass represents a steady-state situation so that on an annual basis the uptake of elements in new growth is approximately equal to the release of these elements by decay (Arthur and Fahey 1990). There are a number of reasons for suspecting that the net effect of biological uptake and release is minimal. The forest in LVWS is mature and has never been logged; tree cores indicate no major fires or other disturbances in the bulk of the forest during the past 450 years (Baron and Bricker 1987; Arthur 1990), and forest vegetation covers less than 7% of the total land area, so that small fluctuations in biomass should not significantly affect solute budgets.

The second assumption is that the release of cations from the soil exchange complex is small compared to the net cation flux from the watershed. Although there is evidence that acid levels in precipitation are higher than historic levels (Baron et al. 1986; see Chapter 4, this volume), precipitation is by no means chronically acidic and it is doubtful that present inputs are sufficient to cause a net loss of cations from exchange sites. Clayton (1988) reported that present levels of atmospheric acidity in Idaho (70 moles $ha^{-1} yr^{-1}$) are insufficient to cause cation stripping in forested watersheds of the Idaho batholith. Acid inputs to Loch Vale average 123 moles $ha^{-1} yr^{-1}$, and although slightly higher than in Idaho, they are 5- to 10 fold lower than hydrogen loading rates in eastern catchments where cation stripping has been documented (Reuss and Johnson 1986). In addition, the combination of a sparse soil cover in LVWS and the release of large volumes of water during spring snowmelt minimizes soil–water contact, thus reducing the influence of soil exchange reactions on net cation fluxes.

The third assumption is that the watershed is hydrologically tight: in other words, precipitation should be the only source of water eventually leaving the drainage as streamflow, and groundwater losses should be minimal. In the headwater drainage of LVWS, it is fairly obvious that precipitation is the only source of input, but groundwater losses are more difficult to evaluate. The bedrock is largely impermeable, and although no major faults have been mapped on the valley floor (Cole 1977), small seeps were observed along joints and fractures in the valley walls. A seismic refraction study revealed shallow deposits of till overlying bedrock, suggesting a minimal loss of water to either deep or shallow reservoirs (Baron and Bricker 1987). The hydrology of LVWS is discussed in more detail in Chapter 3 (this volume).

Four years of input-output budgets (see Chapter 10, this volume) were used to develop a weathering model with an approach similar to that used by Garrels and MacKenzie (1967). The contributions from individual weathering reactions were combined with precipitation inputs in such a way as to yield the obseved concentrations at the outlet of the watershed. The mineral compositions and weathering reactions used for this exercise, shown in Table 6.4 and 6.5, are consistent with the mineralogic and chemical data presented earlier. Because the chemistry and mineralogy of the granite and gneiss are essentially identical, the same weathering reactions apply for both bedrock units.

The results of the mass-balance calculations using average input-output budgets for 1984, 1986, 1987, and 1988 are presented in Table 6.6. The first

Table 6.4. Chemical Composition of Bedrock and Soil Minerals Measured by Electron Microprobe and Bulk Chemical Analysis

Type	Composition
Oligoclase	$Ca_{.27}Na_{.73}Al_{1.27}Si_{2.73}O_8$
Biotite[a]	$(K_{.98}Mg_{1.00}Fe_{1.33}Ti_{.18}Al_{.33})(Al_{1.35}Si_{2.65})O_{10}(OH)_2$
Chlorite	$(Mg_{1.81}Fe_{2.72}Al_{1.39})(Al_{1.23}Si_{2.77})O_{10}(OH)_8$
Calcite	$CaCO_3$
Smectite-illite[b]	$(K_{.32}Ca_{10})(Fe_{.25}Mg_{.39}Al_{1.47})(Al_{.46}Si_{3.54})O_{10}(OH)_2 \cdot nH_2O$
Kaolinite[b]	$Al_2Si_2O_5(OH)_4$

line in Table 6.6 lists chemical inputs calculated using mean annual water year concentrations and moderate estimates of precipitation to the catchment (see Chapters 3 and 10, this volume). In second and third lines, nonprecipitation-derived magnesium and potassium were assigned to the weathering of biotite and chlorite to a mixed-layer smectite clay containing approximately 30% illite layers. The amounts of bicarbonate produced and silica consumed were calculated using reaction stoichiometries from Table 6.5. Both biotite and chlorite reactions were written so that aluminum and iron are conserved in a solid phase. In the fourth line, enough plagioclase

Table 6.5. Weathering Reactions Used in Mass-Balance Calculations

1. Biotite to Mixed-Layer Smectite-Illite
   $1.15\,Bio + 0.10\,Ca^{2+} + 0.49\,H_4SiO_4 + 1.21\,O_2 + 2.13\,CO_2 + (0.73+n)H_2O \, - \!\!> $
   $Smect + 0.76\,Mg^{2+} + 0.81\,K^+ + 1.28\,FeO(OH)_{(s)} + 2.13\,HCO_{13}^- + 0.21\,TiO_{2(s)}$
2. Chlorite to Mixed-Layer Smectite-Illite
   $1.39\,Chlor + 0.10\,Ca^{2+} + 0.32\,K^+ + 0.31\,H_4SiO_4 + 1.78\,O_2 + 3.74\,CO_2 + nH_2O \, - \!\!>$
   $Smect + 2.13\,Mg^{2+} + 3.53\,FeO(OH)_{(s)} + 3.74\,HCO_3^-$
3. Oligoclase to Kaolinite
   $Olig + 1.27\,CO_2 + 4.82\,H_2O \, - \!\!>$
   $0.64\,Kaol + 0.27\,Ca^{2+} + 0.73\,Na^+ + 1.46\,H_4SiO_4 + 1.27\,HCO_3^-$
4. Kaolinite to Aluminum Hydroxide
   $Kaol + 5\,H_2O \, - \!\!> 2\,Al(OH)_{3(s)} + 2\,H_4SiO_{4(aq)}$
5. Pyrite Dissolution
   $FeS_{2(s)} + 3.75\,O_2 + 3.5\,H_2O \rightarrow Fe(OH)_{3(s)} + 2\,SO_4^{2-} + 4\,H^+$
6. Nitrogen Assimilation
   $X\,NH_4^+ + Y\,NO_3 + (X-Y)HCO_3^- \, - \!\!> Organic\text{-}N$
7. Calcite Dissolution
   $Calcite + H_2O + CO_2 \, - \!\!> Ca^{2+} + 2\,HCO_3^-$

Table 6.6. Mass-Balance Calculations Using Average Chemical Budgets for 1984, 1986, 1987, and 1988[a]

Reaction	Ca	Mg	Na	K	SiO_2	HCO_3	H^+	NH_4	NO_3	SO4	Cl	Weathering Rate
Precipitation input	43	11	35	5	0	0	123	69	104	74	23	
Biotite to smec-illite	−2	22		23	−14	61						29 biotite
Chlorite to smec-illite	−1	15		−2	−2	27						7 chlorite
Feldspar to kaolinite	27		72		144	125						99 feldspar
Kaolinite to $Al(OH)_3$					37							18 kaolinite
Pyrite dissolution							24			12		6 pyrite
Nitrogen assimilation						−55		−59	−4			
$H^+ + HCO_3^- = H_2CO_3$						−140	−140					
Calcite disolution	106					212						106 calcite
Calculated output	173	48	107	26	165	230	7	10	100	86	23	
Measured output	173	48	107	26	165	219	7	10	100	87	21	

[a] Dissolved constituents produced (+) or consumed (−) and mineral weathering rates calculated using a watershed area of 880 ha and reported in moles $ha^{-1} yr^{-1}$.

was altered to kaolinite to yield the observed sodium output. Sodium was used as a limiting factor because feldspar is the only significant source of this element in the watershed. In the next line, a small amount of kaolinite was altered to gibbsite to account for the total silica observed in the measured output. Although aluminum hydroxides have been reported to form from the weathering of silicate rocks in alpine environments (Reynolds 1971; Marcos 1977; Giovanoli et al. 1988), the additional silica needed from this reaction may be in part an artifact of the smectite composition used in reactions 1 and 2 (Table 6.5). In line 6, the additional sulfate was assigned to the dissolution of pyrite. Although small amounts of pyrite were observed in the bedrock, some of this sulfate may be introduced as dry deposition from the atmosphere. Next, corrections were made for biological assimilation of ammonium and nitrate according the the net reaction given in Table 6.5, and in lines 8 and 9, the excess hydrogen was reacted with bicarbonate and the remaining calcium was assigned to the dissolution of calcite. No reactions involving chloride were necessary because the measured input balanced the measured output. The behavior of chloride in LVWS should be relatively conservative as there is a negligible contribution from weathering (the concentration in biotite is less than 0.01%) and chloride is relatively inactive with respect to soil exchange reactions and vegetative cycling (Likens et al. 1977; Creasey et al. 1986). Finally, all these reactions were summed to yield the calculated output shown at the bottom of Table 6.6.

The results of this weathering scheme raise two important inconsistencies with the observed bedrock and soil mineralogy. First, this model suggests significant amounts of kaolinite are forming in the drainage, although smectite was identified as the dominant clay mineral in the soils. It is possible that much of the kaolinite is actually present as a poorly crystalline clay not detected by x-ray analysis. In fact Na_2CO_3 extractions indicate the < 2-μm fraction of soils contains a considerable amount of amorphous material with a molar SiO_2/Al_2O_3 ratio about 2. This suggests the presence of an amorphous aluminosilicate in soils with a composition similar to kaolinite.

The second rather unexpected result is that nearly 40% of the total cations derived within the basin are apparently contributed by the dissolution of calcite. Weathering of a carbonate mineral is not only implied by the mass-balance calculations but is supported by the observation of calcite in thin sections using CL microscopy (see Figure 6.2). It is surprising that the weathering of calcite, which makes up much less than 1% of the bedrock (see Table 6.3), is nearly as important as the dissolution of plagioclase, which is considerably more abundant.

From the success of mass-balance calculations made by Garrels and Mackenzie (1967) to interpret the chemistry of ephemeral springs in the Sierra Nevada, it is often assumed that calcium and sodium in waters draining granitic rocks are derived from the weathering of plagioclase. In LVWS, the plagioclase composition necessary to account for the observed CA/Na ratio in stream water would have to be 60% anorthite, which is inconsistent with the measured value of 27%. Several studies of weathering in silicate terrains have also reported higher Ca/Na ratios in surface waters than the observed ratio in the bedrock plagioclase. The excess calcium has been attributed to incongruent feldspar dissolution or the the presence of trace amounts of highly weatherable minerals in the local bedrock. In LVWS, mass-balance calculations indicate the most likely source for the calcium is a reaction that is stoichiometrically similar to calcite dissolution releasing only calcium and bicarbonate to solution.

Although calcite was observed in LVWS, there are other possible sources of calcium that satisfy this stoichiometric constraint. Two recent studies suggest high Ca/Na ratios in surface waters draining granitic terrains are the result of incongruent release of cations with respect to the average feldspar composition of the bedrock (Clayton 1988; Velbel and Romero 1989). Incongruent release refers to the obsevation that cation ratios in surface waters are different from those in the dissolving minerals. Incongruent release of cations may result from heterogeneities within the mineral grains, such as exsolution intergrowths or chemical zonation in crystals. Clayton (1986, 1988) found plagioclase crystals to be commonly zoned in watersheds of the Idaho batholith and hypothesized preferential dissolution of the calcic-rich cores was controlling the Ca/Na ratio in surface waters. In LVWS it is doubtful this mechanism supplies additional calcium to surface waters, as chemical zoning was not observed in the plagioclase.

Another mechanism of incongruent dissolution involves a preferential release of cations from the solid phase as is observed in laboratory studies when freshly ground minerals first contact solution (Busenberg and Clemency 1976; Chou and Wollast 1985; Holdren and Speyer 1985; Schott and Berner 1985; Mast and Drever 1987). This initial incongruency results from an exchange reaction between cations on the mineral surface for hydrogen ions in solution. As the reaction progresses, the release rate of dissolved components eventually reaches a steady state and the dissolution reaction becomes congruent. This initial transient state is rather short (lasting on the order of minutes to days), and it is generally assumed that the steady-state dissolution step is the main process controlling the weathering of feldspars in natural environments (Wollast and Chou 1985).

The dissolution of carbonate minerals introduced with eolian dust is another plausible source of dissolved calcium in surface waters draining LVWS. The transport of sedimentary materials from arid lowlands to high-elevation watersheds in the Sangre de Cristo Mountains of New Mexico has been documented by Gosz et al.(1983) and by Graustein and Armstong (1983). Litaor (1987) proposed that eolian inputs were an important soil-forming factor on Niwot Ridge, and that particulate calcite in the dust strongly influenced solution chemistry, although only four samples of eolian material during 2 years of sampling contained calcite. It is possible that eolian inputs may be episodic (Turk and Spahr 1989) and important in periods of strong winds and drought. Such material should contain readily weatherable calcite, dolomite, and gypsum from the Basin areas west of the Rocky Mountains (Turk and Spahr 1989). At the present time there is no information regarding the flux of windblown carbonate to LVWS, although the very low base saturation levels and high acidity of the soils in the watershed (Walthall 1985; see Chapter 7, this volume) suggest this may not be a continuous or dominant source of calcium in surface waters.

In light of the previous discussion, it is more reasonable to explain the excess calcium by the dissolution of trace amounts of calcite present in the local bedrock. A similar study of weathering in the South Cascade Glacier area (Drever and Hurcomb 1986) reached the same conclusion that the major source of calcium was not from the weathering of plagioclase but rather from minor amounts of calcite found in veins and along joints in the local bedrock. In both the Cascades and LVWS, the importance of carbonate weathering is the result of two factors. First and more obvious is the fact that calcite dissolution rates are typically six orders of magnitude faster than rates observed for common aluminosilicate minerals under neutral pH conditions (Plummer et al. 1979; Lasaga 1984). Second, high rates of mechanical erosion in alpine areas are sufficient to continually expose fresh bedrock material to snowmelt and rain, whereas in lower elevation sites there is often no calcite remaining in near-surface flowpaths and the weathering of silicate minerals dominates.

Annual cationic denudation rates calculated for LVWS between 1984 and 1988 ranged from 230–525 eq ha^{-1} yr^{-1} with a mean annual rate of 390 eq ha^{-1} yr^{-1} (Table 6.7). Cationic denudation rates in LVWS are related to annual water yield and generally increase with increasing water yield from the catchment (Clayton and Megahan 1986). These rates are comparable to denudation rates calculated for another alpine watershed in the Rocky Mountains (Rochette et al. 1988) and to long-term rates in the Adirondack

Table 6.7. Comparison of Cationic Denudation Rates (eq ha^{-1} yr^{-1}) from Different Watersheds

Rate	Location	Reference
390	Front Range, CO	This study
444	Snowy Range, WY	Rochette et al. 1988
9300	Northern Cascades, WA	Reynolds and Johnson 1972
3800	North American continent	Reynolds and Johnson 1972
500–600	Adirondacks, NY[a]	April et al. 1986
1527	Idaho Batholith, ID	Clayton 1988
337	Pond Branch, MD	Cleaves et al. 1970
2200	Northeastern United States	Johnson et al. 1981
450–1500	Scotland	Creasy et al. 1986
650–1000	Sweden	Sverdrup and Warfvinge 1988
1490	British Columbia	Feller and Kimmens 1979

[a]Long-term denudation rates.

Mountains where surface water acidification has occurred (April et al. 1986). This suggests that current weathering rates may not be sufficient to prevent surface water acidification in LVWS. However, predicting the effect of acid precipitation on the chemistry of surface waters requires not only information on present-day rates of weathering, but also information on the effect of an increase of acidity on these rates.

The dissolution rates of various feldspar minerals (Chou and Wollast 1985; Holdren and Speyer 1985; Mast and Drever 1987) and biotite (Acker and Bricker 1989) measured in the laboratory are only weakly dependent on solution pH in the range of most natural waters. In LVWS it is reasonable to assume that rates of silicate mineral weathering would not change significantly if acid levels in precipitation should increase. There is some evidence, however, that if aluminum concentrations in soil and surface waters increase in response to acid deposition, the weathering rates of some silicate minerals may actually decrease (Chou and Wollast 1985; Sverdrup and warfvinge 1988).

The effect of acidic deposition on the rate of calcite weathering is more difficult to evaluate. Calcite dissolution rate is a complex function of pH and depends on other solute concentrations in addition to solution pH (Drever and Hurcomb 1986). However, because the absolute rate is so rapid, the limiting factor controlling calcite weathering rates in LVWS is more likely to depend on the availability of calcite through physical erosion rather than on a change in the absolute rate of mineral weathering as a function of solution pH.

References

Acker JC, Bricker OP (1989) pH influence on dissolution rate and cation release from biotite. EOS Trans. 70(15):328.

April R, Newton R, Coles LT (1986) Chemical weathering in two Adirondack watersheds: past and present-day rates. Geol. Soc. Am. Bull. 97:1232–1238.

Arthur MA (1990) The Effects of Vegetation on Watershed Biogeochemistry at Loch Vale Watershed, Rocky Mountain National Park, Colorado. Ph.D. Dissertation, Cornell University, Ithaca, New York.

Arthur MA, Fahey TJ (1990) Mass and nutrient content of decaying boles in an Englemann spruce/subalpine fir forest, Rocky Mountain National Park, Colorado. Can. J. For. Res. 20:730–737.

Baron J, Norton SA, Beeson DR, Herrmann R (1986) Sediment diatom and metal stratigraphy from Rocky Mountain lakes with special reference to atmospheric deposition. Can. J. Fish. Aquat. Sci. 43:1350–1362.

Baron J, Bricker OP (1987) Hydrologic and chemical flux in Loch Vale Watershed, Rocky Mountain National Park. In: McKnight D, Averett RC, eds. Chemical Quality of Water and the Hydrologic Cycle, pp. 141–156. Lewis Publishers, Chelsea, Michigan.

Braddock WA (1969) Geology of the Empire Quadrangle Grand, Gilpin, and Clear Creek Counties. U.S. Geol. Surv. Prof. Paper 616. USGS, Denver, Colorado.

Busenberg E, Clemency CV (1976) The dissolution kinetics of feldspars at 25°C and 1 atm CO_2 partial pressure. Geochim. Cosmochim. Acta 40:41–49.

Chou L, Wollast R (1985) Steady-state kinetics and dissolution mechanisms of albite. Am. J. Sci. 285:963–993.
Clayton JL (1986) An estimate of plagioclase weathering rate in the Idaho Batholith based upon geochemical transport rates. In: Colman SM, Dethier DP, eds. Rates of Chemical Weathering of Rocks and Minerals, pp. 453–467. Academic Press, Orlando, Florida.
Clayton JL (1988) Some observations of the stoichiometry of feldspar hydrolysis in granitic soils. J. Environ. Qual. 17:153–157.
Clayton JL, Megahan WF (1986) Erosional and chemical denudation rates in the southwestern Idaho batholith. Earth Surface Processes and Landforms 11:389–400.
Cleaves, ET, Godfrey AE, Bricker OP (1970) Geochemical balance of a small watershed and its geomorphic implications. Geol. Soc. Am.•Bull. 81:3015–3032.
Cole JC (1977) Geology of East-Central Rocky Mountain National Park and Vicinity, with Emphasis on the Emplacement of the Precambrian silver Plume Granite in the Longs Peak-St. Vrain Batholith. Ph.D. Dissertation, University of Colorado, Boulder.
Creasey J, Edwards AC, Reid JM, MacLeod DA, Cresser MS (1986) The use of catchment studies for assessing chemial weathering rates in two contrasting upland areas in northeast Scotland. In: Colman SM, Dethier DP, eds. Rates of Chemical Weathering of Rocks and Minerals, pp. 468–502. Academic Press, Orlando, Florida.
Drever JI, Hurcomb DR (1986) Neutralization of atmospheric acidity by chemical weathering in an alpine drainage basin in the North Cascade Mountains. Geology 14:221–224.
Feller MC, Kimmens JP (1979) Chemical characteristics of small streams near Haney in southwestern British Columbia. Water Resour. Res. 15(2):247–258.
Garrels RM, MacKenzie FT (1967) Origin of the chemical compositions of some springs and lakes. In: Gould RF, ed. Equilibrium Concepts in Natural Water Systems, pp. 222–242. Am. Chem. Soc. Adv. Chem. Ser. 67, ACS, Washington, D.C.
Giovanoli R, Schnoor JL, Sigg L, Stumm W, Zorbrist J (1988) Chemical weathering of crystalline rocks in the catchment area of acidic Ticino Lakes, Switzerland. Clays Clay Miner. 36(6):521–529.
Gosz JR, Brookins DG, Moore DI (1983) Using strontium isotope ratios to estimate inputs to ecosystems. BioScience 33(1):23–30.
Graustein WC, Armstrong RL (1983) The use of strontium-87/strontium-86 ratios to measure atmospheric transport into forested watersheds. Science 219:289–292.
Holdren GR, Jr., Speyer PM (1985) Stoichiometry of alkali feldspar dissolution at room temperature and various pH values. Am. J. Sci. 285:994–1026.
Johnson NM, Driscoll CT, Eaton JS, Likens GE, McDowell WH (1981) "Acid rain," dissolved aluminum and chemical weathering at the Hubbard Brook Experimental Forest, New Hampshire. Geochim. Cosmochim. Acta 45:1421–1437.
Kopp OC (1981) Cathodoluminescence petrography: a valuable tool for teaching and research. J. Geol. Educ. 29:108–113.
Lasaga AC (1984) Chemical kinetics of water-rock interactions. J. Geophys. Res. 89(B6):4009–4025.
Lee CM, Macalady DL (1989) Towards a standard method for the measurement of organic carbon in sediments. Int. J. Environ. Anal. Chem. 35:219–225.
Likens GE, Borman FH, Pierce RS, Eaton JS, Johnson NM (1977) Biogeochemistry of a Forested Ecosystem. Springer-Verlag, New York.
Litaor MI (1987) The influence of eolian dust on the genesis of alpine soils in the Front Range, Colorado. Soil Sci. Soc. Am. J. 51:142–147.
Marcos GM (1977) Geochemical Alteration of Plagioclase and Biotite in Glacial and Preglacial Deposits. Ph.D. Dissertation, University of Colorado, Boulder.

Mast MA, Drever JI (1987) The effect of oxalate on the dissolution rates of oligoclase and tremolite. Geochim. Cosmochim. Acta 51:2559–2568.

Mast, MA, Drever JI, Baron J (1990) Chemical weathering in the Loch Vale Watershed, Rocky Mountain National Park, Colorado. Water Resour. Res. 26:2971–2978.

Plummer LN, Parkhurst DL, Wigley TML (1979) Critical review of the kinetics of calcite dissolution and precipitation. In: Jenne EA, ed. Chemical Modeling in Aqueous Systems, pp. 537–573. Am. Chem. Soc. Symp. Ser. 93, ACS, Washington, D.C.

Reuss JO, Johnson DW (1986) Acid Deposition and the Acidification of Soils and Waters. Ecological Studies 59, Springer-Verlag, New York.

Reynolds RC (1971) Clay mineral formation in an alpine environment. Clays Clay Miner. 19:361–374.

Reynolds RC, Johnson NM (1972) Chemical weathering in the temperate glacial environment of the northern Cascade Mountains. Geochim. Cosmochim. Acta 36:537–554.

Rochette E, Drever JI, Sanders F (1988) Chemical weathering in the West Glacier Lake drainage basin, Snowy Range, Wyoming: implications for future acid deposition. Contrib. Geol. 26(1):29–44.

Schott J, Berner R.A. (1985) Dissolution mechanisms of pyroxenes and olivines during weathering. In: Drever JI, ed. The Chemistry of Weathering, pp. 35–54. NATO Adv. Study Inst. Ser., Ser. C, Vol. 149. D. Reidel, Dordrecht, Germany.

Sverdrup H, Warfvinge P (1988) Weathering of primary silicate minerals in the natural soil environment in relation to a chemical weathering model. Water Air Soil Pollut. 38:387–408.

Turk JT, Spahr NE (1989) Chemistry of Rocky Mountain Lakes. In: Adriano DC, Havas M, eds. acid Precipitation, Vol. 1: Case Studies, pp. 181–208. Springer-Verlag, New York.

Velbel AM (1985) Geochemical mass balances and weathering rates in forested watersheds of the southern Blue Ridge. Am. J. Sci. 285:904–930.

Velbel AM, Romero NL (1989) Amphibolite weathering and cation budgets in southern Blue Ridge and Colorado Front Range, USA. In: Programs and Abstracts, 1989 International Geological Congress, 3:291.

Walthall PM (1985) Acidic Deposition and the Soil Environment of Loch Vale Watershed in Rocky Mountain National Park. Ph.d. Dissertation, Colorado State University, Fort Collins.

Wollast R, Chou L (1985) Kinetic study of the dissolution of albite with a continuous flow-through fluidized bed reactor. In Drever JI, ed. The Chemistry of Weathering, pp. 75–96. NATO Adv. Study Inst. Ser., Ser. C, Vol. 149. D. Reidel, Dordrecht, Germany.

7. Soils

Jill Baron, P. Mark Walthall, M. Alisa Mast, and Mary A. Arthur

Forested and meadow soils make up 6% of the land area of Loch Vale Watershed (LVWS) (Figure 7.1). Because these soils are located in the valley floor adjacent to waterways, precipitation has a high possibility of passing through soils as it flows downhill. This creates the potential for these soils to have a large influence on the lakes and streams of LVWS. We explored the soil environment of LVWS in great detail for two reasons: First, soil is extremely reactive material. The kinetics of intensity factors such as anion adsorption or retention, cation exchange, and alkalinity generation by carbonic acid dissolution are rapid enough that they can and do exercise a disproportionate influence on surface water quality (Cosby et al. 1985). Many investigators, including Cosby et al. (1985), Reuss and Johnson (1986), and Christopherson et al. (1982), have emphasized the role of soils in mediating acidic atmospheric deposition.

The second reason is that the processes of soil development are fascinating. This thin layer separating the geosphere from the atmosphere is the cornerstone of the biosphere. Soils, in fact, are a combination of geosphere, atmosphere, and biosphere, being composed of the products of rock weathering, atmospheric gases, and organic material. Without soils there would have been few terrestrial plants or animals, and certainly no evolution of humanity and human civilization.

One can view soils from a descriptive point of view, as in the next section, where we describe the soils, their distribution, mineralogy, and chemical

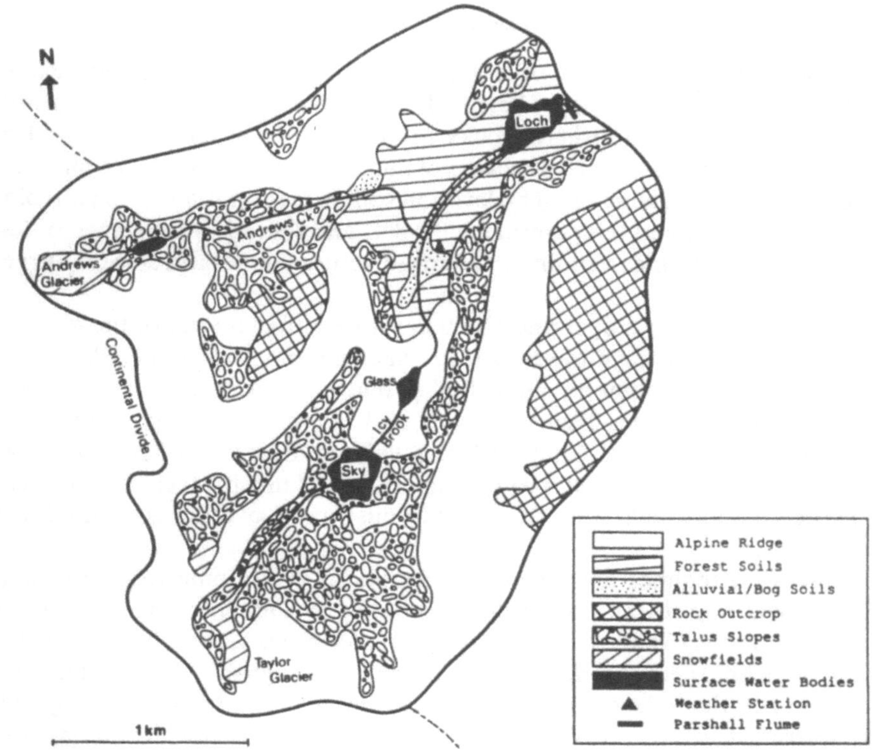

Figure 7.1. Distribution of major soil types and landforms of Loch Vale Watershed.

composition. Soils viewed thus operate in essentially geologic time, and descriptions in this context are historic: the combination of soil-forming factors (climate, organisms, parent material, relief, time) have operated at some time in the past to produce existing soils (Jenny 1941). One can also view soils as dynamic, as in the section on soil–water interactions: here we characterize those parameters that directly influence downstream surface waters. These include exchange capacity, soil acidity, organic matter, sulfur cycling within the soil, and soil water dynamics. Soils viewed this way represent an active biogeochemical environment along a drainage continuum. In actuality, both the static and the dynamic views of soils are necessary to the study of soils. To paraphrase Schumm (1977), the soil system is a biogeochemical system with a history. Our objective is to understand not only the physical and chemical properties of the LVWS soil system, but the importance of soils to the overall alpine/subalpine environment and, in particular, to surface water.

Description of LVWS Soils

Soils of LVWS are similar to other high-elevation soils of the southern Rocky Mountains (Walthall 1985). These soils are coarse textured and often high in coarse fragments. Southern Rocky Mountain soils are not dominated by any specific mineralogic class because of the heterogeneity of parent material. The temperature regime of high-elevation soils is predominantly Cryic (cold); the mean annual soil temperature is 0°–8°C and there is a difference of less than 5°C between mean summer and mean winter temperatures (Soil Survey Staff 1975).

Soils of the valley slopes and floors developed after retreat of the glaciers. Soils in the forest have developed in Pinedale and Bull Lake tills of the Pleistocene age (Richmond 1980). Soil parent material in the cirque valleys above treeline includes Neoglacial tills and taluses, including those of the Temple Lake stade (see Chapter 2, this volume; Richmond 1980; Walthall 1985). The broad alpine slopes of Thatchtop are much older, having been exposed, unglaciated, since the Cretaceous era (Lovering and Goddard 1959).

Three soil regimes are recognized within LVWS: alpine ridge, forested, and alluvial/bog soils. Alpine ridge is confined to Thatchtop Mountain and the ridge above Andrews Creek, with smaller areas above treeline in the basin that holds Sky Pond and Glass Lake (see Figure 7.1). Forested soils are found in the valley floor below Timberline Falls. Alluvial and bog soils are restricted to a narrow band adjacent to Icy Brook and in one area of low relief adjacent to Andrews Creek. Alpine ridge soils cover 11%, forested soils 5%, and alluvial/bog soils approximately 1% of the watershed. The remaining 83% of LVWS is bare rock, boulder fields, surface water, or fields of permanent snow over boulder fields.

Alpine Ridge

The alpine ridge area of LVWS is mostly boulder fields and thin rocky soils drained by intermittent springs and streams. Only 10% to 15% of the ridge is estimated to have had significant soil development or accumulation (Walthall 1985). This contrasts with other alpine areas within Rocky Mountain National Park and in the southern Rockies, such as Niwot Ridge (Benedict 1970). Somewhat gentler slopes on Niwot Ridge may prevent fine material from being lost through mass wasting and erosion. The gentlest slopes on Thatchtop are approximately 30%. The alpine ridge and its soils are physically disjunct from the valley floor of LVWS. Sheer cliffs or steep slopes of talus, scree, and snow separate the ridge from the floor by 300 to 600 m.

Both Cryochrepts and Cryohemists are found on alpine slopes. The Cryochrepts, found on the northeast face of the ridge and in saddles, are shallow, stony, and covered by a thin (~2 cm) organic surface horizon (Table 7.1). A 3-cm A horizon and a minimally developed (20 cm) Bw horizon

Table 7.1. Descriptions of Alpine Ridge Soils (Walthall 1985)

Cryochrept

Parent Material: Gneiss and granite
Parent Material Source: Boulder field
Topography: Saddle position on alpine ridge
Slope: 7%, concave
Aspect: S40°W
Elevation: 3840 m
Vegetation: Alpine forbs and lichens
Soil Profile: (Colors are for moist soil)

Oe	2–0 cm	Moderately decomposed organic matter; many fine and very fine roots; 20% coarse fragments; abrupt smooth boundary.
A	0–3 cm	Black (10 YR 2/1); cobbly sandy loam; weak granular structure; friable; many fine and very fine roots; 20% coarse fragments; pH 4.95; clear, wavy boundary.
Bw	3–30 cm	Brown (10 Yr 4/3); cobbly sandy loam; weak platy structure; friable; common fine roots; 30% coarse fragments; pH 4.66; abrupt irregular boundary.
R	30 + cm	Granite and gneiss boulders

Cryohemist

Parent Material: Forbs and sedges
Parent Material Source: Organics
Topography: Hummocky bog
Slope: 5%, plane
Aspect: N50°W
Elevation: 360 m
Vegetation: *Carex* spp. and alpine forbs
Soil Profile: (Colors are for moist soil)

Oi	0–11 cm	Slightly decomposed organic material; many fine and very fine roots; 20% to 30% coarse fragments; pH 3.98; clear smooth boundary.
Oe	11–24 cm	Moderately decomposed organic material; many fine and very fine roots; 75% coarse fragments; pH 4.20; abrupt irregular boundary.
R	24 + cm	Granite and gneiss boulders.

(subsoil) rest directly on boulders. The percentage of clay-sized particles in these soils is about 7%. Soil development on the ridge top is minimal because of the arid environment. Although measured precipitation at the weather station lower in the watershed is about 100 cm per year, effective precipitation is much less. Intense wind velocities on these ridges cause evaporation of summer precipitation and sublimation or physical removal of winter snow (see Chapters 2 and 3, this volume). The wind also limits vegetative cover to sparse cushion plants and lichens and enhances erosional processes. Small areas of shallow Cryochemists are found in concave areas where springs emerge (Table 7.1). These are characterized by organic mats (24 cm) directly underlain by boulders.

Forested Soils

In contrast to the aridity of the alpine environment, conditions are much more favorable for soil development on the valley floor. Effective precipitation of about 100 cm per year, lower vertical relief (0%–20% slope), partial protection from wind by the canyon walls, and development of a mature stand of coniferous trees have accelerated chemical weathering rates. Inceptisols (Cryochrepts) and Entisols (Cryorthents) occur where talus slopes merge into moraine deposits of the valley floor. These grade into Cryoboralfs lower down where the forest is well developed.

The downslope gradient on the valley floor affects all aspects of soil development. The catastrophic influx of debris material, resulting from debris flows and avalanches, is common. One example of debris influx appears in a Cryorthent with a buried A horizon (Table 7.2). Downslope, and further from the valley walls, vegetation stabilizes this material. The most mature soils are found in sites furthest from talus slopes and sheltered by the wind. Extensive weathering is evident in the presence of approximately 40% saprolite observed in one of the BC horizons (Table 7.2; Walthall 1985). Clay content was greater in all but one of the forest samples, compared with samples from the toe slopes of talus fields.

As of this writing, the forested soils are technically classified as Cryoboralfs. However, they have all the characteristics of frigid Ultisols (Hapludults) and are under consideration for reclassification. The base saturation of less than 35% and argillic horizons (Figure 7.2) are more characteristic of Ultisols than Alfisols (Flach, personal communication; Walthall 1985). Ultisols are characterized by extensive leaching of bases, high acidity, clay formation as a result of in situ weathering in the B horizon, and podzolization (movement of iron downprofile), although in LVWS the argillic horizons are weakly developed (Buol et al. 1980). In contrast with the other soils of LVWS, these soils are highly mature and are the product of a favorable environment for soil development.

Bog/Alluvial Soils

Organic and alluvial soils occur in areas of low relief and poor drainage adjacent to the streams of LVWS. Cryofluvents are found under forested vegetation on isolated terraces where the stream gradient is high. Thick organic horizons (9–10 cm) lie on top of a series of stratified, alluvial C horizons in the Cryofluvents (Table 7.3).

Two other highly organic soils, Cryaquents and Cryohemists, occur in flatter areas that support wet sedge meadows. Both have a series of organic horizons ($\sim$33 cm in thickness) in which the extent of decomposition

Table 7.2. Descriptions of Forested Soils (Walthall 1985)

Cryorthent

Parent Material: Gneiss and granite
Parent Material Source: Colluvium
Topography: Toe slope of talus rock field
Slope: 8%, convex
Aspect: N52°W
Elevation: 3350 m
Vegetation: Alpine forbs, alpine willows, *Carex* spp., *Poa* spp.
Soil Profile: (colors are for moist soil)

Oe	5–0 cm	Moderately decomposed organic material, principally forbs and sedges; many fine to very fine roots; ph 4.51; abrupt smooth boundary.
A	0–12 cm	Very dark gray (10 YR 3/1); sandy loam; common, medium, distinct mottles; weak medium subangular blocky structure; friable; many fine roots; pH 4.71; clear smooth boundary.
C	12–25 cm	Very dark grayish brown; (10 YR 3/2) and reddish brown (5YR 4/3); sandy loam; many fine distinct mottles; structureless, massive; friable; many fine roots; pH 4.73; clear smooth boundary.
Ab/C	25–36 cm	Approximately 70% A material that appears to have been dissected and buried by overlying C material, which makes up the remaining 30%. The Ab material is very dark brown (10 YR 2/2); loam; common, medium, distinct mottles; structureless, massive; friable. The C material is yellowish red (5 YR 5/8); loamy sand; structureless, massive; very friable. As a whole, the horizon contained approximately 5% coarse fragments; common, fine roots; pH 4.64; clear smooth boundary.
Cgb	36–71 cm	Very dark brown (10 YR 3/2); very gravelly sandy loam; common, fine, faint mottles; very weak, coarse, subangular blocky structure; friable; common fine roots; 45% coarse fragments; pH 4.99; gradual wavy boundary.
Cg	71–89 cm	Dark yellowish brown (10 YR 4/4); extremely gravelly silt loam; common fine mottles; very weak, coarse structure; friable; common fine roots; 70% coarse fragments; pH 5.12; abrupt irregular boundary.
R	89 + cm	Fractured boulders of gneiss.

Cryochrept

Parent Material: Granite, gneiss, and schist
Parent Material Source: Colluvium
Topography: Toe slope of of talus rock field
Slope: 29%, convex
Aspect: N10°W
Elevation: 3140 m
Vegetation: Forbs, *Carex* spp., and grasses
Soil Profile: (Colors are for moist soil)

Table 7.2 (*Continued*)

Oe	3–0 cm	Moderately decomposed organic matter, mostly grasses and sedges; many fine and very fine roots; pH 4.56; abrupt wavy boundary.
A	0–11 cm	Dark brown (7.5 YR 3/2); cobbly sandy loam; weak, medium, subangular blocky structure; friable; common fine roots; 20% coarse fragments; pH 4.73; clear wavy boundary.
Bwl	11–17 cm	Brown (7.5 YR 4/4); cobbly loam; weak, medium, subangular blocky structure; friable; commone fine roots; 25% coarse fragments; pH 4.83; clear wavy boundary.
Bw2	17–31 cm	Strong brown (7.5 YR 5/6); very cobbly sandy loam; very weak, medium, subangular blocky structure; very friable, common fine and few medium roots; 45% coarse fragments; pH 4.94; gradual wavy boundary.
BC/R	56 + cm	Massive granite and gneiss boulders.

Cryoboralf

Parent Material: Granite and gneiss
Parent Material Source: Glacial till
Topography: Moraine veneer
Slope: 9%, convex
Aspect: S40°E
Elevation: 3195 m
Vegetation: Fir, spruce, *Vaccinium* spp.
Soil profile: (Colors are for moist soil unless specified otherwise)

Oi	9–5 cm	Slightly decomposed organic matter; many medium, fine, and very fine, and common coarse roots; 60% coarse fragments; pH 4.84; clear smooth boundary.
Oe	5–0 cm	Decomposed organic matter; many medium, fine and very fine, and common coarse roots; 60% coarse fragments; pH 4.84; abrupt wavy boundary.
E	0–19 cm	Pinkish gray (7.5 YR 6/2) and brown (7.5 YR 5/2); pinkish gray (7.5 YR 7/2) dry; very cobbly silt loam; weak medium subangular blocky structure; friable; many medium, fine, and very fine, and common coarse roots; 60% coarse fragments; pH 3.78; clear wavy boundary.
Bt	19–32 cm	Dark brown (7.5 YR 3/4); brown (7.5 YR 4/4) dry; extremely cobbly sandy loam; moderate medium subangular blocky structure; friable; common coarse, medium and fine roots; 75% coarse fragments; pH 3.72; gradual wavy boundary.
BC	32–56 cm	Dark yellowish brown (10 YR 5/4) and dark brown (7.5 YR 3/4); yellowish brown (7.5 YR 5/4) dry; extremely stony loamy sand; weak coarse subangular block structure; friable; few medium and fine roots; 85% coarse fragments; this horizon contains approximately 40% saprolite; pH 3.74; abrupt wavy boundary.
C	56 + cm	Massive granite and gneiss boulders.

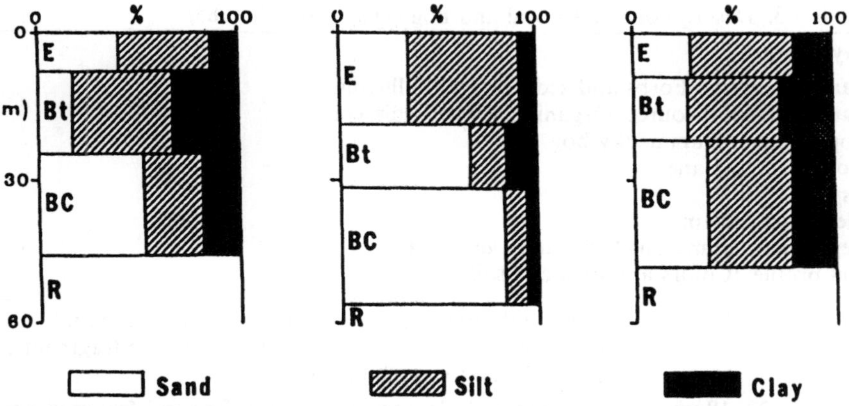

Figure 7.2. Particle-size distribution of three Cryoboralfs depicting weakly developed argillic horizons.

increases with depth. The Cryaquents are underlain by unconsolidated stream alluvium, which exhibit a gleyed C horizon, evidence of a highly reduced environment. Organic horizons of the Cryohemists rest directly on massive boulders of gneiss and granite.

Clay Mineralogy

The percent of clay in LVWS soils varies from a low near 4% in the A horizon of the Cryofluvents to 30% or greater in the E and Bt horizons of the Cryoboralfs. The secondary clay products that form are the result of weathering processes acting on primary minerals. Once formed, their stability is determined, to a large extent, by equilibrium conditions between the clay mineral and the soil solution (Lindsay 1979).

Smectite is the dominant clay mineral in LVWS soils. Formation and maintenance of smectite clays depends on high ionic concentrations of silica and magnesium. Silica is normally very soluble (Lindsay 1979), but high ionic concentrations can be maintained by slow movement or stagnation of water (Buol et al. 1980). Other commonly found clays were kaolinite, mica, and chlorite (Figure 7.3). Less abundant minerals identified in the clay-sized fraction ($< 2\,\mu m$) included vermiculite and hydro-biotite (mixed-layer vermiculite-biotite). Amorphous or poorly crystalline clay minerals were also found by selective extractions of the $< 2\,\mu m$ fraction (Follett et al. 1965) to be abundant in the soils. The molar silica-to-aluminium ratio between 2 and 3 suggests the presence of a poorly crystalline clay with a composition similar to kaolinite (Mast 1989). Quartz and feldspar were also present.

The occurrence of primary clays biotite and chlorite in the alpine soils and their absence in the forested soils suggests these micaceous minerals are destroyed by weathering. The young soils at the base of the talus (represented

Table 7.3. Descriptions of alluvial and bog soils (Walthall 1985)

Cryaquent

Parent material: Forbs and sedges, recent alluvium
Parent material source: Organics, stream sediment
Topography: Hummocky bog
Slope: 1–2%, plane
Aspect: N80°W
Elevation: 3110 m
Vegetation: Numerous forbs and *Carex* spp.
Soil profile: (Colors are for moist soil)

Oi	33–29 cm	Slightly decomposed organic matter; many fine and common very fine roots; less than 5% coarse fragments; pH 3.83; clear wavy boundary.
Oel	29–18 cm	Decomposed *Carex* spp.; common fine and few medium roots; less than 5% coarse fragments; pH 3.96; gradual wavy boundary.
Oe2	18–8 cm	Decomposed *Carex* spp.; pH 4.12; abrupt smooth boundary.
Oa	8–0 cm	Highly decomposed *Carex* spp. with some mineral material; pH 4.30; abrupt smooth boundary.
Cg	0–10 cm	Dark gray (5 YR 4/1); loam; structureless massive; very firm; less than 5% coarse fragments; pH 4.36; abrupt smooth boundary.
R	10 + cm	Massive granite and gneiss boulders.

Cryofluvent

Parent material: Recent alluvium
Parent material source: Stream sediment
Topography: Flood plain
Slope: 1%, plane
Aspect: S30°E
Elevation: 3110 m
Vegetation: Spruce, fir, forbs, and sedges
Soil profile: (Colors are for moist soil)

Oe	10–0 cm	Moderately decomposed spruce and fir needles; many fine and very fine roots; pH 3.91; abrupt smooth boundary.
C	0–22 cm	Very dark grayish brown (10 YR 3/2); sandy loam; common fine faint and few medium distinct mottles; structureless massive; friable; common fine roots; pH 4.38; clear wavy boundary.
Cg	22–39 cm	Very dark gray (10 YR 3/1); cobbly sandy loam; structureless massive; friable; few fine roots; 45% coarse fragments; pH 4.50; abrupt irregular boundary.
R	39 + cm	Coarse boulders of granite and gneiss.

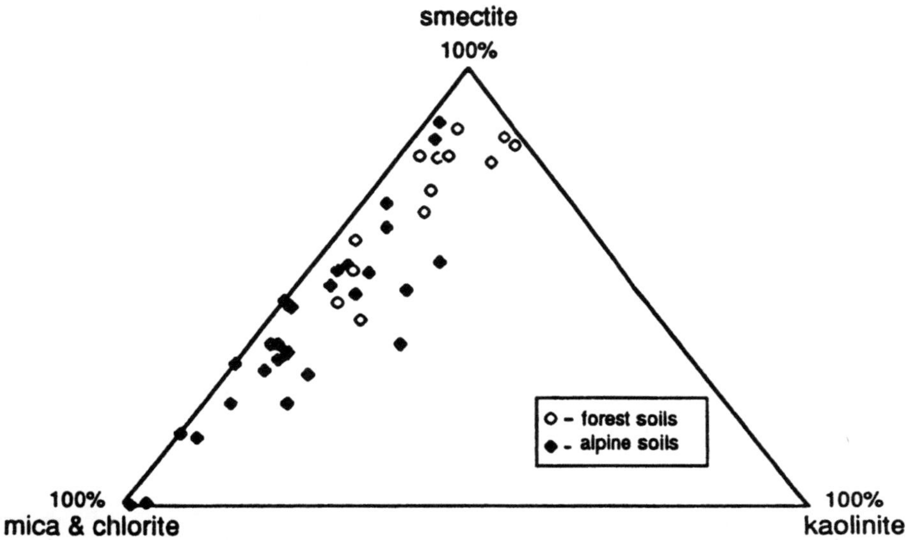

Figure 7.3. Relative abundance of commonly occurring clay minerals in the <2-μm size fraction of Loch Vale Watershed soils.

by a Cryochrept in Table 7.4) contain higher contents of illite, vermiculite, and chlorite, whereas mature soils of the forest floor are dominated by smectite (Table 7.4). The composition of the smectite was rarely pure, however (Mast 1989). X-ray diffraction patterns exhibited broad, poorly defined reflections with a high degree of scattering between 2° and 9° 2θ (Mast 1989). These broad reflections may indicate a poorly crystalline smectite or more likely the presence of a mixed-layer smectite-illite mineral (Reynolds 1980; Mast 1989). A series of extractions to determine the presence of amorphous clay material (Follett et al. 1965; Jackson 1967) suggested that most of the smectite was actually a mixed-layer smectite mineral containing roughly 30% illite layers (Mast 1989). This is supportive of initial stages of a weathering sequence suggested by Stoch and Sikora (1976) for granite and gneiss in Poland to weather to illite, then to illite/smectite, to smectite, and finally to kaolinite. The sequence of mineral weathering may not be straight-forward, however. In LVWS there are a number of starting parent minerals, including biotite, chlorite, illite, and feldspar, each with its own weathering sequence. Clay-sized chlorite in soils was iron rich, as was the chlorite in the bedrock, suggesting that chlorite is not a pedogenic material (Mast 1989). A possible weathering sequence for smectite may include transformation of primary biotite and chlorite, with vermiculite as an intermediate phase. With continued leaching, kaolinite may form from the smectite (Walthall 1985; Mast 1989), although kaolinite may also precipitate from solution.

Table 7.4. Semiquantitative Estimates of Coarse and Fine Clay Fractions (Walthall 1985)

		Smectite		Vermiculite		Chlorite		Illite		Kaolinite	
Site	Horizon	Coarse	Fine	Coarse	Fine	Coarse	Fine	Coarse	Fine	Coarse	Fine
Cryoboralfs											
	1 Bt	xxxx[a]	xxxx			x		x		x	
	2 E	xxx	xxxx	x		xx		x		xx	x
	Bt	xxxx	xxxx	x		x		x		x	
	3 E	xxxx	xxxx	xx						xx	
	Bt	xxxx	xxxx	x						xx	
	BC	xxx	xxxx	x				x		xx	x
	7 E	xxxx	xxxx					x		xx	x
	Bt	xxx	xxxx	xx		x		x		x	x
Cryochrept											
	10 A			xxx	xxx	x		xx	xxx	xx	x
	Bw	xx	xxxx	xxx	x			xx	x	xx	x

[a] Symbols: xxxx, > 60%; xxx, 30–60%; xx, 10–30%; x, < 10%.

Smectite is not thought to be stable in well-drained, acidic environments (Karathanasis and Hajek 1984). Flushing of soil water normally leaches out silica and base cations necessary for the maintenance of smectite (Karathanasis and Hajek 1984). Kaolinite is the dominant clay mineral in most Ultisols (Buol et al. 1980), although smectite is known to dominate in some (Flach, personal communication; Karathanasis and Hajek 1984), including those of LVWS. One theory to account for this phenomenon suggests that at some point in the weathering enough smectite forms to become self-sustaining. Because smectite retains water, silica is less likely to be lost, unlike other soils in which leaching of solution silica occurs readily (Karathanasis and Hajek 1984). The retention of water creates favorable conditions for formation of more smectite instead of more weathered clays, and smectite persists in this environment as a stable clay mineral.

There is no unique interpretation to account for LVWS clay mineralogy, but the greater dominance of what are commonly thought to be more mature clay minerals in the forested soils compared to a more heterogenous assemblage of clay minerals in the alpine soils supports the idea that forested soils represent the most highly weathered soils of the watershed.

Chemical Properties

Soil Acidity

All the soils of LVWS are acidic (Table 7.5). The pH (1:1 soil/water suspension) for soils of the valley floor range from 3.3 to 5.0. The soils above treeline have a somewhat higher pH range, 4.0–5.9, and these are similar to reported values from nearby Niwot Ridge (Burns 1980; Litaor 1987). The acidity of the forest soils results from the presence and persistence of organic acids formed from decomposition of litter and root exudation (Buol et al. 1980). The most acidic soils are the Cryoboralfs. This extreme acidity is attributed to (1) a mature weathering state in which further neutralization of acidity by weathering is minimal, (2) regular supplies of organic matter from coniferous forest litterfall, which serve as a source of organic acids, and (3) a relatively gentle topography, which increases water residence time and effective subsoil leaching. In the Inceptisols and the Entisols of the canyon walls and alpine zones, these conditions are not met. Instead, soils are less developed, steep slopes minimize contact time of runoff with soils, and there is less vegetative cover to supply organic matter. Additionally, these soils reflect a more active physical weathering environment in which calcite from microveins of the bedrock (Mast 1989; see Chapter 6, this volume) might be more frequently exposed. Higher pH values are expected in soils where calcite is present. This is in contrast to the more stable weathering environment of the valley floor, where leaching has exhausted the calcite supply.

Table 7.5. Exchangeable Cations (cmol kg^{-1}) of LVWS Soils (values are averages for similar horizons)

Site	Horizon	BUFCEC (cmol kg^{-1})	NSCEC (cmol kg^{-1})	BUFBS (%)	NSBS (%)	Organic Carbon (%)	Clay (%)	pH (H_2O)	pH (KCl)
Cryboralf									
	Oi + Oe	48.4	17.5	23.0	64.5	13.1	—	4.2	3.6
	E	30.0	12.8	15.4	30.6	2.7	19.5	3.7	3.4
	Bt	27.6	13.5	9.3	23.1	2.5	28.3	3.3	3.5
	BC	23.3	8.3	8.1	29.2	1.3	16.2	4.0	3.4
Cryochrept									
	Oe	—	23.9	—	91.2	11.6	—	5.0	5.1
	A	—	10.4	—	64.8	2.8	9.5	4.8	4.0
	Bw	—	7.5	—	44.4	0.9	12.6	4.7	4.0
	BC	—	3.9	—	25.6	0.2	6.1	5.0	3.5
Cryorthent									
	Oe	—	12.2	—	85.2	9.8	—	4.4	4.0
	A	—	6.4	—	45.3	2.3	8.2	4.0	3.3
	C	—	4.6	—	32.6	2.3	6.8	4.3	3.4
	Ab/C	—	5.9	—	30.5	2.1	8.4	4.3	3.3
	Cgb	—	6.5	—	53.8	1.2	9.6	4.6	3.6
	Cb	—	7.9	—	54.4	1.6	16.1	4.8	3.7
Cryofluvent									
	Oi + Oe	39.8	13.9	14.3	41.0	—	3.8	—	—
	A	36.9	13.7	5.2	13.9	6.7	3.8	3.8	3.8
	Cg	30.2	10.4	9.8	33.3	9.0	17.7	4.1	3.9
	Oeb	38.8	9.1	14.7	63.7	—	14.0	4.4	4.0
Cryaquent									
	Oi	—	16.8	—	63.2	16.8	—	4.0	4.1
	Oe	—	10.9	—	44.7	22.1	—	4.1	3.8
	Oa	—	10.5	—	32.8	17.0	—	4.1	3.7
	Cg	—	7.3	—	47.9	3.0	17.0	4.2	3.7
Cryohemist									
	Oi	—	15.8	—	89.9	18.1	—	4.0	4.2
	Oe	—	11.1	—	41.4	17.2	—	4.2	3.7

Exchange Capacity

The cation exchange capacity (CEC) of a soil is the capacity for soils to sorb and hold cations and then to release them back to the soil solution in reversible chemical reactions (Buol et al. 1980). Cation exchange values can be used to interpret the relative degree of weathering of a soil as well as the types of clay minerals that should be present (Buol et al. 1980). Total CEC is determined by summing the exchangeable base cations (Ca, Mg, Na, K) and the exchange acidity (Al and H) (Soil Survey Staff 1975). The exchangeable base cations and the exchange acidity are also useful for interpreting relative degree of weathering related to leaching, climate, parent material, and vegetation (Buol et al. 1980). We estimated exchange acidity by two different methods, resulting in different CEC values. The first method (buffered cation exchange capacity, or BUFCEC) considered the exchange acidity associated with both permanent charge exchange sites (independent of pH) and variable charge or pH-dependent exchange sites extracted in a buffered salt solution at pH 7.0 (see Table 7.5; Thomas 1982). This is the standard methodology used for the taxonomic characterization of soils (Soil Survey Staff 1975), and its use allows for comparisons at a standard or reference pH. In the second approach (neutral salt cation exchange capacity, or NSCEC), exchange acidity (Al and H; Table 7.6) was extracted with a neutral salt solution of 1 *M* KCl (McLean 1965). This extract determines the permanent charge but differs in that pH-dependent charge is measured at the pH of the soil. In the case of an acid soil the exchange acidity can be substantially less than that determined at pH 7.0 (Table 7.5). The NSCEC is more valuable for simulating exchange reactions in acid soils because it estimates effective exchange capacity (Kalisz and Stone 1980; Reuss 1983).

The greatest differences between BUFCEC and NSCEC in LVWS soils were found in the organic surface horizons (see Table 7.5); here the majority of the exchange sites were expected to be organic and pH dependent, resulting in extreme estimates of the pH 7.0 BUFCEC values. The difference in exchange capacity decreased in the mineral horizons of the soils where the influence of organic matter decreased. Permanent charge originating from secondary clays played a more important role in the CEC of mineral soil horizons.

Aluminum was the dominant cation on the exchange phase in subsoil horizons of LVWS (Table 7.6). The highest concentrations of base cations were found in the surface organic horizons from nutrient cycling driven by litterfall and decomposition. Neutral salt base saturation averaged 69.5% in organic surface layers of the Cryoboralfs. Underlying horizons were about 30.0% neutral salt base saturation (see Table 7.5). Some of these values approach a critical neutral salt base saturation level of 15%–20% at which soils become unable to supply base cations to surface waters under the influence of acidic deposition (Reuss and Johnson 1986). Soils with lower

Table 7.6. Exchange Chemistry and Related Soil Properties of LVWS Soils (values are averages for similar horizons)

Site	Horizon	Ca	Mg	K	Na	Al	H
		(cmol kg^{-1})					
Cryoboralf							
	Oi & Oe	9.2	1.6	0.9	0.1	3.5	1.8
	E	2.5	0.5	0.4	0.1	6.3	1.0
	Bt	1.8	0.4	0.2	0.1	8.4	0.7
	BC	1.3	0.3	0.1	0.1	5.4	0.4
Cryochrept							
	Oe	18.3	3.6	0.9	0.1	0.5	0.5
	A	6.5	1.5	0.2	0.1	1.7	0.5
	Bw	2.4	1.1	0.1	0.1	4.2	0.2
	BC	0.7	0.1	0.1	0.1	2.3	0.6
Cryorthent							
	Oe	8.0	1.5	0.8	0.1	1.0	0.8
	A	2.2	0.4	0.2	0.1	3.0	0.5
	C	1.2	0.1	0.1	0.1	2.8	0.3
	Ab/c	1.4	0.2	0.1	0.1	3.7	0.4
	Cgb	2.8	0.4	0.1	0.2	2.6	0.4
	Cb	3.5	0.5	0.2	0.1	3.4	0.2
Cryofluvent							
	Oi & Oe	3.5	1.3	0.7	0.2	5.4	2.0
	A	1.3	0.3	0.1	0.2	10.8	1.5
	Cg	2.5	0.6	0.1	0.2	6.4	0.7
	Oeb	4.2	1.2	0.1	0.2	2.6	0.7
Cryaquent							
	Oi	3.3	0.8	0.6	0.3	7.2	1.7
	Oe	1.6	0.2	0.2	0.1	6.8	1.1
	Oa	2.8	0.3	0.1	0.1	4.7	0.9
	Cg	2.7	0.6	0.1	0.1	3.1	0.7
Cryohemist							
	Oi	11.2	2.0	0.7	0.3	0.6	1.0
	Oe	3.6	0.7	0.2	0.1	5.8	0.7

neutral salt base saturation values are unable to exchange base cations for H that enters in precipitation. Instead, H and Al are leached from these soils when SO_4 and NO_3 flush through from precipitation to surface waters (Cosby et al. 1985).

Aluminum

Aluminum was the most abundant exchangeable cation in the forested soils and was also found in high concentrations in the other soils (see Table 7.6). The activity of aluminum in soil solutions was related to the solubility of a solid mineral phase according to the distribution of the total concentration of inorganic aluminum between its inorganic, monomeric form, Al^{3+}, and a number of associated complexes (Lindsay 1979; Walthall 1985). Most of the aluminum found in LVWS soils was found as an organoaluminum complex, as indicated by solution concentrations three orders of magnitude higher than the theoretical solubility of amorphous aluminum hydroxide (Drever 1988; Denning et al. in press). For those forested soil samples where organic carbon did not interfere with the analytical techniques (Walthall 1985), Al^{3+} activities appeared to be controlled by the solubility of smectite. The less mature soils had higher Al^{3+} activities, perhaps controlled by a range of unstable alumino-silicates (Walthall 1985).

Organic Matter

Organic matter in the soils exerts a dominant influence on almost every soil property. Its effect on LVWS CEC, pH, and Al solubility have already been mentioned. Soil organic matter also is an important parameter controlling soil fertility (Buol et al. 1980), complexation of metals such as Fe and Al (Reuss and Johnson 1986), accumulation and adsorption of SO_4 in litter and soil (Watwood et al. 1988), mineral weathering rates (Antweiler and Drever 1983; Drever 1988), and long-term pedogenesis (Mulder et al. 1989). Soil organic matter may influence many surface water properties, including pH, acid-neutralizing capacity (Sullivan et al. 1989; Denning et al., in press), Al mobilization (Driscoll 1985; Cozzarelli et al. 1987), and water clarity and light attenuation (McKnight and Feder, in press; Baron et al. in manuscript).

The organic carbon content varied greatly among soil horizons and sample locations. Organic carbon was highest in the upper organic horizons of all soils and decreased rapidly in the mineral layers. In the Cryoboralfs organic carbon ranged from 7.6% to 29.9% in the Oe horizons (high value from Watwood, personal communication; low from Table 7.5). In the E horizons and below, the organic carbon was much lower and less variable, with a range of 0.8%–3.7%. In bog soils of both the streamsides and alpine ridge, the organic carbon of organic horizons was less variable at 13.3%–22.3% (high value, Watwood, personal communication; low, Table 7.5).

Sulfur Fractions

Sulfate was most concentrated in the upper soil horizons (Table 7.7). A two-stage biochemical process is probably responsible for this concentration. First, sulfate from precipitation is incorporated into organic matter of the forest soil and litter through processes of microbial metabolism (Fitzgerald et al. 1983; Strickland and Fitzgerald 1984; Fuller et al. 1986; Watwood et al. 1988). The primary source of LVWS sulfur is wet deposition in the form of SO_4 ($\sim$8–10 kg SO_4 ha^{-1} yr^{-1}). Second, organic S is rapidly mineralized back to sulfate (Freney et al. 1975; Strickland and Fitzgerald 1984; Watwood et al. 1988). This breakdown and mineralization process is believed to result from the presence of preformed enzymes in the soil (Strickland and Fitzgerald 1984; Strickland et al. 1986). These enzymes may cause precipitation sulfate to cycle rapidly in and out of organic phases.

Table 7.7. Sulfur Fractions from Two Soils of Loch Vale Watershed[a]

			HI Reducible S		Carbon-Bonded S	
Soil	Total S	SO_4^{-2}	Total Ester S	Total HI Reducible S	Amino Acid S	Sulfonate S
OE	1039.6	7.7	317.0	324.7	124.9	589.9
E	87.6	2.6	28.1	30.7	19.0	38.0
Bt	166.6	7.8	45.1	52.9	18.1	95.6
Bog	1965.0	5.7	842.8	848.5	285.3	831.3

[a] Collected in August 1984. Values are given as $\mu g\ g^{-1}$ dry weight (Watwood unpublished data).

On a dry weight basis, 95%–99% of the sulfur in LVWS soils is organically bound and is present as either ester-bonded S or carbon-bonded S (Table 7.7). Only the small remaining fraction is present as SO_4. This finding is in common with other forested sites where studies have shown the largest pools of S to be organically bound. The most rapid cycling of S, however, occurs as inorganic SO_4–S (Fitzgerald et al. 1983; Johnson 1984).

Sulfate retention and accumulation in soils has been recognized as an important means of postponing surface water acidification and soil base cation depletion in areas that receive acidic deposition (Reuss and Johnson 1986). Accumulation of S can occur via two processes: (1) adsorption onto clay minerals (Chao et al. 1962; Johnson 1984), and (2) formation of organic S (Freney 1979; Strickland and Fitzgerald 1984). Both processes were examined for LVWS soils.

Sulfate adsorption isotherms were developed for an Inceptisol (alpine), Histisol (meadow), and an Alfisol (forested) soil by equilibrating soils with solutions of varying $MgSO_4$ concentrations (Cappo et al. 1987). Sulfate adsorption capacities of all the LVWS soils (Figure 7.4) were very low compared with capacities suggested for other forest soils. Harrison et al. (1989) reported adsorption of as much as 7.2 mmol kg^{-1} in East Tennessee soils, while Watwood et al. (1988) determined (by salt extraction; Fitzgerald et al. 1983) the amount of SO_4 already adsorbed to be 3.6 mmol kg^{-1} for the A1 horizon of a coastal plain pine forest in South Carolina, 18.0 mmol kg^{-1} for the A1 horizon of a hardwood forest at Coweeta Experimental Forest in North Carolina, and 15 mmol kg^{-1} for the A1 horizon of a mixed spruce, fir, pine, and aspen forest in the Santa Fe National Forest, New Mexico. By contrast, the LVWS Inceptisol and Alfisol soils adsorbed less than 1.2 mmol kg^{-1} when supplied with almost 1 mM SO_4 solution (Figure 7.4 A, B). The low observed levels are partially due to low amounts of amorphous aluminum oxide minerals, such as gibbsite (sesquioxide), which provide most anion adsorption (Chao et al. 1962; Johnson 1984).

The Histisol isotherms (Figure 7.4C) desorbed SO_4 with an increase in SO_4 loading, probably because of the release of water soluble SO_4 from organics. Other investigators have reported the negative influence of organic matter on SO_4 adsorption (Johnson and Todd 1983; Johnson 1984; Watwood et al. 1988), possibly because of blockage of adsorption sites by organic ligands or the influence of organic matter on clay minerals.

The rates of formation and retention of organic S in LVWS soils were examined in the laboratory with 48-hour incubations with ^{35}S (Strickland et al. 1986). The resulting sulfur-processing capacities indicate that 9.7 and 10.0 nmol organic S 48 hr^{-1} g^{-1} soil were formed in the O2 Alfisol horizon and the Histisol, respectively, while far less was formed in the E and Bt2 of the Alfisol (Table 7.8). The rate of organic S formation in the O2 horizon is about twice as high as that reported for O2 horizon soils of a lower elevation Alfisol in New Mexico, while the rates for formation in E and Bt2 horizons in LVWS were actually lower than that reported from New Mexico (1.9 and 2.1 nmol organic S 48 hr^{-1} g^{-1} soil for the E and Bt2 Alfisol horizons in LVWS, respectively, compared with 8.0 nmol organic S 48 hr^{-1} g^{-1} soil in the Santa Fe National Forest; Watwood et al. 1988). The potential for mobilization of organic S in LVWS soils was quite high, and between 18.9% and 40.3% of the organic S formed during the incubations was mobilized within a subsequent 24-hour period (Table 7.8). This is comparable to rates of mobilization reported from New Mexico (29.9.% for the O2 horizon and 31.6% for the E horizon; Watwood et al. 1988). Mobilization rates were lower than reported from eastern low-elevation forests (Watwood et al. 1988).

Table 7.8. Sulfur-Processing Capacities for an Alfisol and a Histisol of Loch Vale Watershed[a], and one from the Santa Fe National Forest[b]

Soil	pH	Moisture (%)	SO_4^{-2} Adsorption	Organic S Formation	Organic S Accumulation	Organic S Mobilization
			nmoles 48 h^{-1} g^{-1}			% 24 h^{-1}
Loch Vale Watershed						
Oe (2 cm)	4.1	84.0	3.6	9.7	7.8	18.9
E (0–15 cm)	3.7	30.7	4.3	1.9	1.2	36.1
Bt (15–30 cm)	3.8	38.8	7.7	2.1	1.3	40.3
Bog (0–20 cm)	3.6	268.5	25.5	10.0	6.9	30.9
Santa Fe National Forest						
Oe	—	—	3.1(0.3)	4.8(0.6)	3.4	29.9(4.2)
A	—	—	5.0(0.4)	3.0(0.2)	2.1	31.6(2.6)

[a] Collected in August 1984 (Watwood, unpublished data),
[b] n = 15 ± 1S.D.; (Watwood et al. 1988, used with permission).

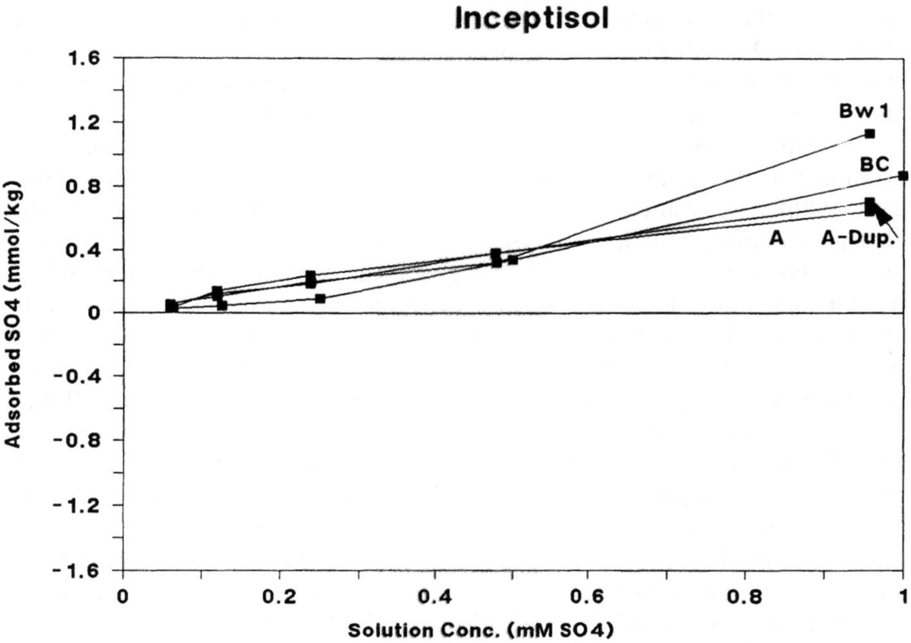

A

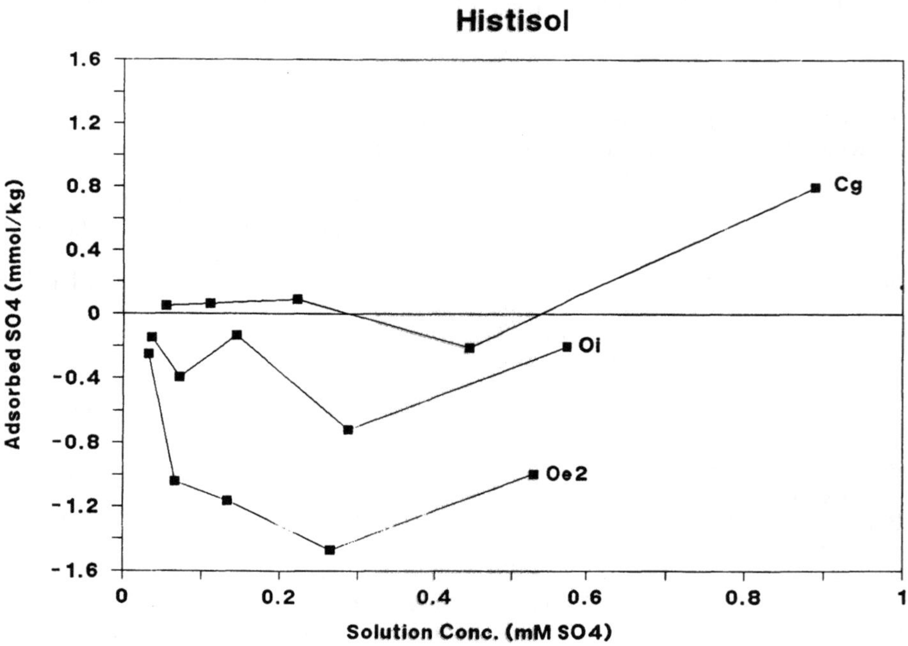

B

Figure 7.4. Sulfate adsorption isotherms for three Loch Vale Watershed soils: (A) Inceptisol, (B) Histisol, and (C) Alfisol. Isotherms were determined according to methodology of Cappo et al. (1987).

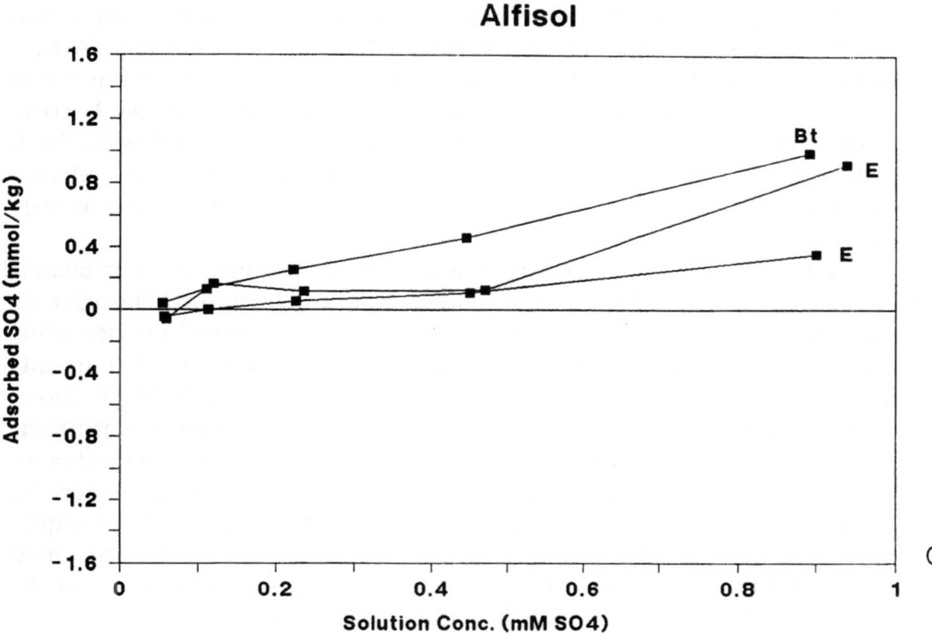

C

Figure 7.4 (*Continued*)

Soil Water Dynamics

Soil Solution Chemistry

In subalpine and alpine ecosystems, much of the annual precipitation accumulates in the snowpack and then quickly flushes through the soil during snowmelt. In LVWS soil water was readily available only during the time period May, June and July. Only rarely do summer rains infiltrate the soil sufficiently to leach through to surface waters. Elevated concentrations of solutes in soil solutions during snowmelt have been attributed to the accumulation of atmospheric inputs of solutes to the snowpack and to the flushing of by-products of N mineralization and nitrification during the winter (Johnson et al. 1969; Rascher et al. 1987; Fahey and Yavitt 1988). Snowmelt has been shown to cause short-term changes in the composition of surface water (Schofield et al. 1985; Rascher et al. 1987; Denning et al., in press). Soil water was interpreted to determine the pathways and influences on water passing through the forested soil during and after snowmelt.

Gravity flow lysimeters were installed in October 1985 at 10 randomly chosen locations in forested plots (Arthur 1990). The forested plots were among 20 permanent vegetation plots placed in LVWS; plot selection was based on accessibility and sufficient soil depth for lysimeter placement (Arthur 1990). The fritted-glass-plate lysimeters (8.5-cm diameter, Corning Glass Works, Inc.) were placed in pairs, with one directly below the Oie horizon (forest floor lysimeters) and the other approximately 20 cm below in the B horizon (root-zone lysimeters). Solutions from the lysimeters were collected for 3 years, twice weekly throughout the period of snowmelt, starting in early May of 1986 (Arthur 1990).

Interpretation of the chemistry of soil solutions collected by lysimeters is difficult for a number of reasons, including spatial and temporal heterogeneity within soils (Fahey and Yavitt 1988), differences in pH and p*e* from degassing and aeration (Lindsay 1979), and microbial consumption of nutrients after sample collection but before sample analysis (Arthur 1990). Volume-weighting was employed when calculating mean soil solution concentrations to minimize some of the variability caused by samples with very low volumes. Volume-weighted concentrations for each lysimeter were calculated by summing the total quantity of each solute for all the soil solution samples for a given year and dividing by the total water flux. The volume-weighted mean (VWM) concentration was then calculated as the mean value of the volume-weighted concentrations for all the lysimeters.

Snowpack depth and snowmelt were highly nonuniform, resulting in large differences both in amount and timing of flow into lysimeters. One pair of lysimeters flowed continuously from early May through July during 1986–1988; several others had barely flowed at all; and the rest of the lysimeters varied in the amount that could be collected. To test differences in soil solutions resulting from snowmelt, solutions were arbitrarily divided into "early" and "late" snowmelt, with late snowmelt defined as solutions collected after the date at which all the lysimeters had collected soil water at least once. By this definition, early solutions in 1986 were collected before June 14, whereas early solutions in 1987 and 1988 were collected before May 18. The differences between years is a direct result of variability in snowpack depth and onset of snowmelt in different years.

Seasonal trends in forest floor and root zone soil solution chemistry were observed, with significantly higher VWM concentrations of nearly all solutes at the beginning of snowmelt (Figure 7.5). Differences between early and late snowmelt were detected in all years but were most dramatic in 1986, when snowpack was deepest and snowmelt started the latest of the 3 years of study. After the first flush of solutes early in the snowmelt period, concentrations in all lysimeters declined rapidly to minimum values of the late samples. Minimum values were very similar among years and were also not very different between soil solutions and surface waters.

The VWM H concentration of soil solutions was 7 μeq L^{-1} (pH of 5.2) in the forest floor solutions and 8 μeq L^{-1} (pH of 4.9) in the root zone solutions (Table 7.9). Although H appeared to increase slightly with depth, no significant conclusion can be reached because there were large differences

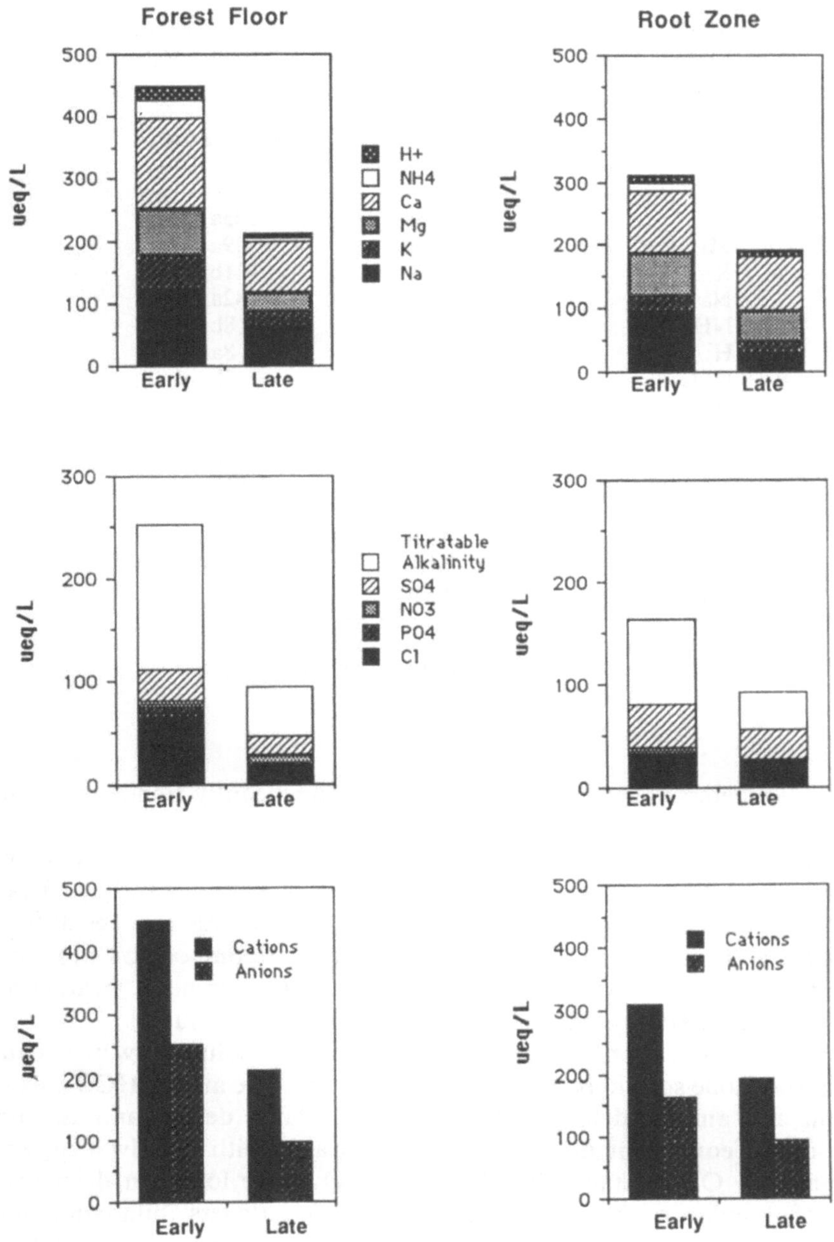

Figure 7.5. Concentrations of major cations (top), major anions (middle), and (bottom) sum of cations versus sum of anions in "early" and "late" forest floor and root zone soil solutions (left and right columns, respectively) from Loch Vale Watershed for the 1986 melt season. (Arthur 1990).

Table 7.9. Volume-Weighted Mean Concentrations of Major Solutes in Forested Areas of Loch Vale Watershed[a]

Solute	Forest Floor	Rooting Zone
	(μeq L^{-1})	
Ca	142a	105a
Mg	63a	49a
K	66a	21b
Na	96a	82a
NH_4	25a	8b
H	7a	8a
NO_3	11a	5b
SO_4	31a	34a
Cl	48a	20a
PO_4	9a	4b
Alkalinity	86a	81a
	(mg L^{-1})	
Si	0.95a	1.70a
Al	0.18a	0.43a
DOC	26.0a	25.0a

[a] Values for forest floor and root zone solutes are for 1986–1988. In columns, values with different letters are significantly different ($p = 0.10$, Wilcoxon signed rank test). (Arthur 1990).

among lysimeters. There were no statistically significantly trends in acidity over time.

Total alkalinity was not significantly different between forest floor and root-zone solutions but was dramatically different between early and late solutions. Alkalinity declined as snowmelt progressed. This may result from dissolution of respiratory CO_2 accumulated from decomposition processes occurring during the winter (Arthur 1990), as well as mineral weathering, which produces HCO_3 (Mast 1989; see Chapter 6, this volume).

The dominant cations in forest floor and root zone solutions were Ca and Na. In root zone solutions, Mg was also important. The anions HCO_3, weak organic acid anions [determined from charge balance deficits and assumptions of the equivalent negative charge associated with dissolved organic carbon from Oliver et al. (1983)], Cl, and SO_4 were found in decreasing order of abundance. Nitrate and PO_4 were significantly less abundant in late solutions at both depths and had much lower solution concentrations than the other anions. Statistically significant differences in solute concentrations between root zone lysimeters and forest floor lysimeters were found for NH_4, NO_3, PO_4, and K, only (Table 7.9).

Concentrations of dissolved organic carbon (DOC) were similar in forest floor and root zone solutions. Correlations between DOC and most other solutes were high, and were especially high for total Al (Arthur 1990; Denning et al., in press). Large charge balance deficits (estimated as sum of cation equivalence minus sum of anion equivalence) were observed in all forest soil solutions. High correlations between DOC and the charge balance deficit ($r = 0.65$, $n = 78$ for forest floor; $r = 0.73$, $n = 115$ for root zone samples) suggest that many of the unmeasured anions in solution may be attributed to anions of weak organic acids (Oliver et al. 1983). DOC concentrations were much higher during the early stages of snowmelt, probably because of the dissolution of soluble by-products of decomposition that occurred over winter.

Soluble organic anion concentrations have been observed in other forested systems to peak in the forest floor and be lost from solution in the mineral soil (Cronan 1980; Sollins and McCorison 1981). In LVWS, however, concentrations of DOC were similar between soil depths. In spodic horizons, deposition of organic matter (and thus loss of DOC from solution) is attributed to precipitation and adsorption reactions involving iron, aluminum, and organic matter (Schnitzer 1969). The solubility of organic carbon is dependent on several factors, including soil clay content (Oades 1988). Fahey and Yavitt (1988) found that coarser soils had higher DOC concentrations than clayey soils in lodgepole pine forests, probably because of lower adsorption of organics on mineral colloids. Mineral soils at LVWS are moderately coarse (Walthall 1985), which may contribute to the relatively high concentrations of DOC in the subsoil solutions and lack of adsorption of organics with depth.

Concentrations of total N were similar between the two lysimeter depths, but N was dominated by inorganic forms (64%) in forest floor solutions, whereas the reverse was true in root zone solutions (NO_3-N and PO_4-P were subtracted from total N and P values determined by standard methods to yield amounts of inorganic and organic N and P, respectively; Arthur 1990). Seventy-five to 91% of total N was organic in the root zone solutions (Table 7.10). DOC concentrations were similar at both depths, resulting in a much higher organic C/N ratio in the forest floor solutions than in the root zone solutions. A similar trend of inorganic to organic forms was observed for P, with nearly all of total P accounted for as PO_4 in forest floor solutions, but only 40%–45% in root zone solutions (Table 7.10). Concentrations of NH_4, NO_3, and PO_4 were all significantly hiher at the onset of snowmelt, decreasing to undetectable levels after the first several weeks. Thus, differences between forest floor and root zone solutions in the proportion of inorganic to organic N and P resulted from much higher concentrations of inorganic forms of these elements in early solutions.

Concentrations of Al and Si increased with depth (see Table 7.9). There was a high correlation between total Al and DOC ($r = 0.76$, $n = 114$), again suggesting that Al occurred primarily as an organic aluminum complex.

Table 7.10. Volume-Weighted Mean Concentrations of N and P in Forested Soil Solutions of Loch Vale Watershed[a]

Species	Forest Floor	Root Zone
pH	5.2a	5.1a
NH_4-N	0.35a	0.11b
NO_3-N	0.15a	0.07a
OrgN	0.28a	0.59a
Total N	0.78a	0.77a
PO_4-P	0.06a	0.05a
OrgP	0.00a	0.06b
Total P	0.06a	0.11a
DOC	26.00a	25.00a
C:N	93	42

[a] Values are for 1986–1988. In columns, values with different letters are significantly different ($p = 0.10$, Wilcoxon signed rank test). All N, P, and DOC values are in mg L^{-1}. (Arthur 1990).

A Solution Continuum: Snowmelt, Soil Solutions, and Surface Waters

The importance of soil solutions to surface waters was evaluated by comparing solute concentrations between snowmelt water, soil solutions, and surface waters for the early and late periods of analysis. If surface waters (interpreted from samples collected at The Loch outlet) were simply snowmelt water piped to surface water either as overland flow or through macropores, they should not be enriched in DOC, total Al, Si, H, base cations, or alkalinity. We tested this with a comparison of VWM snowmelt water and soil water for 1987. A continuum was constructed of the first samples (April 27–May 2) from snowmelt, soil water, and surface waters (Figure 7.6). The lysimeters were enriched in H, Ca, Mg, Na, Cl, total Al, and Si. They were depleted in NO_3 when compared to snowmelt water, equal in K, and variable with depth for SO_4. Ammonium in soil water was below detection limits. Orthophosphate, DOC, and alkalinity in snowmelt were not analyzed during this period. Clearly the soil solutions reflected soil processes in addition to inputs from snow.

▶

Figure 7.6. Volume-weighted mean concentrations of snowmelt, forest floor solutions from forest floor and root zone solutions and Loch outlet stream water during snowmelt period in 1987. Early snowmelt represents period 4/27/87–5/2/87, and late snowmelt represents period 5/3/87–6/13/87 for snowmelt and stream water samples. Lysimeter values represent first four sample dates for early period and remainder of 1987 sample period for late period.

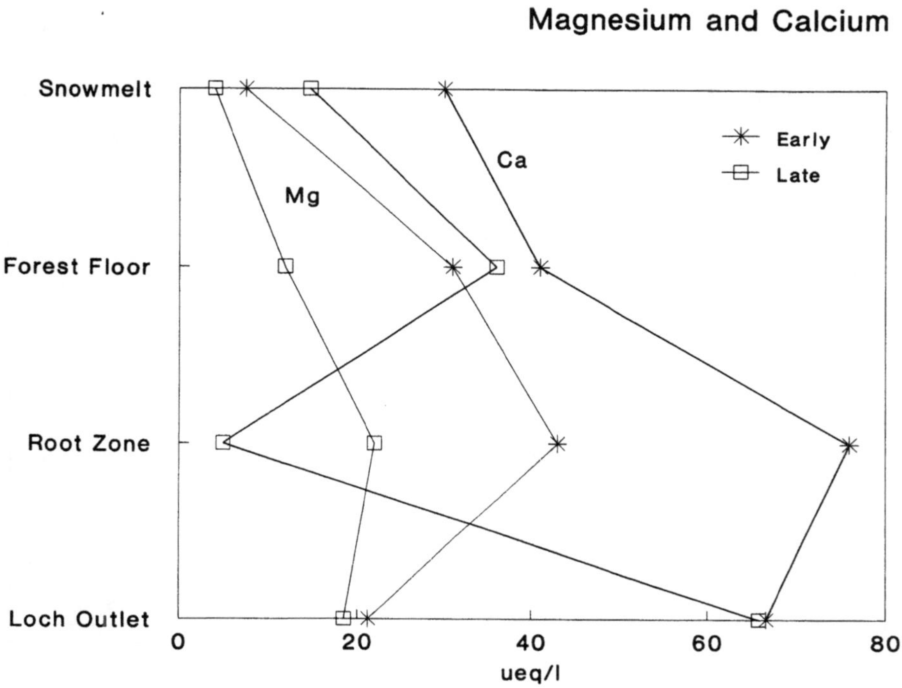
Magnesium and Calcium
Snowmelt
Forest Floor
Root Zone
Loch Outlet
Ca
Mg
Early
Late
0
20
40
60
80
ueq/l

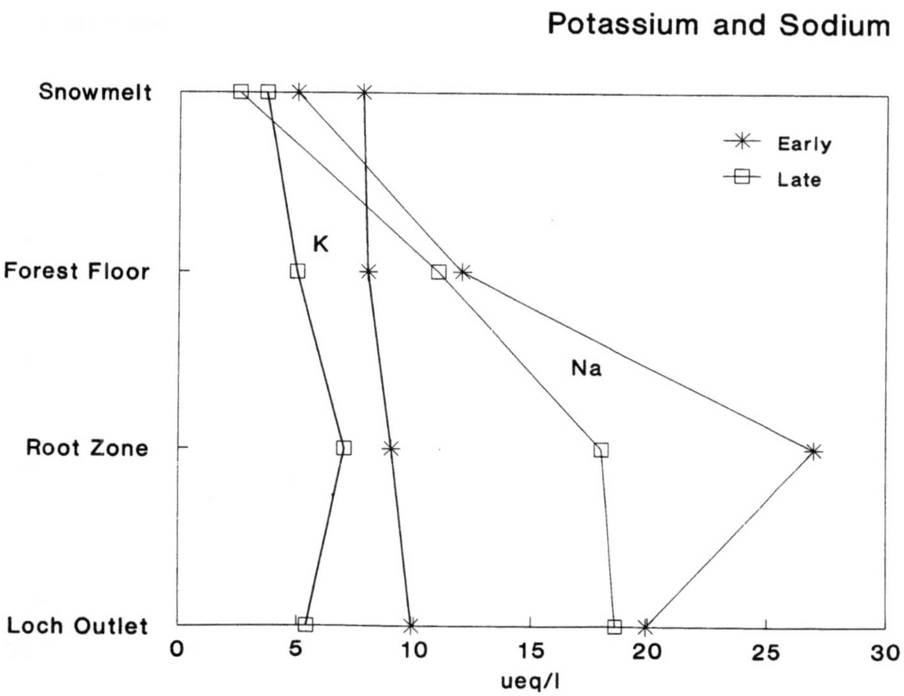
Potassium and Sodium
Snowmelt
Forest Floor
Root Zone
Loch Outlet
K
Na
Early
Late
0
5
10
15
20
25
30
ueq/l

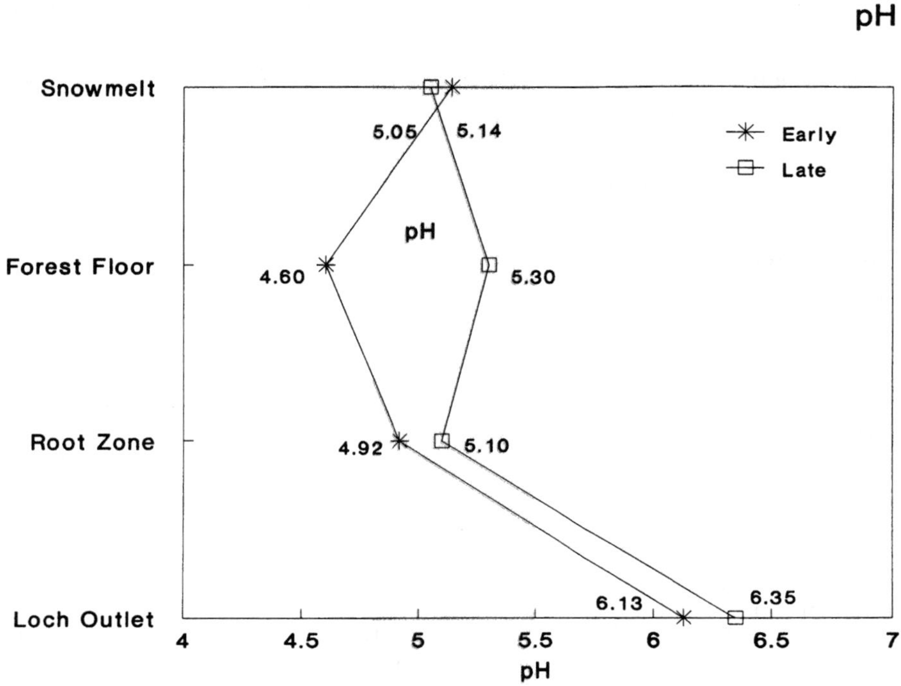

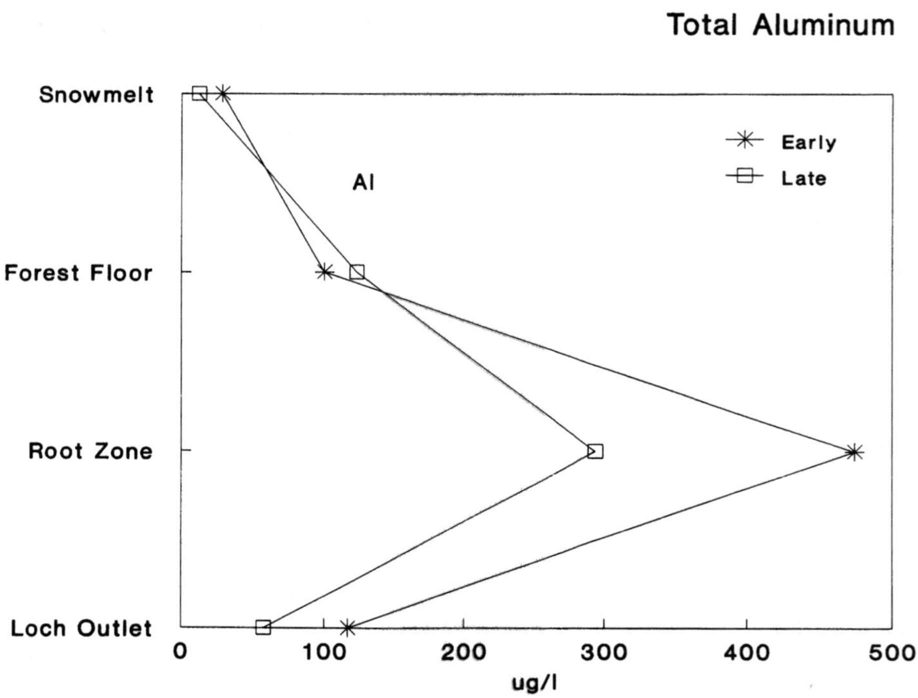

Figure 7.6 (*Continued*)

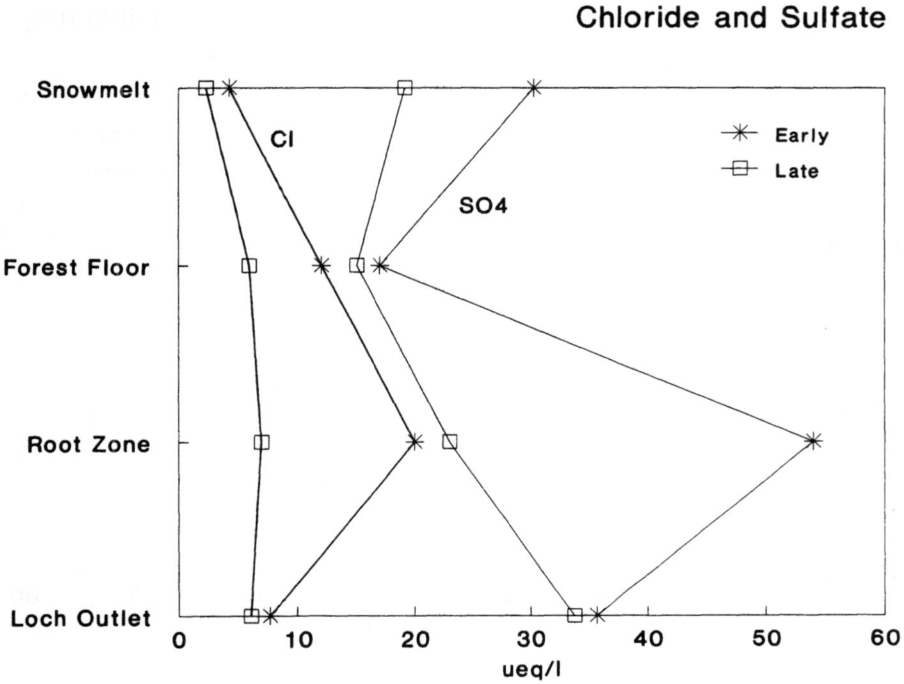

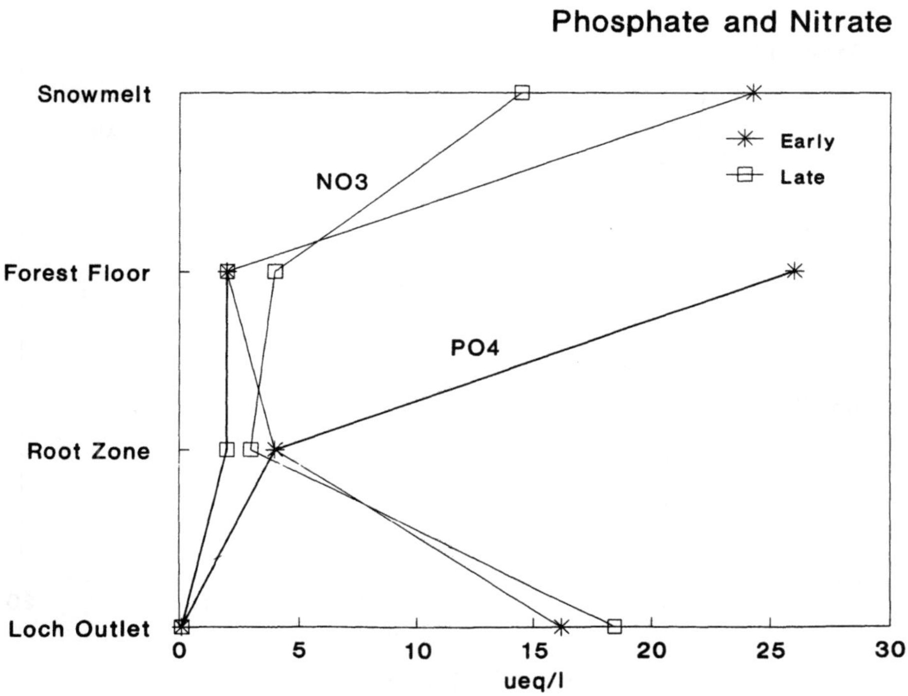

Figure 7.6 (*Continued*)

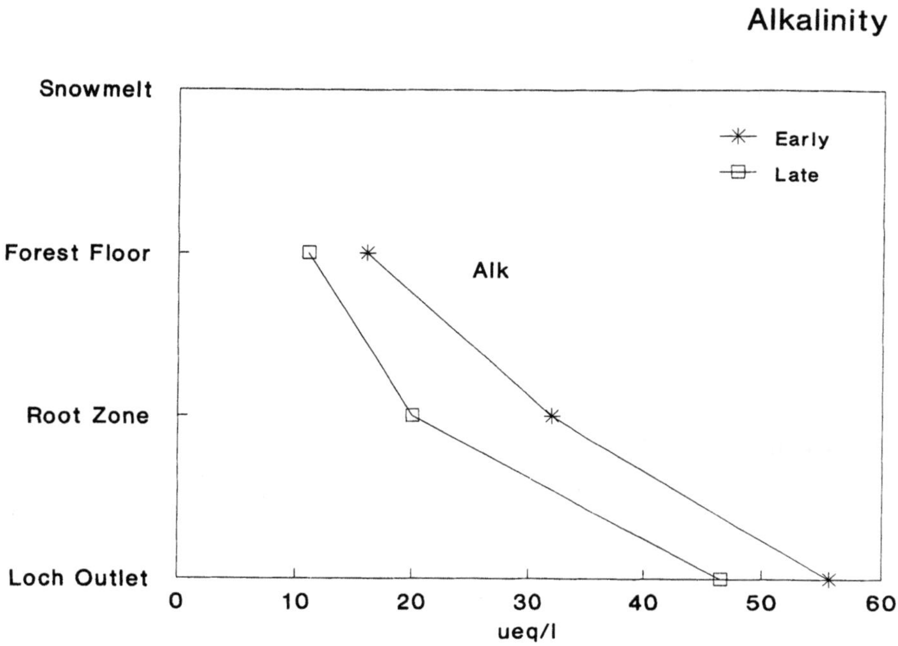

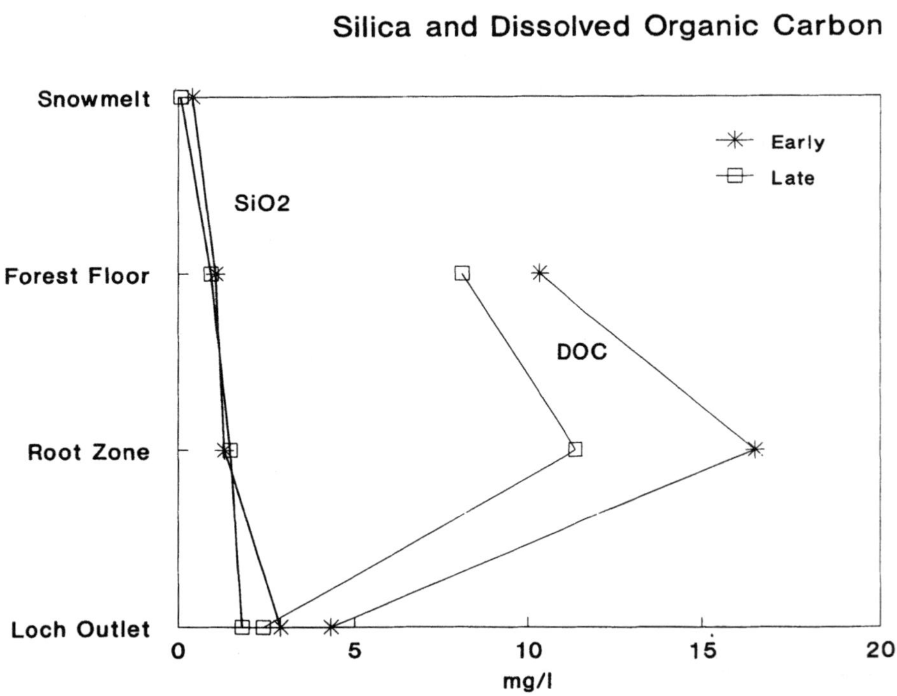

Figure 7.6 (*Continued*)

Table 7.11. Root Zone Leaching of Cations ($eq\ ha^{-1}\ yr^{-1}$) Minus Precipitation Inputs[a]

Cation	$Eq\ ha^{-1}\ yr^{-1}$
Ca	528
Mg	329
K	120
Na	400
Total	1377
Minus precipitation input	66
Total	1311
Total cationic denudation rate for watershed	382

[a] Compared to cationic denudation rate calculated for Loch Vale Watershed (Mast 1989; Chapter 6) (Arthur 1990).

Forest floor and root zone solutions had higher concentrations of most cations during the early snowmelt period, most likely because of mineralization of organic matter. Rascher et al. (1987) observed similar trends in snowmelt and forest floor flux in the Adirondack Mountains of New York, with the exception that the forest floor there was an important source for NO_3 whereas in LVWS it was not.

Surface waters showed increased concentrations of some solutes, including total Al, Na, and Ca, over their concentrations in snowmelt water, indicating a soil source. Concentrations of H were lower in the stream than in the soils. These same species had their greatest concentrations in soil solutions over surface waters and snowmelt waters, suggesting the soils serve as a pool for these ions.

A similar comparison was made with late season (May 3–June 13) samples for the 1987 snowmelt period. Lysimeters were again enriched in Ca, Mg, Na, Si, and total Al. Concentrations of H, K, Cl, and SO_4 in snowmelt and soil waters did not differ (see Figure 7.6). Nitrate had lower concentrations in soils than in snowmelt.

To sum, while the soil solutions clearly reveal the influence of soil processes, this influence is not totally mirrored in stream water. Concentrations of solutes in stream water were lower than those in the soils, suggesting that much of the snowmelt bypasses soils. This is not surprising considering that only 6% of LVWS has either forested or meadow soils. Increases in the concentrations of alkalinity, total Al, and some of the base cations during the earliest snowmelt (Figure 7.6) indicate that, at this time at least, there is an influence; this is discussed further in Chapter 8 (this volume).

On a molar basis, the ratio of Ca to Na in root zone solutions (1.50) was similar to the ratio measured in stream water (1.55; Mast 1989). This suggests that weathering reactions, rather than decomposition of organic matter, dominate the cationic concentrations in the soil solution (Arthur 1990). Higher unit area rates of cationic denudation from the forest than the watershed (taken as a whole) were expected, because of H exudation from roots during nutrient uptake and the generation of organic acids in forested soils. Based on soil solution chemistry and snowmelt flux, base loading was calculated by Arthur (1990) to be 1311 eq $ha^{-1} yr^{-1}$ (Table 7.11). This was about 3.5 times as high as the cationic denudation rate calculated for the whole watershed using a mass-balance approach (380 eq $ha^{-1} yr^{-1}$; Chapter 6, this volume; Mast 1989). Although precipitation inputs and mineral weathering (ignoring any contribution from soils) can explain the total solute flux from LVWS on an annual basis, during the snowmelt period, moderate changes in surface water chemical composition reflect the input of solutes, especially cations (Ca, Mg, Na), DOC, and total Al from the soils (see Chapter 8, this volume; Denning et al. in press).

Conclusions

The soil environment of the LVWS valley floor is not extensive, covering only 6% of the total watershed area. Alpine soils, which cover another 11% of LVWS, are physically disjunct from the valley floor, separated by bedrock cliffs and talus. Alpine soils are poorly developed because of the harsh physical conditions of extreme winds and cold temperatures, which create almost desert-like conditions in much of the alpine. A thin band of relatively young, immature soils occurs at the base of steep talus slopes. Bog and alluvial soils are found in low-lying areas next to streams and lakes. Forested soils are the most extensive soil type. They are characteristically highly acid, with pH values between 3.0 and 5.0. While currently classified as Cryoboralfs, these soils have characteristics of frigid Ultisols, including low base saturation, argillic horizons, and high acidity. Mean base saturation percentages range from 21% in forest Bt horizons to 37% in C horizons underlying the bog soils. These soils are unusual in that their dominant clay mineral is smectite, thought to be not highly stable in acid environments, and their soluble Al concentrations are very high (between 4.0 and 5.0 log molar). Most of this Al is complexed with organic matter. Solid-phase controls of the remainder were in keeping with extant clay minerals and the maturity of the soils. The younger and more immature soils, with higher concentrations of transitional clay minerals such as illite, vermiculite, and chlorite, have Al^{3+} activities poised above the aluminum solubility of smectite. In the forested soils, a substantial decrease in aluminum potential is observed, and Al^{3+} shows good alignment with the solubility of smectite.

Currently, soil water undergoes a dramatic temporal change during snowmelt. The earliest snowmelt flushes the accumulated by-products of 8 months of soil weathering and decomposition as soil water infiltrates horizons. Differences in N and K between the forest floor root zone solution chemistry reflect the importance of root uptake in the higher horizons. Other important soil processes influencing soil solution chemistry include litter mineralization, soil cation exchange and weathering, and the lack of adsorption of organics and SO_4. Mobility of SO_4 and organics control the leaching of cations at this site when there is freely moving water (May, June, and July only). The lack of adsorption of soluble organics with depth may be the result of coarse soil texture. The lack of SO_4 adsorption is further affected by low sulfate adsorption capacity and low degree of sulfate retention by organic matter.

These soils will not provide much buffering to surface waters from increased acidic deposition. Base saturation is low, offering limited ability to exchange base cations for acidity from deposition. On the basis of sulfate adsorption isotherms and rates of organic S formation, it appears the forested and organic soils of LVWS have limited capacity for retaining incoming sulfate. If acid deposition increases, sulfate or nitrate can easily deplete the base cation supply within the soil. Soil will then become a source of acidification for surface waters. This may not cause lake and stream acidification on an annual basis, since the primary source of acid-neutralizing capacity in LVWS is produced from physical exposure of fresh rock surfaces. It may, however, cause an acidic pulse to the lower watershed during the first weeks of snowmelt when the soils have their greatest influence on lake and stream composition.

References

Antweiler RC, Drever JI (1983) The weathering of a late tertiary volcanic ash: importance of organic acids. Geochim. Cosmochim. Acta 47:623–629.

Arthur MA (1990) The Effects of Vegetation on Watershed Biogeochemistry at Loch Vale Watershed, Rocky Mountain National Park, Colorado. Ph.D. Dissertation, Cornell University, Ithaca, New York.

Benedict JB (1970) Downslope soil movement in a Colorado alpine region: rates, processes and climatic significance. Arct. Alp. Res. 2:165–226.

Buol SW, Hole FD, McCracken RJ (1980) Soil Genesis and Classification. The Iowa State University Press, Ames, IA.

Burns SF (1980) Alpine Soil Distribution and Development, Indian Peaks, Colorado Front Range. Ph.D. Dissertation, University of Colorado, Boulder (Diss. Abstr. 81–13948).

Cappo KA, Blume LJ, Rayab G, Bartz JK, Engels JE (1987) Analytical Methods Manual for Direct/Delayed Response Soil Survey. Environ. Mon. Sys. Lab. ORD, Section 13: Sulfate Adsorption Isotherms. U.S. Environmental Protection Agency, Las Vegas, Nevada.

Chao TT, Harward ME, Fang SC (1962) Adsorption and desorption phenomena of sulfate in soils. Soil Sci. Soc. Am. Proc. 26:234–237.

Christopherson N, Seip HM, Wright RF (1982) A model for streamwater chemistry at Birkenes, Norway. Water Resour. Res. 18:977–996.

Cosby BJ, Wright RF, Hornberger GM, Galloway JN (1985) Modelling the effects of acid deposition: estimation of long-term water quality responses in a small forested catchment. Water Resour. Res. 21:1591–1601.

Cozzarelli IM, Herman JS, Parnell, RA, Jr. (1987) The mobilization of aluminum in a natural soil system: effects of hyrologic pathways. Water Resour. Res. 23:859–874.
Cronan CS (1980) Solution chemistry of a New Hampshire subalpine ecosystem: a biogeochemical analysis. Oikos 34:272–281.
Denning AS, Baron J, Mast MA, Arthur MA Hydrologic pathways and chemical composition of runoff during snowmelt in Loch Vale Watershed, Rocky Mountain National Park, Colorado. Water Air Soil Pollut. 55.
Drever JI (1988) The Geochemistry of Natural Waters, 2d Ed. Prentice-Hall, Englewood Cliffs.
Driscoll CT (1985) Aluminium in acidic surface waters: chemistry, transport and effects. Environ. Health Perspect. 63:93–104.
Fahey TJ, Yavitt JB (1988) Soil solution chemistry in lodgepole pine (*Pinus contorta* ssp. *latifolia*) ecosystems, southeastern Wyoming, USA. Biogeochemistry 6:91–118.
Fitzgerald JW, Ash JT, Strickland TC, Swank WT (1983) Formation of organic sulfur in forest soils: a biologically mediated process. Can. J. For. Res. 13:1077–1082.
Follett E, McHardy AC, Mitchell WJ, Smith BFL (1965) Chemical dissolution techniques in the study of soil clays: Part I. Clays Clay Miner. 6:23–34.
Freney JR (1979) Sulfur transformations. In Fairbridge RW and Finkl CW, eds. The Encyclopedia of Soil Science. Part I. Physics, Chemistry, Biology, Fertility and Technology, Pennslyvania. Dowden, Hutchinson & Ross, Stroudsburg, pp. 536–544.
Freney JR, Melville GE, Williams CH (1975) Soil organic matter fractions as sources of plant available sulphur. Soil Biol. Biochem. 7:217–221.
Fuller RD, Driscoll CT, Schindler SC, Mitchell MJ (1986) Modelling of sulfur transformation in forested spodosols. Biogeochemistry 2:313–328.
Harrison RB, Johnson DW, Todd DE (1989) Sulfate adsorption and desorption: reversibility in a variety of soils. J. Environ. Qual. 18:419–426.
Jackson ML (1967) Soil Chemical Analysis–Advanced Course. Department of Soil Science, University of Wisconsin, Madison (unpublished).
Jenny H (1941) Factors of Soil Formation: A System of Quantiative Pedology. McGraw-Hill, New York.
Johnson DW (1984) Sulfur cycling in forests. Biogeochemistry 1:29–43.
Johnson DW, Todd DE (1983) Relationships among iron, aluminium, carbon and sulfate in a variety of forest soils. Soil Sci. Soc. Am. J. 47:792–800.
Johnson NM, Likens GE, Bormann FH, Pierce RS (1969) A working model for the variation in streamwater chemistry at the Hubbard Brook Experimental Forest, New Hampshire. Water Resour. Res. 5:1353–1363.
Kalisz PJ, Stone EL (1980) Cation exchange capacity of acid forest humus layers. Soil Sci. Soc. Am. J. 44:407–413.
Karathanasis AD, Hajek BF (1984) Evaluation of aluminum-smectite stability equilibria in naturally acid soils. Soil Sci. Soc. Am. J. 48:413–417.
Lindsay WL (1979) Chemical Equilibrium in Soils. Wiley-Interscience, New York.
Litaor MI (1987) The influence of eolian dust on the genesis of alpine soils in the Front Range, Colorado . Soil Sci. Soc. Am. J. 51:142–147.
Lovering TS, Goddard EN (1959) Geology and Ore Deposits of the Front Range, Colorado. U.S.G.S. Prof. Paper 223, U.S. Geological Service, Denver, Colorado.
Mast MA (1989) A Laboratory and Field Study of Chemical Weathering with Special Reference to Acid Deposition. Ph.D. Dissertation, University of Wyomng, Laramie.
McKnight D, Feder J Ecological aspects of humic substances in the environment. In: McCarthy P, Malcolm RL, Swift R, Hayes M, eds. Humic Substances: Environmental Interactions. Wiley, London (in press).

McLean EO (1965) Aluminum. In: Black CA, ed. Methods of Soil Analysis, Agronomy, Vol. 9, pp. 978–998. American Society of Agronomy, Madison, Wisconsin.

Mulder J, van Breeman N, Eijk HC (1989) Depletion of soil aluminum by acid deposition and implications for acid neutralization. Nature (London) 337:247–249.

Oades JM (1988) The retention of organic matter in soils. Biogeochemistry 5:35–70.

Oliver BG, Thurman EM, Malcolm RL (1983) The contribution of humic substances to the acidity of colored natural waters. Geochim. Cosmochim Acta 47:2031–2035.

Rascher CM, Driscoll CT, Peters NE (1987) Concentration and flux of solutes from snow and forest floor during snowmelt in the West-Central Adirondack region of New York. Biogeochemistry 3:209–224.

Reuss JO (1983) Implications of the calcium-aluminium exchange system for the effect of acid percipitation on soils. J. Environ. Qual. 12:591–595.

Reuss JO, Johnson DW (1986) Acid Deposition and the Acidification of Surface Waters. Ecological Studies, Vol. 59, Springer-Verlag, New York.

Reynolds RC (1980) Interstratified clay minerals. In: Brindley GW, Brown G, eds. Crystal Structures of Clay Minerals and their X-Ray Identification, pp. 249–304. Monograph No. 5, Mineralogical Society, London.

Richmond GM (1980) Glaciation of the east slope of Rocky Mountain National Park, Colorado. In: Ives JD, ed. Geoecology of the Colorado Front Range: A study of Alpine and Subalpine Environments, pp. 24–34. Westview Press, Boulder, Colorado.

Schnitzer M (1969) Reactions between fulvic acid, a soil humic compound, and inorganic soil constituents. Soil Sci. Soc. Am. Proc. 33:75–81.

Schofield CL, Galloway JN, Hendry GR (1985) Surface water chemistry in the ILWAS basins. Water Air Soil Pollut. 26:403–423.

Schumm SA (1977) The Fluvial System. Wiley, New York.

Soil Survey Staff (1975) Soil Taxonomy. USDA Handbook No. 436, U.S. Government Printing Office, Washington, D.C.

Sollins P, McCorison FM (1981) Nitrogen and carbon solution chemistry of an old-growth coniferous forest watershed before and after cutting. Water Resour. Res. 17:1409–1418.

Stoch L, Sikora W (1976) Transformation of micas in the process of kaolinization of granites and gneisses. Clays and Clay Miner. 24:156–162.

Strickland TC, Fitzgerald JW (1984) Formation and mineralization of organic sulfur in forest soils. Biogeochemistry 1:79–95.

Strickland TC, Fitzgerald JW, Swank WT (1986) *In situ* mobillization of ^{35}S-labelled organic sulphur in litter and soil from a hardwood forest. Soil Biol. Biochem. 18:463–468.

Sullivan TJ, Driscoll CT, Gherini SA, Munson RK, Cook RB, Charles DF, Yatsko CP (1989) Influence of aqueous aluminum and organic acids on measurement of acid neutralizing capacity of surface waters. Nature (London) 338:408–410.

Thomas GW (1982) Exchangeable cations. In: Page AL, ed. Methods of Soil Analysis. Part 2-Chemical and Microbiological Properties. American Society of Agronomy, Madison, Wisconsin.

Walthall PM (1985) Acidic Deposition and the Soil Environment of Loch Vale Watershed in Rocky Mountain National Park. Ph.D. Dissertation, Colorado State University, Fort Collins.

Watwood ME, Fitzgerald JW, Swank WT, Blood ER (1988) Factors involved in potential sulfur accumulation in litter and soil from a coastal pine forest. Biogeochemistry 6:3–20.

8. Surface Waters

Jill Baron

The surface waters of Loch Vale Watershed (LVWS) integrate the effects of the upstream processes discussed in previous chapters. Surface waters include lakes, streams, ponds, and temporary melt pools, although we have studied only lakes and streams. Physical factors that govern the surface water system include the annual cycle of accumulation and melting of the mountain snowpack, the steep topographic gradient, the size and aspect of each drainage subcatchment, and the presence of glaciers and icefields in the upper cirques.

Surface waters are closed by ice to direct precipitation for more than 6 months each year. During this time, precipitation is stored in cold snowpacks that do not melt until spring. The rapid release of stored water dominates the seasonal cycles of dissolved constituents in the lakes and streams (Baron and Bricker 1987). Steep topographic gradients, coupled with low watershed to water body ratios in high mountain systems, cause rapid flow rates in the streams and short hydrologic retention times. This steepness has important implications for chemical budgets and is discussed here.

LVWS has a northeast-facing aspect, but the 9-ha drainage of Little Loch Creek faces south, which gives the creek very different chemical and flow characteristics from the rest of the drainage. Glaciers and permanent snow at the top of the watershed provide a direct source of water, particularly during the summer months when temperatures are warm enough to melt this water. As surface drainage slows with the onset of winter, slow seepage drains the lakes below their outlets in some years. The lakes then become closed basins of water under ice cover, sealed from atmospheric precipitation and from replenishment by water flowing downstream.

There is a strong seasonal difference in lake characteristics that is as or more important than the spatial differences between alpine and subalpine lakes. During snowmelt, all water bodies are similar to each other and are dominated by rapid water movement; meltwater flushing overwhelms all other lake processes. As snowmelt ends and flushing rates decrease, within-lake processes become progressively more important. During the winter, when the lake basins are closed or nearly closed, maximum expression of biological and chemical dynamics occurs.

Physical Characterization

There are four drainage lakes within Loch Vale Watershed, and many very shallow ponds of less than 1 m depth. These shallow ponds range in area from Embryo Pond at 1000 m^2 to depressions less than a meter square. Only three of the lakes, Sky Pond, Glass Lake, and The Loch, have been studied in detail, and none of the ponds (Figure 8.1). Andrews Tarn, an alpine lake at the foot of Andrews Glacier, is inaccessible for most of the year and has not been studied. All the lakes of LVWS are cirque lakes that were formed during Pinedale age advances of Taylor and Andrews Glaciers.

Icy Brook originates at the base of the snowfields in the cirque above Sky Pond (Figure 8.1). It flows through the three major lakes, dropping 450 m along its course, and leaves the watershed at The Loch outlet. The overall gradient of Icy Brook over its 3 km course in LVWS is about 15‰. Two tributaries, Andrews Creek and Little Loch Creek, have been studied. Many small intermittent streams drain portions of the basin.

Morphology of the Lakes

Sky Pond is an alpine lake (Pennak 1963) and the highest lake studied, at 3320 m (Table 8.1). The lake is surrounded on three sides by the steep slopes of Thatchtop, Powell and Taylor Peaks, and the Cathedral Spires. Boulder fields of moraines and talus surround and penetrate the lake except where bedrock forms the periphery. A few areas above the lake with relatively flat topography support tundra vegetation. These range from wet meadows to dry tundra. Krummholz grows up to the northern shore of the lake. Two permanent snowfields and Taylor Rock Glacier occupy the cirque headwalls to the southwest.

The surface area of Sky Pond is 3 ha; its volume is $1.2 \times 10^5 m^3$, its maximum depth is 7.3 m, and its average depth of 4.5 m (Figure 8.2A). Lake sediments are highly organic gyttja interspersed with rocks and boulders. A poorly defined inlet flows through the talus field below Taylor Rock Glacier; this inlet drains approximately 20% of the Sky Pond catchment.

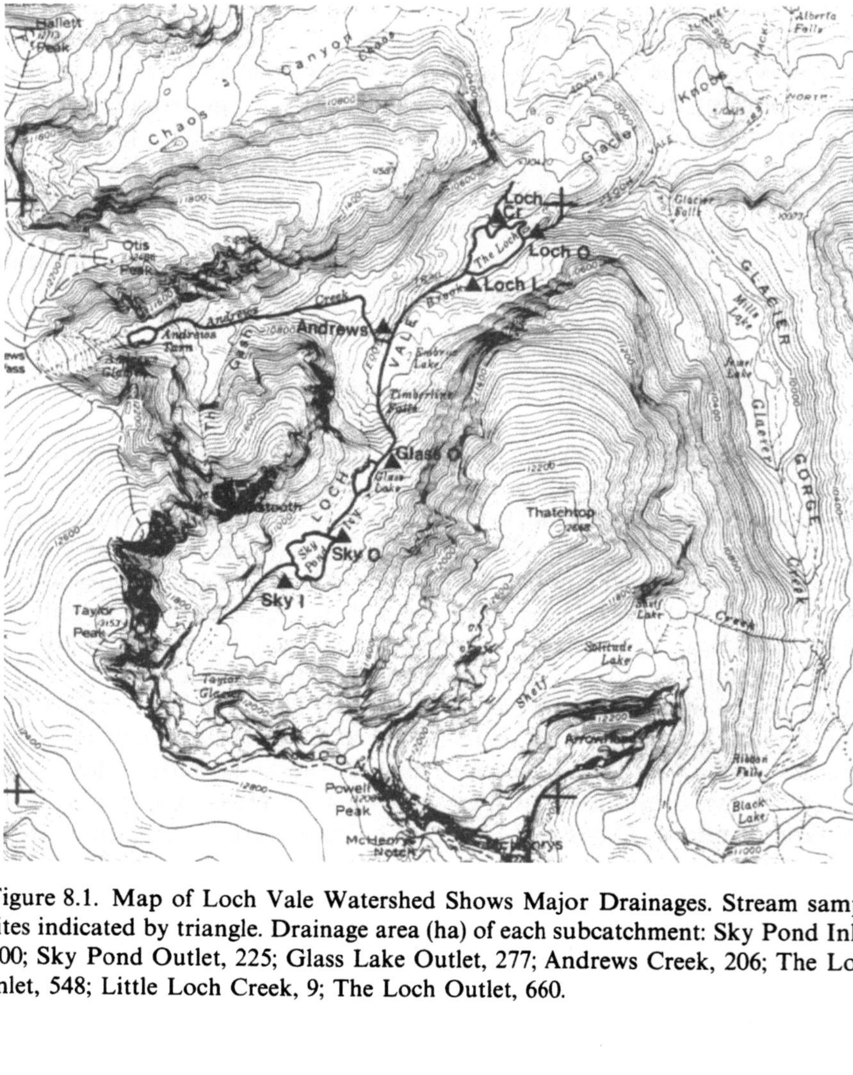

Figure 8.1. Map of Loch Vale Watershed Shows Major Drainages. Stream sample sites indicated by triangle. Drainage area (ha) of each subcatchment: Sky Pond Inlet, 100; Sky Pond Outlet, 225; Glass Lake Outlet, 277; Andrews Creek, 206; The Loch Inlet, 548; Little Loch Creek, 9; The Loch Outlet, 660.

Table 8.1. Morphometric Characteristics of the Study Lakes

Lake	Elevation (m)	Watershed Area (ha)	Suface Area (ha)	Lake Volume (m^3)	Average Depth (m)	Maximum Depth (m)
Sky Pond	3,322	204	3.03	121,684	4.5	7.3
Glass Lake	3,292	257	1.01	25,690	2.8	4.7
The Loch	3,048	660	4.98	61,100	1.5	4.7

Unchannelized flow through boulder fields, particularly along the south and southeastern shores, supplies most of the input water.

Glass Lake, at 3300 m, is smaller than Sky Pond, with a surface area of 1 ha and a volume of $2.5 \times 10^4 m^3$ (Table 8.1); its maximum depth is 4.7 m and its average depth is 2.8 m (Figure 8.2B). Glass Lake sits at the top of a hanging valley between steeply sloping rock and boulder walks. Cliffs and boulder fields make up most of the shoreline, and there are patches of krummholz vegetation along the western and eastern shores. Willows and wet meadows are found at both inlet and outlet. The inlet supplies most of the water to the lake, because most of the Glass Lake watershed includes the watershed area of Sky Pond. As in Sky Pond, the lake bottom is made up of gyttja and rocks.

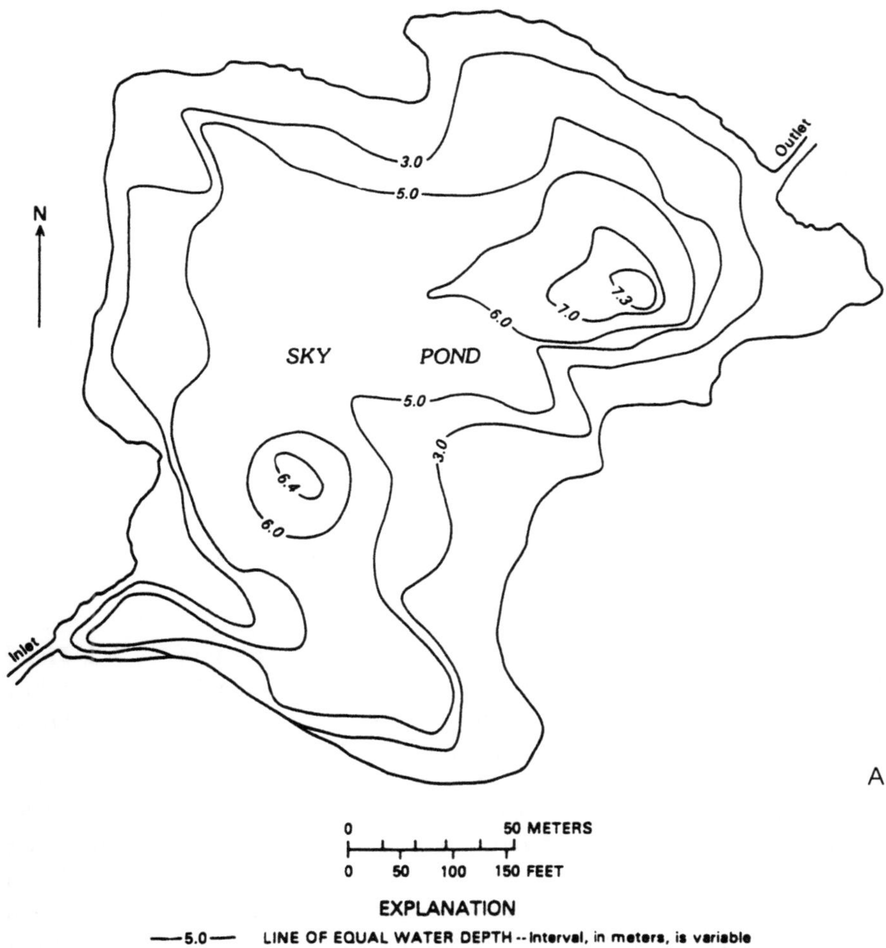

Figure 8.2. Bathymetric maps of Loch Vale Watershed lakes: (A) Sky Pond; (B) Glass Lake; (C) The Loch.

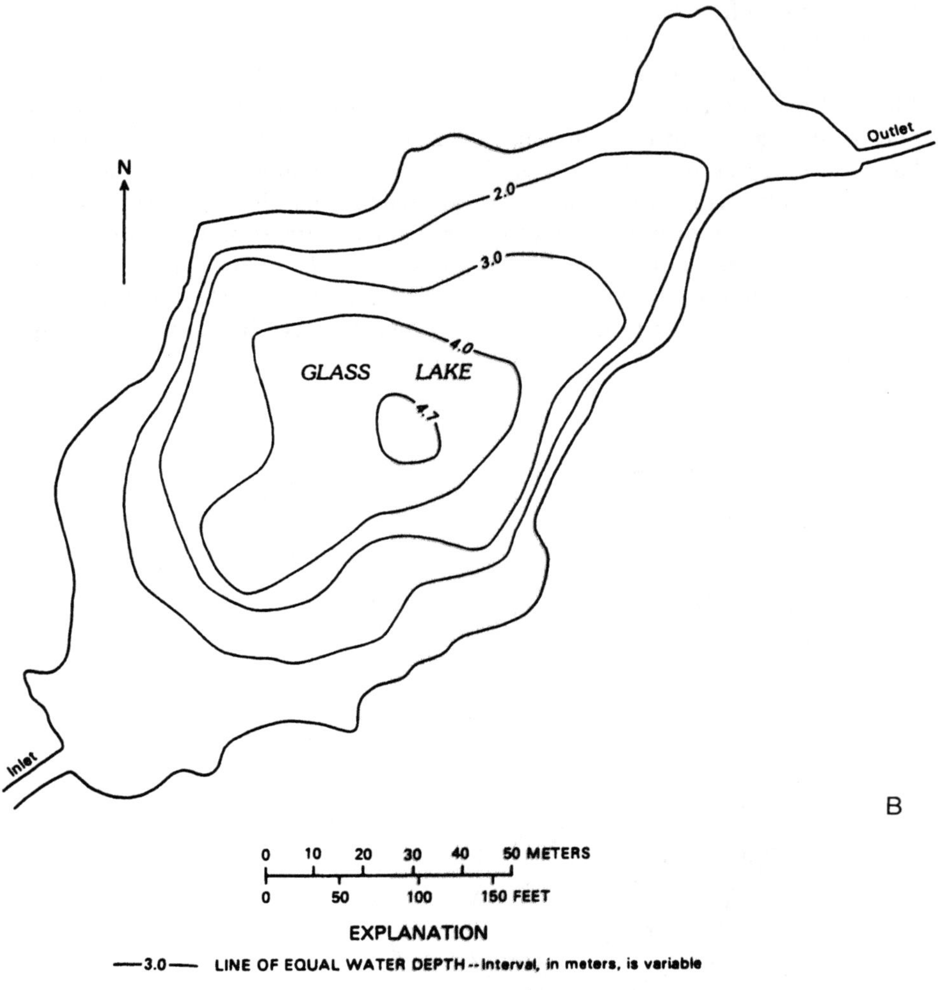

Figure 8.2 (*Continued*)

The Loch, at 3050 m, drains the entire subalpine portion of LVWS. There are boulder fields and cliffs along its shores, but more than half of the shoreline is boggy or forested. The Loch has the largest surface area of the study lakes, 5 ha. It is very shallow (Figure 8.2C), and its volume, at $6.0 \times 10^4\ m^3$, is only half the volume of Sky Pond (see Table 8.1). The average depth is 1.5 m, and the maximum depth of 4.7 m is found in one depression near the outlet. The sediments are fine, made up of gyttja and sand. Most of the water that flows through The Loch enters from Icy Brook because more than 90% of Loch Vale Watershed lies above The Loch (Figure 8.3). The path of this stream appears to follow a channel along the south shore of The Loch. Several other intermittent streams enter The Loch and may contribute to the circulation of water within the lake, especially during the snowmelt period.

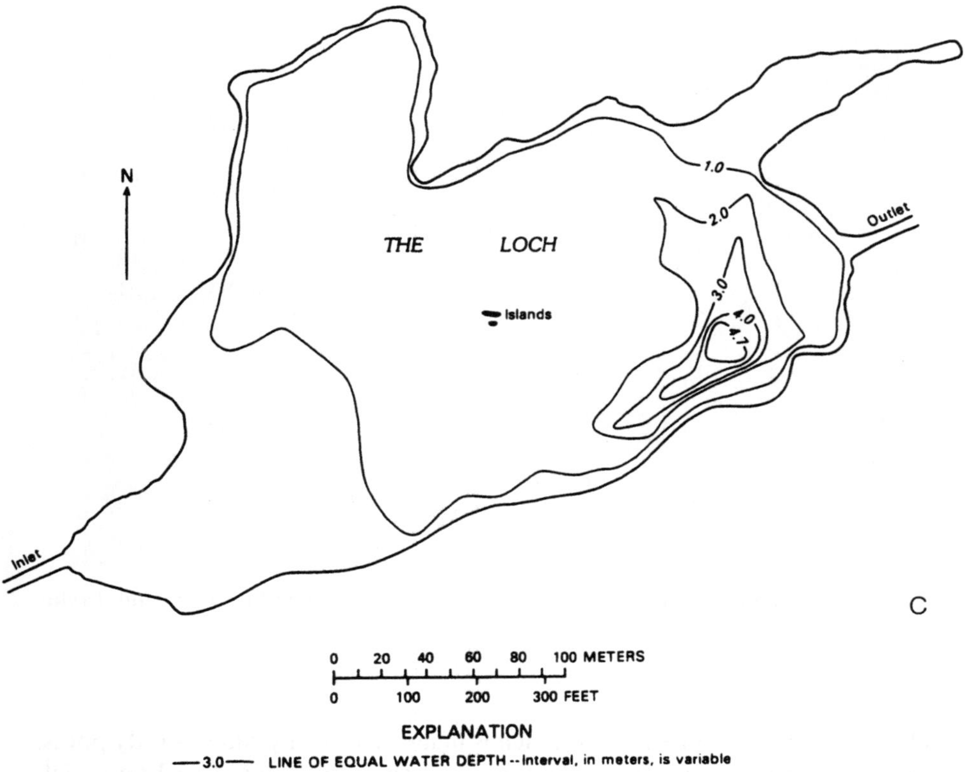

Figure 8.2 (*Continued*)

Description of Streams

Icy Brook originates in the high alpine landscape above Sky Pond. In addition to annual snow and rain, Taylor Glacier, several smaller snowfields, and Taylor Rock Glacier supply water to Icy Brook. The water at the Sky Pond inlet is very cold, with an average September temperature of 1°C (Table 8.2). The highest reach of Icy Brook is extremely steep, with a gradient of about 40% between the base of Taylor Glacier (3540 m) and Sky Pond inlet (3320 m).

The lower alpine reach of Icy Brook between Sky Pond and Glass Lake flows through boulders and alpine meadows with a gradient of approximately 7%. Solar radiation to this reach and to Sky Pond warms the water so that the average September stream temprature of this stretch of Icy Brook is 7°C.

A transition occurs between alpine and subalpine reaches of Icy Brook at Timberline Falls, downstream from the Glass Lake outlet. The stream drops 110 m in a series of falls and cascades over 430 m of horizontal distance before leveling out in the forest below. Stream gradient is approximately 7% between the base of Timberline Falls and the Andrews Creek confluence

Figure 8.3. Looking southwest over The Loch toward Timberline Falls and Taylor Glacier.

(3160 m). The streambed in this reach is made up of stony riffles, sandy pools, and extensive wet meadows. Between Andrews Creek and The Loch inlet, the stream steepens to a 20% grade and flows through rocky cascades in a well-constrained channel through dense forest. The inlet to The Loch is a large sandy pool surrounded by meadow. The average September water temperature of 6°C at this site is slightly lower than that of Glass Lake outlet and probably results from shading and the influx of colder water from Andrews Creek.

Andrews Creek is the largest tributary to Icy Brook in the study area. It drains 30% of LVWS, originating in Andrews Tarn at the base of Andrews Glacier (3470 m). During most of the ice-free season Andrews Glacier terminates in Andrews Tarn, so the water is cold and derived almost entirely from melting snow and glacier ice. The creek drops steeply from the tarn through boulder fields and cascades (40% grade) for 500 m, then levels out (15% grade) at the confluence with an intermittent stream flowing from The Gash. At treeline (3230 m along Andrews Creek), the stream passes through a 2-ha meadow before dropping into forest shade. Between the meadow and the Icy Brook confluence the creek flows steeply (15% to 20% grade) over rocky cascades in a well-constrained channel. The average September temperature at the sample site is 4°C.

Little Loch Creek flows directly into The Loch at the bay on the northwest side of the lake. Its 9-ha catchment has no multiyear snow. Little Loch Creek

Table 8.2. Stream Temperature Data for 1983–1988[a]

Site	Average Summer (°C)	Maximum Summer (°C)	*n*	Average Sept. (°C)	Maximum Sept. (°C)	*n*	Comments
Sky Inlet	2.0	5.0	18	1.4	2.5	4	Flows through talus above treeline
Sky Outlet	5.8	9.5	30	6.7	8.0	6	Sunny, shallow
Glass Inlet	6.4	9.0	7	7.5	9.0	2	Sunny, shallow
Glass Outlet	5.7	9.0	26	6.7	8.0	5	There are two locations for this site, above and below Timberline Falls
Andrews Creek	3.6	9.0	27	4.3	7.0	4	Shaded
Loch Inlet	5.3	9.5	32	6.1	9.5	5	Shaded. 1983 data excluded for QC reasons
Loch Outlet	6.2	12.0	110	8.4	11.0	11	Shady summer samples biased by many May and June readings. 1983 data excluded
L. Loch Creek	5.9	11.0	28	11.0	11.0	1	No Sept. samples taken 1984–1988 due to insufficiently flowing water

[a] Average summer values are for May–October.

drains steep slopes of south to southeast aspect consisting of bedrock cliffs and talus, with small patches of trees. Although small seeps occur along these slopes, Little Loch Creek itself exists as surface flow for less than 50 m. The water flows from coarse talus through a wet meadow along a moderate grade (10% to 15%) into The Loch. Snowmelt supplies flow to the creek until late June or early July. Base flow is maintained by groundwater, which can be depleted by late August in some years. The higher late-season temperatures reflect Little Loch Creek's southern exposure, low discharge rate, and lack of upstream snow or ice.

Thermal Characterization of the Lakes

Strong winds, rapid flushing of snowmelt water, and shallow lake depths prevent thermal stratification from occurring during the open water season. The maximum recorded temperature of Sky Pond is 8°C and that of Glass Lake is 9°C (Figure 8.4A, B). Water temperatures reach a maximum of 12°C in August or September in The Loch (Figure 8.4C). Snowmelt water is a major influence on lake temperatures until July for most years. Solar radiation warms the lakes through August and September. Thereafter, until ice formation, wind mixing prevents thermal stratification. Inverse stratification occurs during the winter after ice formation. The Loch Vale lakes do not fall neatly into the stratification classes that are commonly used today unless one stretches the definition of cold monomictic (Wetzel 1983). They can be described as third-order lakes using an older classification (Whipple 1898); third-order lakes never develop thermal stratification. These lakes remain in circulation throughout the summer (Hutchinson 1975).

The timing of ice formation on Loch Vale lakes varies from year to year depending on air temperature and on wind, but the lakes have always been frozen over by November 15. Generally, ice formation occurs about 1 week sooner for Sky Pond, and ice breakup occurs between 2 and 3 weeks later at this higher lake. Ice can form a thin layer over part of The Loch for several days in October, only to be broken up again by strong winds. The first layers of ice on the lakes are frozen from lake water and are transparent. Over time snow is added from above, and the ice becomes translucent. The wind is so strong during the winter months, however, that snow rarely rests on the lakes for more than a few days before blowing off to drifts at the outlets. Lake ice for The Loch and Sky Pond increases in thickness to about 1 m by February; it has been recorded as thick as 2.9 m for Glass Lake.

The weight of snow during the winter can cause the ice to buckle or even cause lake water to melt holes through the ice during some periods of low wind. Water from these holes can flow out onto the ice surface (Knight 1987, 1988), and meltholes have been observed on The Loch (Spaulding 1991). Water spouts have been observed on Sky Pond in midwinter, possibly formed from the increased pressure of snow on ice (Schoepflin, personal communication). Ice remains on Sky Pond between 195 and 220 days, or

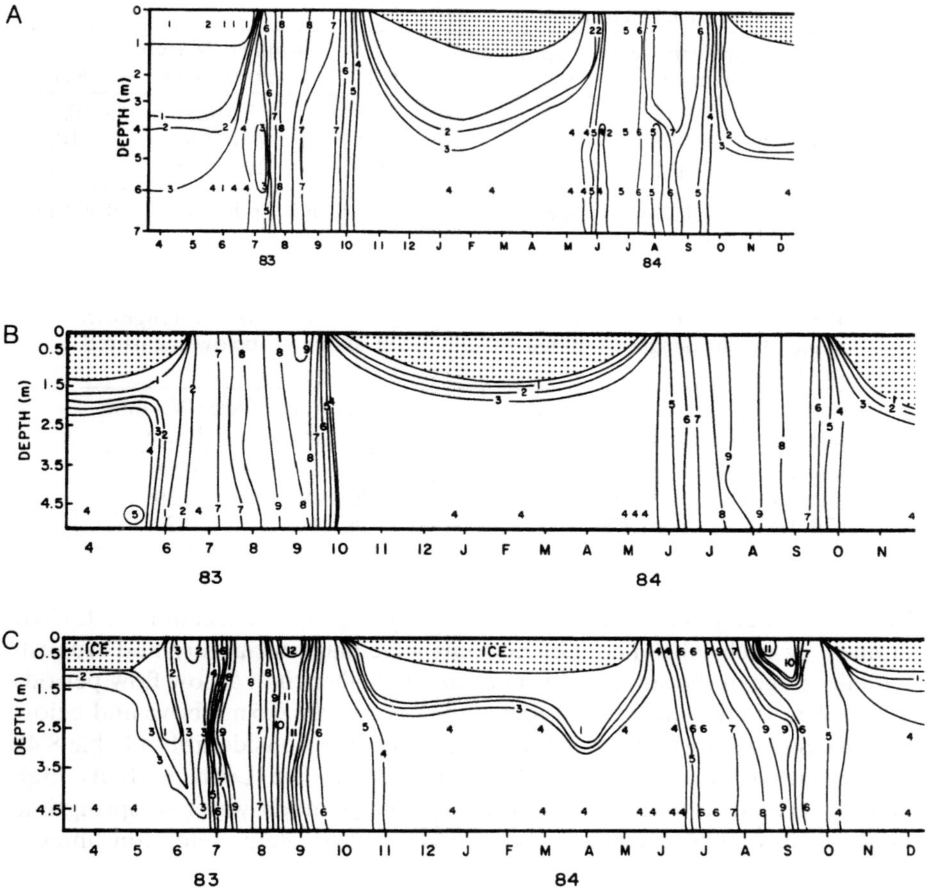

Figure 8.4. Temperature isotherms for Loch Vale Watershed lakes: (A) Sky Pond; (B) Glass Lake; (C) The Loch.

53%–60% of each year. Ice on The Loch can last between 185 and 210 days, or 51%–58% of the year. The breakup of ice in the spring is dependent on air temperature and solar radiation. Lake ice melts from the top, forming a layer of wet slush and sometimes ponded water on top of harder, black ice. As melting progresses, layers of slush become interspersed with layers of harder ice. For a period each spring immediately before ice breakup in Loch Vale, the ice forms vertical columns hexagonal in cross section.

Hydrologic Retention Times

Stream discharge is measured continuously at The Loch outlet between April and December. Estimates of flushing rates (Table 8.3) were obtained by dividing the total discharge at The Loch outlet by lake volumes estimated

Table 8.3. Flushing Characteristics[a]

Lake	#Turnover per yr	April–June turnover/yr	R.T.	August–March turnover/yr	R.T.
Sky Pond	12.7 (2.0)	9.3 (1.3)	39.4	3.5 (0.8)	105.2
Glass Lake	73.1 (15.3)	53.2 (9.7)	6.9	20.0 (5.9)	18.3
The Loch	81.8 (12.9)	59.5 (8.1)	6.1	22.3 (5.4)	16.4

[a] Notes: $n = 5$; standard deviations are in parentheses. Snowmelt is April–June. Residence time (R.T.) is measured in days.

Table 8.4. Calculated Changes in Lake Volume (m^3) with Changes in Stage Height (stage estimated to vary 0.5 m above or below measured values)

Lake	Lake Volume	Maximum Volume	Minimum Volume	Area/Volume Ratio
Sky Pond	121,684	136,834	106,534	0.25
Glass Lake	25,690	30,740	20,640	0.39
The Loch	61,100	86,000	36,200	0.82

from bathymetry measurements and multiplying by the fraction of LVWS drained by each lake. The lake bottoms were mapped in July and August, but lake stage in the spring is as much as 1 m higher than at low flow periods in October. If lake stage is considered to vary betwen 0.5 m above and below the calculated values, lake volumes can fluctuate considerably (Table 8.4). The change in volume is most pronounced for The Loch, with its high area-to-volume ratio; the lake can hold twice as much water in springtime as at low flow. The uncertainty of estimates of hydrologic retention times is therefore large.

Sky Pond, Glass Lake, and The Loch turn over 13 times, 73 times, and 82 times per year, respectively. Most of this water flux occurs during snowmelt. Lake retention times are so short in June for The Loch and Glass Lake that the lakes behave more like stream pools than lakes.

By late July, most annual snow has melted and the longer hydrologic retention reflects the transition to a base flow regime. Stream discharge decreases through autumn and approaches zero by December. From late December through at least February of most years, there is no measurable flow at the stream gauge at The Loch outlet. The lakes form mostly closed basins during these months.

Slow seepage of water through lake sediments is responsible for some water loss during the winter months. In the absence of replenishment from upstream sources, such seepage losses have caused water to drain out from under the ice in The Loch, forming an air space in some winters before the ice collapsed into the dry lake bed in the vicinity of the inlet. This phenomenon has not been observed at the upper lakes, but they are much less frequently visited in the winter. In October 1988, lake stage at Andrews Tarn was observed to fall 2 m below the outlet. Andrews Creek emerged from the surficial material about 100 m downstream, fed by seepage from the tarn.

Hydrologic Pathways and Sources of Water

The ultimate source of all surface waters in LVWS is precipitation, but meteoric water may move through several different pathways before being incorporated into surface water. These pathways include rain, melting of annual snow, meltwater from glaciers and icefields, and drainage from soil moisture. Impermeable crystalline bedrock makes deep groundwater storage unlikely (see Chapters 3 and 6, this volume), and overland flow has never been observed.

The different contributions to streamflow can be distinguished on the hydrograph from The Loch outlet (Figure 8.5). Maximum discharge occurs in June from snowmelt (Figure 8.5A). A strong diurnal period with greatest flow about midnight is observed. The greatest solar radiation and maximum air temperatures are reached in the afternoon, but we estimate an 8- to 11-hour lag time between peak meltwater production in the upper snowfields and the time this water reaches the outlet. Daily evapotranspiration cycles were observed to contribute to diurnal patterns in the hydrograph at Fraser Experimental Forest, Colorado, where the uptake of water by vegetation is greatest during the afternoons and least at night (Kaufmann 1981). Early spring storms tend to depress rather than increase stream discharge because they reduce the input of solar and thermal energy to the snowpack. Clearly, meltwater inputs from annual snowpack are the dominant source of streamflow at this time of the year.

By mid-July (Figure 8.5B) the diurnal cycle is largely absent. If losses of water via evapotranspiration are important to the diurnal cycle in stream discharge observed earlier, the periodicity should continue later in the summer, but it does not. Likens et al. (1977) postulated that water is first lost to evapotranspiration in the heavily vegetated Hubbard Brook Experimental Forest, New Hampshire, and what is left over supplies streamflow. These authors remarked that this holds true so long as there is adequate soil moisture, and this may also be important to Loch Vale water dynamics. Large rain events in midsummer cause an immediate and significant increase in discharge, suggesting soil moisture is high enough to minimize infiltration. By late July, precipitation and drainage of soil moisture dominate streamflow at The Loch outlet, with some unknown contribution from the melting of glaciers and icefields.

The annual snow is completely melted by August, and stream discharge rates decline (Figure 8.5C). This reflects the progressive drawdown of soil moisture from high levels immediately following the melting of annual snow. By mid-September (Figure 8.5D), even large precipitation inputs infiltrate soils and have little effect on streamflow. As the weather cools with the approach of winter, melting of ice and snow at the highest elevations in

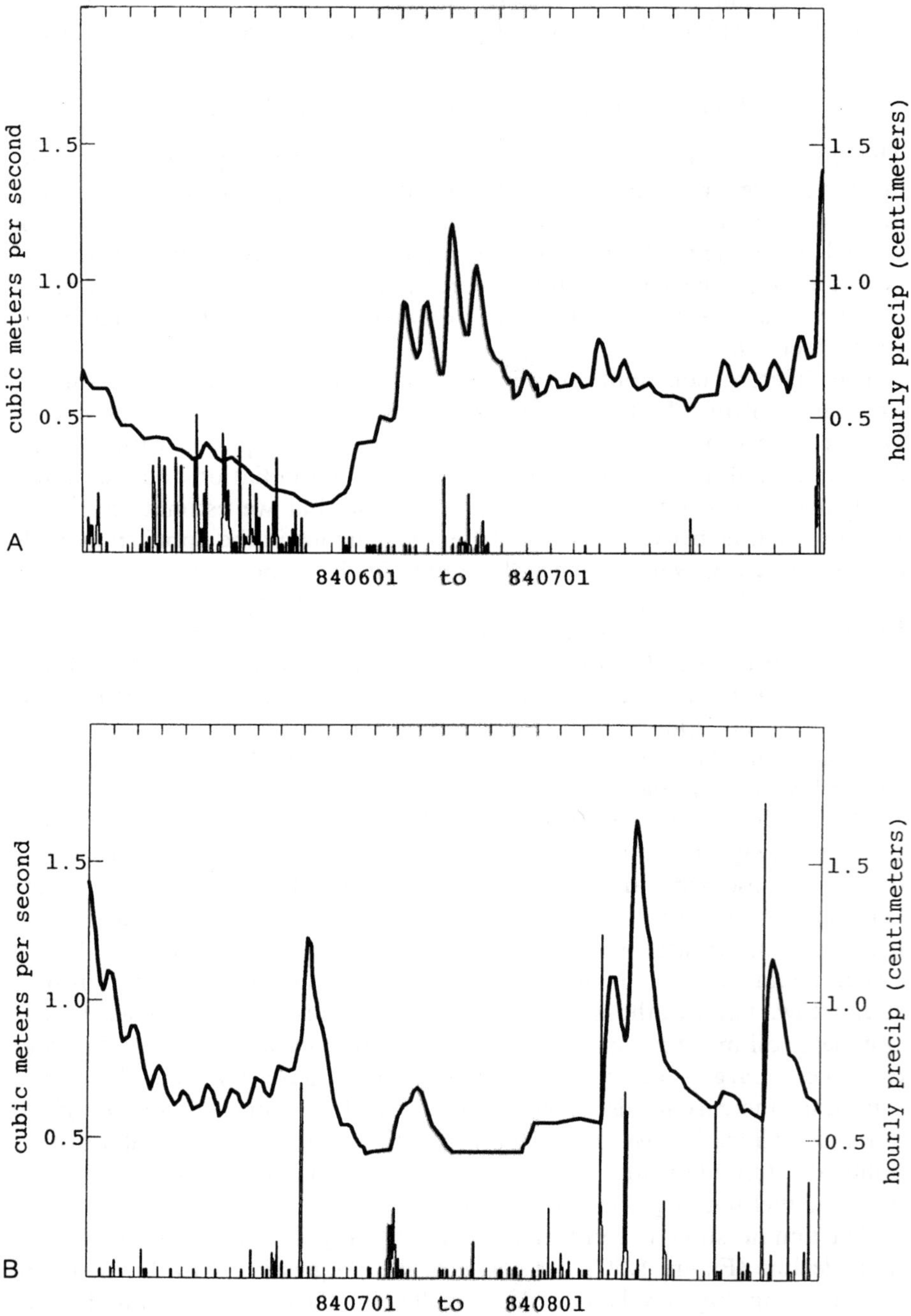

Figure 8.5. Hourly precipitation events (bars cm) and hydrograph (line) from The Loch outlet ($m^3 sec^{-1}$ for 1984: (A) June 1984; (B) July 1984; (C) August 1984; (D) September 1984.

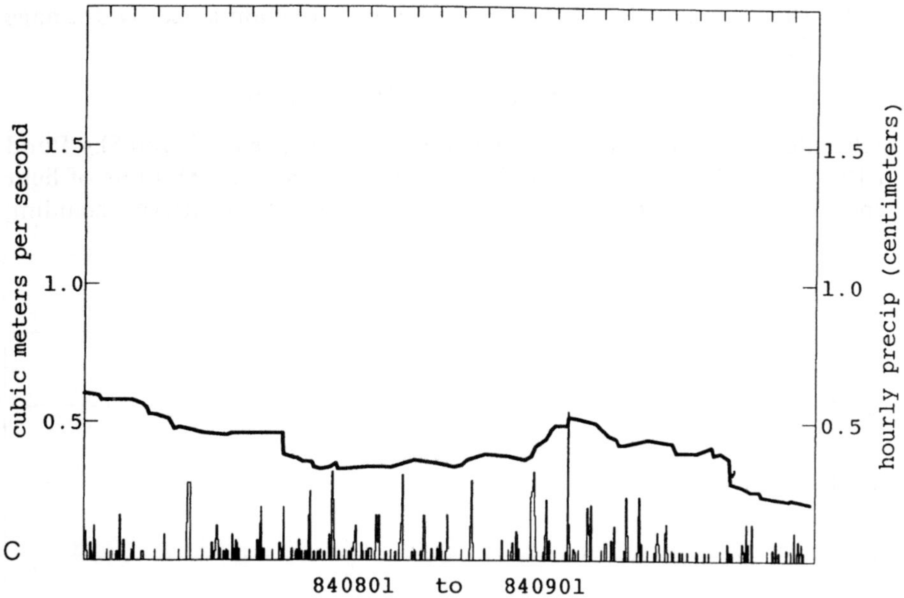

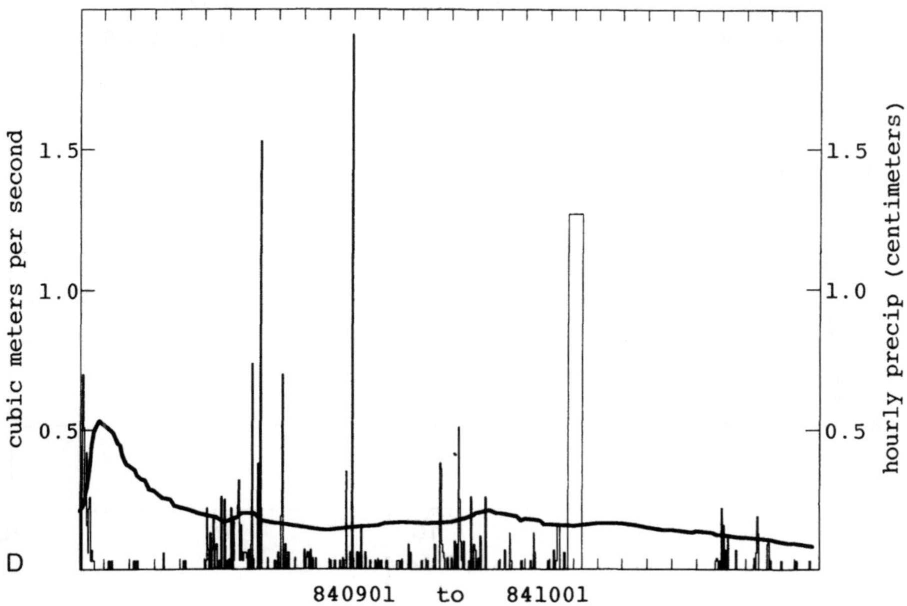

Figure 8.5 (*Continued*)

LVWS slows, and autumn base flow conditions are maintained by drainage from soils.

Visibility and Light Penetration

Light intensity, in lux, was measured periodically in The Loch and Sky Pond in 1984 and 1985 (data are from McKnight et al. 1986, 1988). Plots of light penetration for these data (Figure 8.6A–D) reflect seasonal patterns including

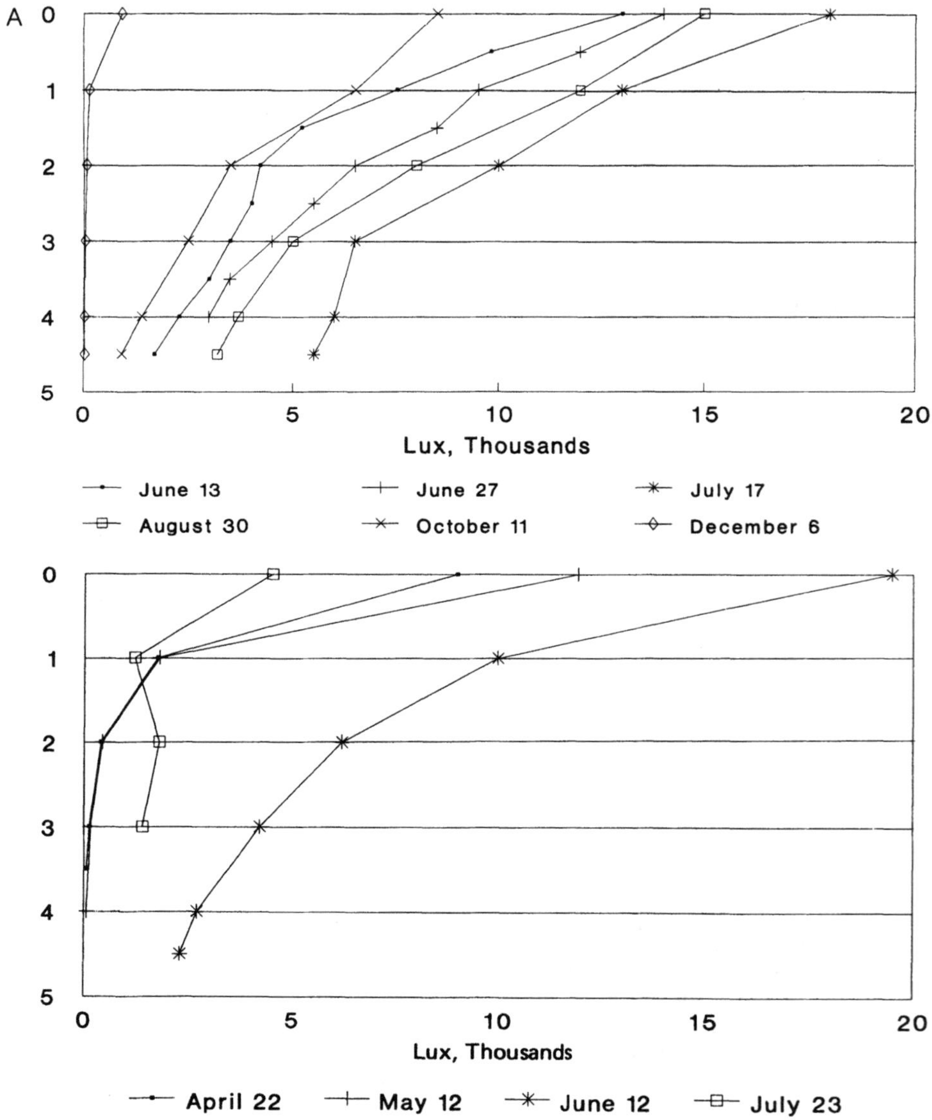

Figure 8.6. Light penetration (lux) at different times of year for The Loch, 1984 (A) and 1985 (B), and for Sky Pond, 1984 (C) and 1985 (D).

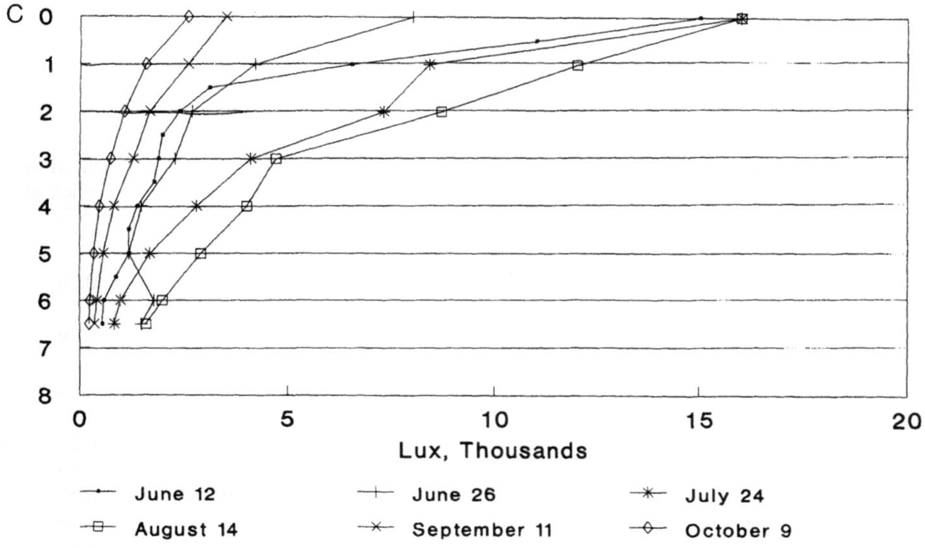

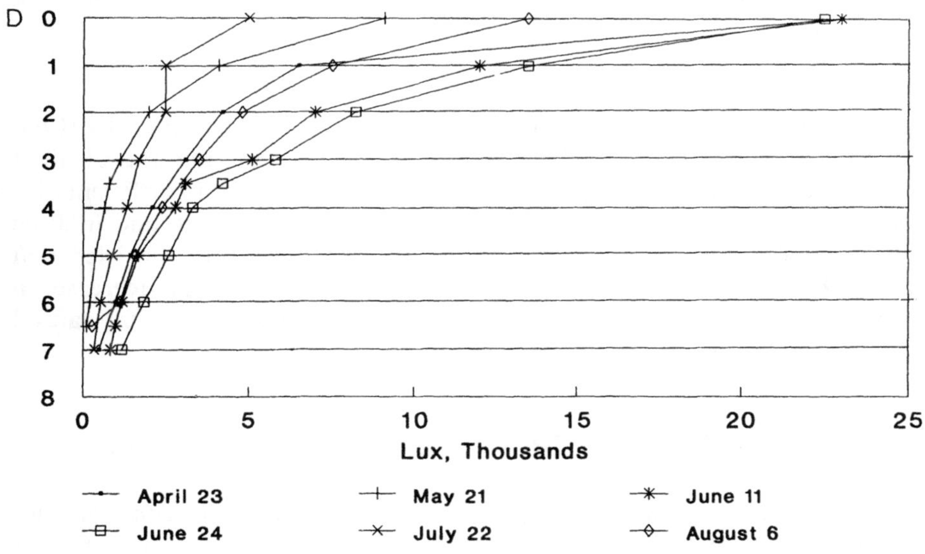

Figure 8.6 (*Continued*)

the presence of ice and springtime suspended particulates, as well as cloudy sample days compared to clear ones. Light intensity curves for Sky Pond showed consistently less light penetration to depth, with the amount of light falling off sharply below the lake surface. Extinction coefficients (Table 8.5) calculated from these data (McKnight et al. 1986, 1988) provide a measure

Table 8.5. Light Extinction Coefficient for The Loch and Sky Pond[a]

The Loch (lux)				Sky Pond (lux)			
1984		1985		1984		1985	
Date	Extinction Coefficient	Date	Extinction Coefficient	Date	Extinction Coefficient	Date	Extinction Coefficient
Jun 13	0.43	Apr 22 (ice)	1.40	Jun 12	0.47	Apr 23 (ice)	0.47
Jun 27	0.38	May 6 (ice)	0.89	Jun 26	0.36	May 6 (ice)	0.58
Jul 3	0.33	May 12 (ice)	1.30	Jul 10	0.27	May 14 (ice)	0.79
Jul 17	0.27	May 30	0.67	Jul 24	0.45	May 21 (ice)	0.64
Aug 15	0.33	Jun 2	0.74	Aug 14	0.35	Jun 5	0.58
Aug 30	0.36	Jun 12	0.46	Sep 11	0.36	Jun 11	0.46
Sep 13	0.34	Jun 27	0.47	Oct 9	0.35	Jun 24	0.41
Oct 11	0.49	Jul 10	0.53			Jul 9	0.34
Dec 6 (ice)	0.88	Jul 23	0.31			Jul 22	0.35
		Aug 8	0.93			Aug 6	0.49

[a] Lakes were ice covered on dates noted as (ice).

of lake transparency by describing the slope of the log of light intensity (Lind 1979) and show that the two lakes were similar in amount of light transparency in 1984 and again in 1985. The high extinction coefficients for both lakes in May 1985 resulted from ice cover on the lakes, and in June 1985 might result from a combination of suspended clay particles and phytoplankton, as suggested by McKnight et al. (1986). A decrease in phytoplankton abundance and turbulence caused by snowmelt in July caused the extinction coefficient in both lakes to decrease.

Chemical Characteristics of Loch Vale Surface Waters

The lakes of LVWS are very dilute, as can be seen from median concentrations of major solutes (Table 8.6). Calcium is the most abundant cation in all three lakes; bicarbonate and sulfate are the dominant anions. Solute concentrations increase slightly downstream through the watershed with total ionic strength (calculated from the medians) of 215 $\mu eq\ L^{-1}$ at Sky Pond, 242 $\mu eq\ L^{-1}$ at Glass Lake, and 256 $\mu eq\ L^{-1}$ at The Loch. Surface water chemistry varies by season (Figure 8.7) and is categorized by hydrologic phase into winter, snowmelt, and summer seasons.

Phase I, October 1 through April 14, corresponds to the beginning of the hydrologic water year and describes the winter period in LVWS. Phase II, April 15 through July 14, is the snowmelt period of LVWS, the time of

Table 8.6. Median Concentrations of Major Solutes in Loch Vale Watershed Waters, 1983–1988[a]

Parameter	Sky Pond	Glass Lake	The Loch
Field pH	6.5	6.5	6.3
Lab pH	6.7	6.8	6.6
Ca ($\mu eq\ L^{-1}$)	59.9	64.9	69.9
Mg ($\mu eq\ L^{-1}$)	14.8	16.5	19.7
Na ($\mu eq L^{-1}$)	16.3	17.4	21.8
K ($\mu eq\ L^{-1}$)	3.8	4.3	5.2
NH_4 ($\mu eq\ L^{-1}$)	1.2	1.2	1.4
Cl ($\mu eq\ L^{-1}$)	2.6	3.3	4.1
SO_4 ($\mu eq\ L^{-1}$)	29.1	32.9	36.0
NO_3 ($\mu eq\ L^{-1}$)	15.0	12.9	16.4
SiO_2 ($mg\ L^{-1}$)	1.2	1.2	1.9
DOC ($mg\ L^{-1}$)	0.8	0.8	1.4
Alkalinity ($\mu eq\ L^{-1}$)	29.8	37.6	42.6

[a] All valid samples for each lake and its inelt and outlet were used for this summary (Sky Pond: $n = 142$; Glass Lake: $n = 96$; The Loch: $n = 361$).

greatest chemical and physical change. Phase III occurs in the summer, July 15 through September 31, and is characterized by dilute water and decreasing flow to less than $0.5\ m^3\ sec^{-1}$ at The Loch outlet. The lowest concentrations of all the major ions occur during Phase III, and the variability is also low. The variability in solute concentrations is greater for the other two phases because they include seasonal transitions.

Spatial Variability in Winter Lakes (Phase I)

Discharge declines and solute concentrations increase as precipitation becomes snow after October 1 each year. The lakes begin to freeze in October. By the end of each December there is usually no surface water flow into or from each of the lakes. During winter, the lakes of LVWS are mostly closed basins, with little exchange of water or gases with the atmosphere or with water flowing downstream. The maximum ice thickness is reached in December or January depending on weather conditions (Spaulding 1991).

Chemical concentrations in The Loch increase as more than half the volume of lake water is converted to ice. Concentrations of most solutes are more than twice as high as in summer samples from the same sites (Figure 8.8; sample sizes are shown in Table 8.7). The process of ice formation excludes chemical solutes, and analyses of two ice cores collected in February 1989 showed them to be very dilute (Spaulding 1991). Freeze concentration causes most ions to reach their maximum concentration in LVWS in the winter, a phenomenon that has also been reported from other small lake systems (Barica 1977; Canfield et al. 1983). More than 60% of the volume of The Loch freezes, but ice accounts for less than 20% of the volume of Sky

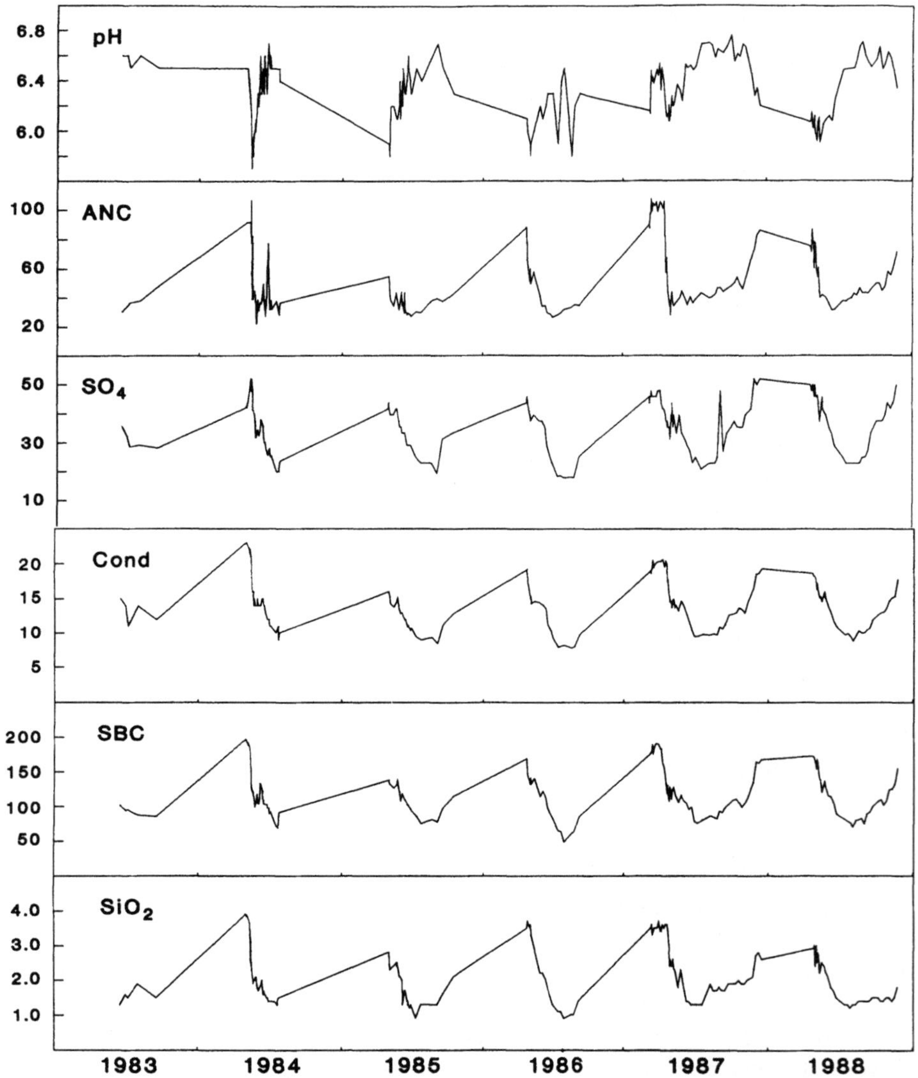

Figure 8.7. Time series of major ions, 1983–1988. ANC, acid-neutralizing capacity; Cond, specific conductance; SBC, sum of base cations Ca, Mg, Na, and K. Values, $\mu eq\,L^{-1}$; except pH, pH units; specific conductance, $\mu S\,cm^{-1}$; SiO_2, $mg\,L^{-1}$.

Figure 8.8. Mean solute concentrations for Phase I, October 1 through April 14. Bars represent 2 standard errors of the mean. Lake locations are surface (Surf), sampled at 0.5 m below water surface, and deep (Deep), sampled between 0.5 and 1.0 m above lake bottom. Sample sizes are found in Table 8.7. Values, $\mu eq\,L^{-1}$; except pH, pH units; Al, Fe, and Mn, $\mu g\,L^{-1}$; SiO_2 and DOC, $mg\,L^{-1}$.

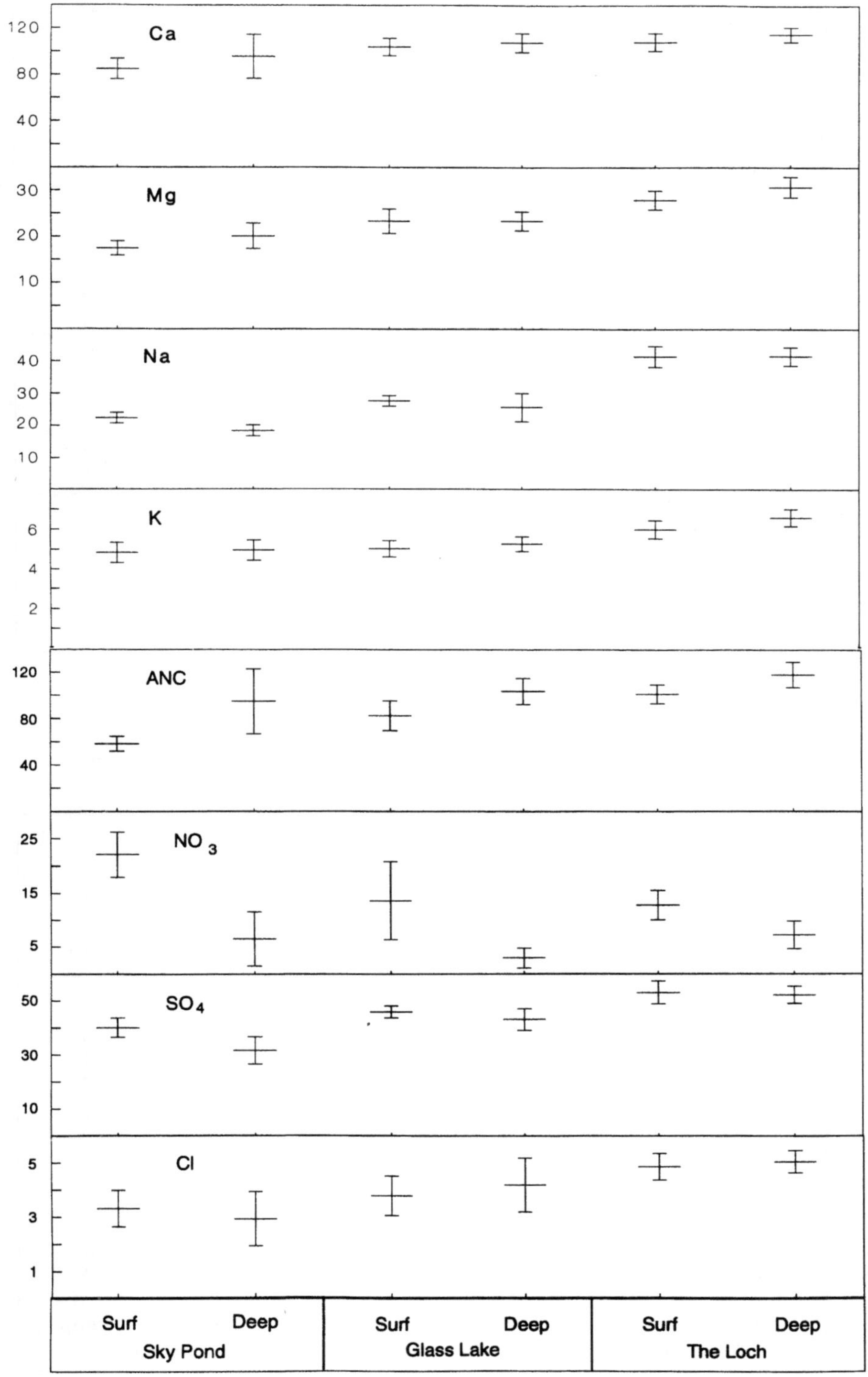

120
80
40
Ca
30
20
10
Mg
40
30
20
10
Na
6
4
2
K
120
80
40
ANC
25
15
5
NO 3
50
30
10
SO4
5
3
1
Cl
Surf
Deep
Sky Pond
Surf
Deep
Glass Lake
Surf
Deep
The Loch

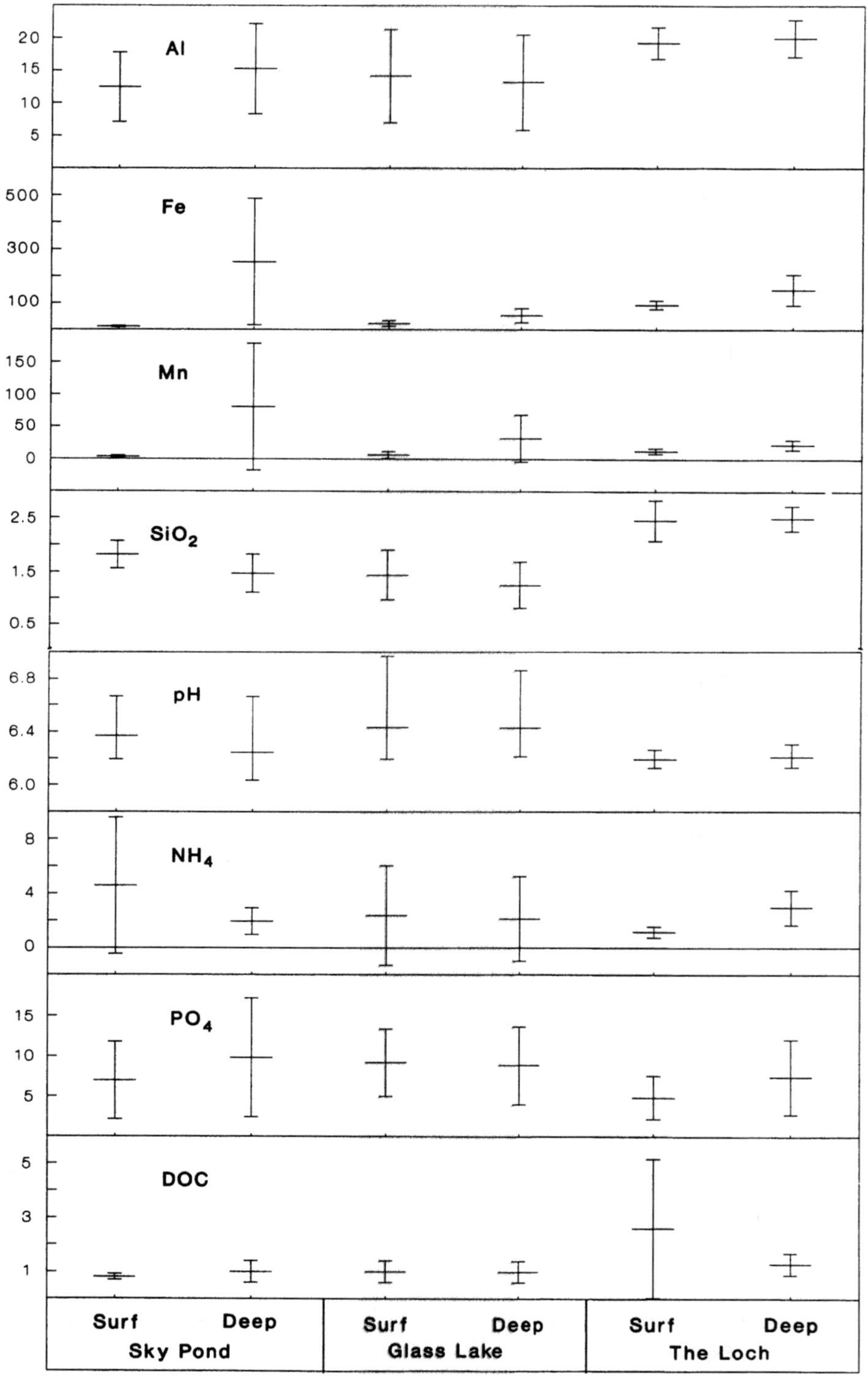

Figure 8.8 (*Continued*)

Table 8.7. Number of Samples Included in Discussion of Solute Chemistry of Phases I, II, and III. 1983–1988

	Phase I	Phase II	Phase III
Sky I	0	4	17
Sky S	13	25	14
Sky B	7	21	13
Sky O	0	17	11
Glass S	11	16	11
Glass B	9	13	12
Glass O	0	13	11
Andrews	0	14	14
Loch I	0	51	20
Lake Loch Creek	0	6	5
Loch S	22	32	14
Loch B	16	26	15
Loch O	0	128	37

Pond, so this process cannot account for all the increase in solute concentrations.

Additional reasons for increased concentrations of solutes in winter lake water include diffusion of solutes from the sediments (Hutchinson 1975; Likens 1985; Schindler 1986; Stoddard 1987; Psenner 1988; Giblin et al. 1990), and an influx of soil drainage water during a period of little or no flushing (Hutchinson 1975; Stoddard 1987). Hydrologic residence of lake waters is too short during the open water period for diffusion from sediments to have an effect on lake concentrations, but the closing of the lakes in winter allows these solutes to build up to higher levels.

Calcium concentrations are not significantly different between surface and bottom waters and between lakes during Phase I (see Figure 8.8). Calcium ranged from 80 to 120 μeq L^{-1} during this winter period. There is a slight trend of increasing concentration of Mg and K from the highest elevation lake to the lowest elevation lake, and almost a doubling in concentration of Na between Sky Pond and The Loch (Figure 8.8). In the Rocky Mountains several workers have found that solutes become progressively more concentrated from higher to lower elevations (Gibson et al. 1983; Turk and Adams 1983; Turk and Campbell 1987). This is because increased drainage area contributes greater amounts of weathering products. Lower elevations support greater soil development, and the vastly increased surface area available from soil particles as compared to rock particle surface area enhances leaching of base cations (Drever 1988; Drever and Zobrist, in manuscript). The inverse relation between elevation and base cation concentration may be apparent for LVWS lakes only during periods of very low flow when hydrographic processes become less important than biogeochemical processes, such as during Phase I. Conversely, the observed differences in cation concentration could result from the greater effect of freeze concentration in The Loch. The elevation difference between Sky Pond

(3320 m), Glass Lake (3300 m), and The Loch (3050 m) is not great, and although we have detected a soil influence to The Loch during snowmelt (discussed later), we have not studied winter cation inputs.

Chloride concentrations during Phase I are consistently between 1.5 and 4.5 $\mu eq\ L^{-1}$ and do not show a significant increase with decreasing elevation (Figure 8.8). Surface waters are about twice as concentrated as Cl measured in wet deposition (Chapter 4, this volume), during Phase I, and this might be due to a combination of evaporation of lake water in late summer before the onset of ice and freeze concentration as ice forms.

Winter sulfate concentrations (28–58 $\mu eq\ L^{-1}$) are similar to those observed during Phase II and about twice as high as during Phase III (Figure 8.8). There is a very slight trend of increasing concentration from Sky Pond to The Loch, and possibly lower concentrations in the bottom waters of Sky Pond. Reduction of sulfate can occur through microbial processes, forming either organic S (Nriagu and Soon 1985; Cook et al. 1986; Rudd et al. 1986) or iron sulfide (Carignan and Tessier 1988; Giblin et al. 1990) Iron and Mn concentrations are elevated in the bottom water of Sky Pond (see following discussion but dissolved organic carbon is not. A sample of Sky Pond sediment interstitial waters collected October 8, 1988, contained 5.3 mg C L^{-1} (McKnight, unpublished data), so the possibility that sulfate reduction occurs within the sediments occurs cannot be ruled out. Giblin et al. (1990) found that most of the reduced sulfur found in sediments of a diverse set of New England lakes was stored as FeS_2 rather than as carbon-bonded S.

Nitrate and ammonium display similar dynamics during Phase I (Figure 8.8). Average surface water concentrations for NO_3 and NH_4 are 15–27 and 2–5 $\mu eq\ L^{-1}$, respectively. Nitrate concentrations are high throughout the year in LVWS, with 1984–1988 average value of 16 $\mu eq\ L^{-1}$ at The Loch outlet. Highest values occur at the highest elevations, and Phase I is no exception. The 1984–1988 average annual concentrations of NO_3 in precipitation were 9–13 $\mu eq\ L^{-1}$, so lake waters are substantially more concentrated than precipitation. Not all the surface water NO_3 can be accounted for by evaporation (see Chapter 3, this volume). Additional sources could include unmeasured dry deposition, nitrification in and resuspension from sediments, bedrock sources, or nitrification of atmospheric N_2 during lightning events. Except for sediment processes, these sources would contribute NO_3 during Phases II and III, so high NO_3 concentrations during Phase I may represent a residual. There is no reason to suppose mineralization of sediment-deposited N is occurring at a greater rate than bacterial immobilization (Rudd et al. 1990). There is also no bedrock source in LVWS for nitrogen (Mast et al. 1990; see Chapter 6, this volume) and dry deposition of anthropogenic N is not considered to be very important (see Chapter 4, this volume). Lightning will be discussed in the Phase III discussion, which follows.

The lowest and least variable concentrations of both NO_3 and NH_4 are found in the bottom waters of Sky Pond and The Loch. The bottom waters of Glass lake also have low concentrations of NO_3. Rudd et al. (1990) found that NO_3 from experimental additions penetrated more deeply into lake sediments during the winter as compared to summer. These authors found a marked increase in microbial activity in the epilimnetic sediments during the summer months. The seasonality of microbial activity in Loch Vale lakes is not known, but seasonality may be less pronounced, at least in Sky Pond and Glass Lakes, because of smaller differences between summer and winter maximum temperatures (8° and 4°C, respectively, in Sky Pond and Glass Lake). It appears to me that the lower N values in Phase I lake bottom waters results from microbial immobilization. Whether bacterial activity is greater during the summer is unknown and a good topic for study. The uptake of N is noticeable during the winter because it is not masked by the overwhelming influence of lake flushing at other times of the year.

Phosphate concentrations average between 5 and 10 $\mu g\ L^{-1}$ during Phase I throughout LVWS (Figure 8.8). The lowest concentrations are found in bottom waters of The Loch, perhaps from algal consumption. Concentrations in LVWS are somewhat higher than those reported from Mirror Lake, New Hampshire, where winter profiles of the water column show PO_4 less than 4 $\mu g\ L^{-1}$. Winter LVWS PO_4 values are less than those reported for more eutrophic lakes such as Clear Lake, California (10 $\mu g\ L^{-1}$), and Lakes Windermere and Esthwaite in the English Lake District (30 $\mu g\ L^{-1}$; Goldman and Horne 1983).

Values for pH through Phase I are not significantly different between lakes, with depth, or with time (Figure 8.8). The Loch pH declines slightly from 6.3–6.5 in December to 6.0–6.3 in January and remains at these values until snowmelt. Average Sky Pond values are 6.3–6.4 during the period of Phase I. A paleolimnological reconstruction of pH over the past 150+ years for several nearby lakes suggested pH values have not changed significantly since about 1800, having remained between 6.2 and 6.9 (Baron et al. 1986).

Acid-neutralizing capacity (ANC) is normally very low in LVWS lakes, with an annual average value of 55 $\mu eq\ L^{-1}$ (Figure 8.8). It reaches its highest concentrations in the winter, with average values about 100 $\mu eq\ L^{-1}$ in bottom waters of The Loch and Sky Pond. Mast et al. (1990) have argued that weathering processes of the parent material are sufficient to account for the ANC in 1984–1988 mass-balance calculations, and on a watershed basis this may be true. During the winter period, however, when there is little or no hydrologic input, minimal inputs from weathering are expected, and in-lake processes take on greater importance for ANC generation (Schindler 1986; Schafran and Driscoll 1987; Psenner 1988). In situ ANC generation can occur from reduction of SO_4, NO_3, Fe, and Mn. Two HCO_3 ions are produced for each SO_4 ion that is biologically reduced, but this alkalinity gain can be reversed by the return of oxidizing conditions in the springtime. For the alkalinity gain to be permanent, the reduced S must react with Fe or organic matter and either be volatilized as H_2S or be buried in lake sediments (Giblin et al. 1990). By assuming the difference between surface and bottom water SO_4

concentrations results from a permanent loss of SO_4 to the sediments, 18, 4, and 4 μeq HCO_3 L^{-1} can be generated for Sky Pond, Glass Lake, and The Loch, respectively, during the winter.

Biological immobilization of NO_3 generates an equivalent amount of HCO_3 (Reuss and Johnson 1986; Dillon and Molot 1990). The degree to which this will occur is a function of water residence time and lake depth (Kelly et al. 1987; Dillon and Molot 1990). These factors, in essence, describe the potential for NO_3 to be "captured" via either algal or microbial consumption. In experimentally acidified lakes in Ontario, Rudd et al. (1990) concluded that bacterial denitrification and sedimentation was the primary means of NO_3 loss because algal populations were limited by phosphorus and could not use available N. In LVWS lakes the amount of alkalinity generated by NO_3 reduction and sedimentation can be estimated by subtracting bottom water from surface water NO_3 concentrations. If we assume this difference is caused by microbial activities and that the loss after reduction is permanent, then approximately 17, 11, and 7 μeq HCO_3L^{-1} are produced for Sky Pond, Glass Lake, and The Loch respectively, via this pathway. The combined alkalinity production from NO_3, SO_4, Fe, and Mn reduction processes can account for approximately one-third of the observed Phase I alkalinity for Sky Pond and 10%–15% of the alkalinity of Glass Lakes and The Loch.

Iron and manganese concentrations increase in the bottom waters of Sky Pond and The Loch during some winters, although not every winter, and are not coincident in the two lakes (Figure 8.8). The yearly and spatial variation in snow distribution on each frozen lake might be responsible for this heterogeneity. The products of redox reactions in the sediments involving solid organic matter, O_2, NO_3, SO_4, and HCO_3, cause iron and manganese to solubilize and diffuse into the water column (Stumm and Morgan 1981). These metals increase in concentration as a result of oxygen reduction in lake bottom waters.

Iron fluctuates in LVWS lakes during Phase I over time and with depth. Values in early winter and surface water samples throughout the winter in Sky Pond average about 10 $\mu g\,L^{-1}$. Bottom water sample values fluctuate during the winter from 20 to 30 $\mu g\,L^{-1}$ in October to values above 100 $\mu g\,L^{-1}$ in February, March, and April (Figure 8.9A). A very high Fe value of 800 $\mu g\,L^{-1}$ was observed in March 1986, accompanied by 360 $\mu g\,L^{-1}$ Mn (Figure 8.9A). Early winter values in The Loch range from 25 to 50 $\mu g\,L^{-1}$ in the surface and outlet waters of The Loch. In the 1984 and 1985 water years, Fe concentrations never rose above 120 $\mu g\,L^{-1}$. In 1987 and 1988 anaerobic conditions must have been moderate, and Fe concentrations reached 190 $\mu g\,L^{-1}$ in the bottom waters of the lake. In 1986 and 1989, values reached 550 $\mu g\,L^{-1}$ and 650 $\mu g\,L^{-1}$, respectively, in bottom waters during the greatest reducing conditions late in the winters (Figure 8.9B).

Dissolved oxygen concentrations are available for 1989, and show O_2 to have decreased to 2 mg L^{-1} in Loch bottom waters by the end of February

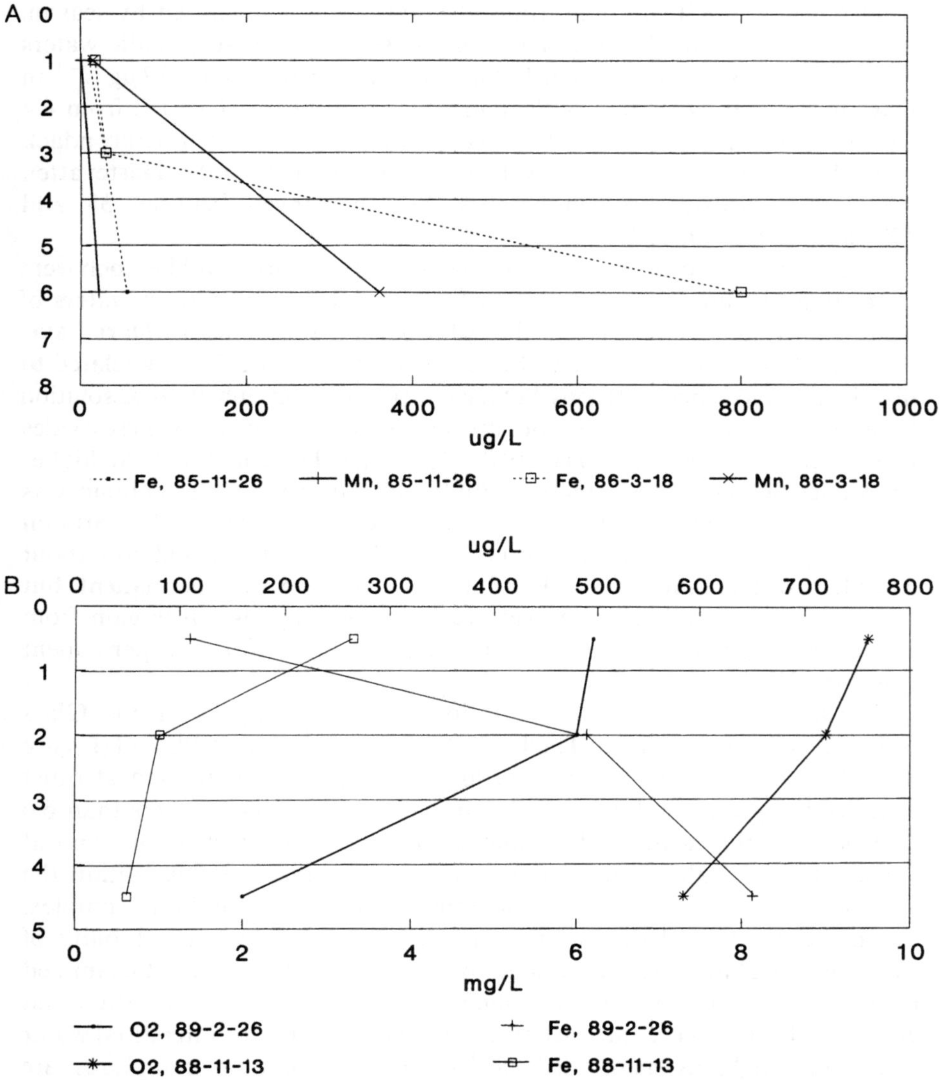

Figure 8.9. Metal concentrations under ice in Sky Pond and The Loch: (A) Iron and manganese concentrations with depth in Sky Pond in November 1985 and later that same winter in March 1986; (B) iron and dissolved oxygen with depth in The Loch in November 1988 and later that same winter in February 1989. Fe and Mn, $\mu g\ L^{-1}$; O_2, $mg\ L^{-1}$.

(Figure 8.9B). Our record for Mn is incomplete, but Mn concentrations in The Loch increased from lows of $1.0\ \mu g\ L^{-1}$ during most of the year to $51.0\ \mu g\ L^{-1}$ in lake bottom waters in December 1989. Surface water samples on February 26, 1989, were $27\ \mu g\ L^{-1}$; bottom samples are not available for that date. A summary of other available Mn concentration data is available from Nilsson (1985), although concentrations are not separated by season. The data show that Mn ranged as high as 400 $\mu g\, L^{-1}$ in very acidic waters in Swedish char lakes (Dickson 1978), varied between 11 and 137 $\mu g\, L^{-1}$ in lakes in the Galloway area of Scotland, where average pH varied from 4.4 to 6.3 (Wright et al. 1980), and ranged as high as 99 $\mu g\, L^{-1}$ in Adirondack lakes (Schofield 1976). Average values over a year in Lake St. Hästevatten and Lake Gårdsjön in southwestern Sweden ranged between 38 and 170 $\mu g\, L^{-1}$ (Nilsson 1985).

The average concentrations of Mn (93 $\mu g\, L^{-1}$) in solution in The Loch were lower than the highest values (120 $\mu g\, L^{-1}$) recorded from bottom waters of acidified (pH 5.1) Dart Lake in the Adirondack Mountains. In Dart Lake, White and Driscoll (1987) found Mn concentrations to be directly related to dissolved oxygen concentrations, but the low pH probably enhanced dissolution because both protons and electrons are necessary for many manganese oxides to solubilize to Mn^{2+} (Lindsay 1979; White and Driscoll 1987). At higher pH values Schindler and others (1980) found Mn in the water column was strongly sorbed onto particulate organic matter (Hutchinson 1975; Stumm and Morgan 1981; White and Driscoll 1987). Iron and Mn could add about 5 $\mu eq\, L^{-1}$ of alkalinity to Sky Pond if their reduction were permanent, but I assume these metals oxidize again each spring. Any alkalinity gain from Fe and Mn applies only to Phase I and does not constitute a permanent gain for LVWS.

The average aluminum concentration is lower in Sky Pond and Glass Lake (13–16 $\mu g\, L^{-1}$) than in The Loch (19–21 $\mu g\, L^{-1}$) during Phase I (Figure 8.8). Phase I Al concentrations are lower during the winter than at other times of the year. The low Al is due in part to pH values greater than 6.0 that occur during winter. Aluminum solubility is controlled by pH, and at values above 5.5 aluminum is mostly insoluble (Lindsay 1979). Within the soils of LVWS aluminum was found either as an organoaluminum complex, or, in the absence of organic matter, was controlled by the solubility of smectite (see Chapter 6, this volume; Walthall 1985). We have not examined lake sediment properties in any detail, but bulk sediment from one Sky Pond grab sample had an organic carbon content of 3.5‰, and the interstitial water DOC was high, as mentioned earlier. Organoaluminum complexes are strongly correlated with both particulate and dissolved organic carbon concentrations (Johnson et al. 1981; Driscoll et al. 1984; Schafran and Driscoll 1987), and it is likely that most of the Al in LVWS waters complexes with organic matter and settles to the lake bottoms.

The highest silica concentrations of the year are found in the winter (Figure 8.8). Sky and Glass concentrations average 1.0–2.0 mg $SiO_2\, L^{-1}$, while The Loch is somewhat higher, approximately 2.5 mg L^{-1}. Diatom populations are quite active during the winter months, with *Asterionella formosa* appearing in great numbers in December and again between February and April. In water year 1988 Spaulding (1991; see Chapter 9, this volume) reported 1100 cells ml^{-1} in surface waters of The Loch in December and later peaks in bottom and middepth waters of as many as 1500 cells ml^{-1}. In December of water year 1989, 6000 cells ml^{-1} were counted at middepths of The Loch, and a later bloom in the bottom waters attained densities of 2000 cell ml^{-1}. Possible evidence of biogenic depletion of silica concentrations is seen in SiO_2 time plots for these 2 years (Figure 8.10). In 1988 silica recovered slightly in January after a December diatom bloom and became depleted to nearly $1.0\,mg\,L^{-1}$ by mid-April. Diatoms were two to three times more numerous the following winter, and the silica concentrations were lower. The correspondence between diatom dynamics and silica concentrations is not as apparent, however, for 1989. Wetzel (1983) categorized concentrations of $< 5\,mg\,SiO_2\,L^{-1}$ as moderate to low and suggested silica can become limiting to algal growth at these levels should other nutrients become abundant. In LVWS lakes SiO_2 may be low but it varies consistently between 0.8 and $4.0\,mg\,L^{-1}$, so it is not likely to be the limiting nutrient, particularly during the winter months.

Unlike most other soluble constituents in winter lakes, dissolved organic carbon (DOC) is equal to or lower in concentration than during the rest of the year (Figure 8.8). An exception occurs in surface waters of The Loch, which average $3\,mg\,L^{-1}$ and have a wide variance. Possible sources for organic matter include phytoplankton, algal exudates, lake sediment sources, soil water, precipitation, and direct plant leaf and litter deposition. During the winter, direct inputs of leaf litter and precipitation do not occur. We did not measure DOC excretion by phytoplankton, but Zlotnik and Dubinsky

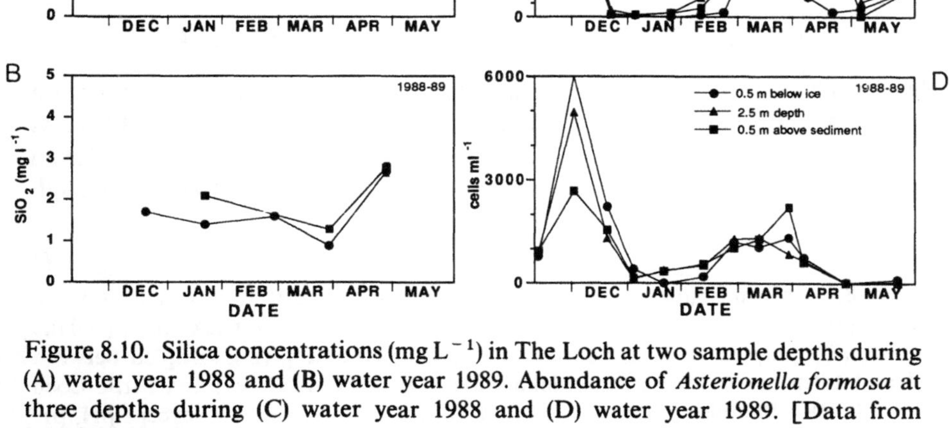

Figure 8.10. Silica concentrations ($mg\,L^{-1}$) in The Loch at two sample depths during (A) water year 1988 and (B) water year 1989. Abundance of *Asterionella formosa* at three depths during (C) water year 1988 and (D) water year 1989. [Data from Spaulding (1991).]

(1989) suggested this may be a small but significant flux into the water column. Glucose concentrations in 1990 [samples were filtered through precombusted glass-fiber filters and analyzed according to Cavari and Phelps (1977)] were generally lower under winter conditions (5 $\mu g\,L^{-1}$ to below detection for The Loch, 10–15 $\mu g\,L^{-1}$ for Sky Pond) than later in the year (10 $\mu g\,L^{-1}$ to below detection for The Loch, 15–20 $\mu g\,L^{-1}$ for Sky Pond; Baron et al., in manuscript). The lower glucose concentrations under the ice suggest greater microbial uptake of available sugars during this time.

Stable carbon isotope ($\delta^{13}C$) analyses of the humic fraction of dissolved organic carbon in the water column gave values of −30.8 for Sky Pond surface, −33.1 for Sky Pond bottom, and −30.0 for The Loch bottom waters in early April 1990 (Baron et al., in manuscript). These $\delta^{13}C$ values are similar to those reported by LaZerte from midlake sediment cores and phytoplankton from Lake Memphremagog in Quebec (LaZerte 1983). LaZerte concluded the major source of organic material to sediments in midlake locations was phytoplankton. Autochthonous organic matter is also well represented in Mirror Lake New Hampshire sediments (Likens 1985). On the basis of the $\delta^{13}C$ values that we observed, phytoplankton, either in the water column or resuspended from the sediments, is the major source of winter DOC in Sky Pond. The light $\delta^{13}C$ ratio from The Loch bottom water on April 10, 1990, also suggested a phytoplankton contribution to DOC in late winter. A linear regression of chlorophyll *a* with DOC for The Loch and Sky Pond for the winter months 1984, 1985, and 1987 did not reveal a significant relationship, so we have postulated algal settling and resuspension from the sediments as the pathway for DOC to the water column. The $\delta^{13}C$ ratio for DOC in surface waters of The Loch on that same date was −26.4. This suggests there is an input of soil drainage water during the winter to this lake, because the value falls within the range of −24.0 to −27.0 expected from terrestrially derived carbon (Degens 1969; Rau 1980; LaZerte 1983; Baron et al. in manuscript).

Chemical Dynamics During Snowmelt (Phase II)

Icy Brook begins to flow during late March or April. For a short period this causes laminar flow under the ice because incoming water at 0°C is less dense than the bottom lake water at 4°C (Bergmann and Welch 1985; Baron and Bricker 1987). Phase II covers the time from initial snowmelt through

▶

Figure 8.11. Mean solute concentrations for Phase II, April 15 through July 14. Bars represent 2 standard errors of the mean. Data presented for lake inlet (I), lake surface (S), sampled at 0.5 m below water surface; lake deep (D), sampled between 0.5 and 1.0 m above lake bottom; lake outlet (O); Andrews Creek; and Little Loch Creek (CRK). Sample sizes are found in Table 8.7. Values, $\mu eq\,L^{-1}$; except pH, pH units; Al, Fe, and Mn, $\mu g\,L^{-1}$; SiO_2 and DOC, $mg\,L^{-1}$.

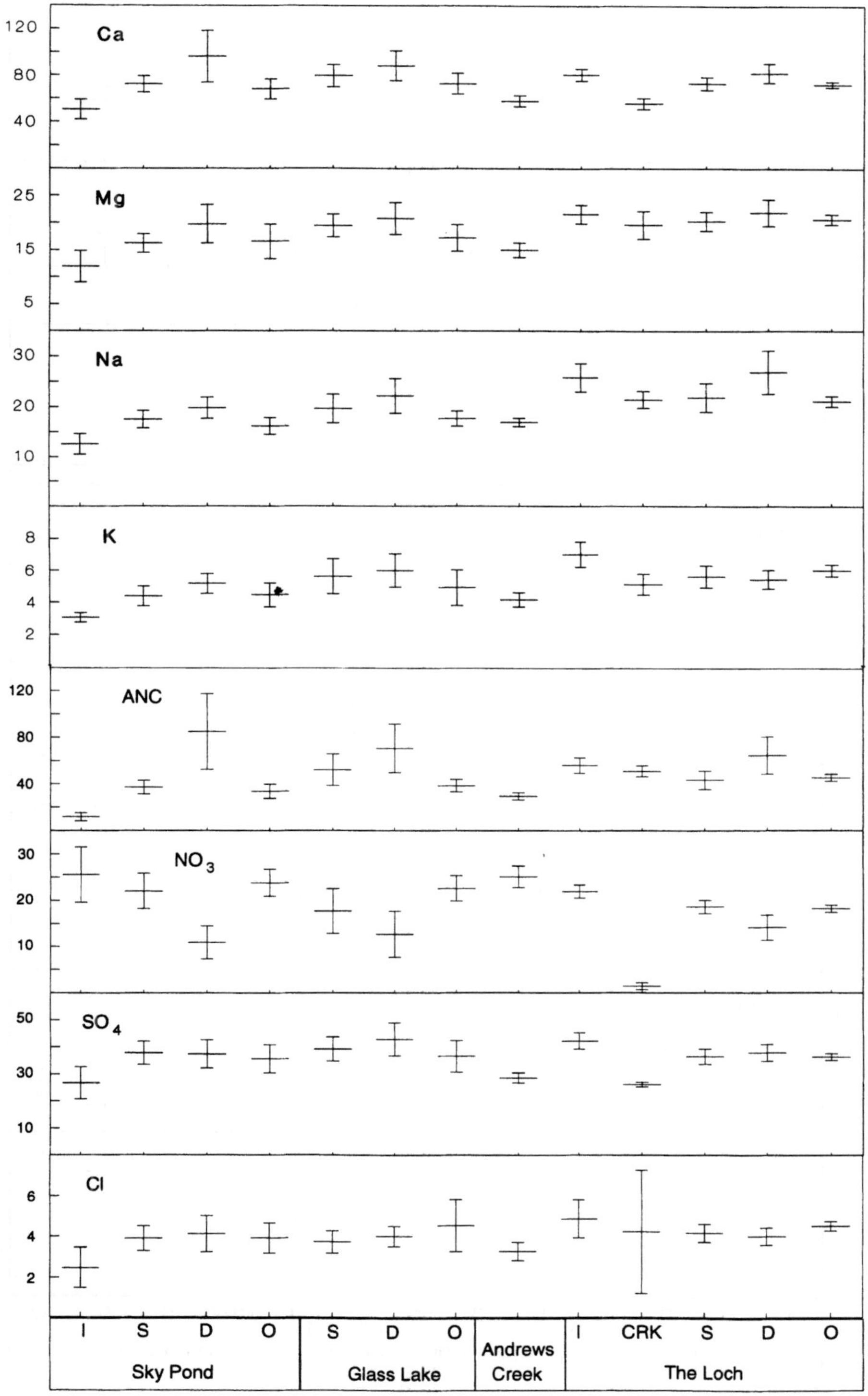
Ca
Mg
Na
K
ANC
NO_3
SO_4
Cl
I S D O
Sky Pond
S D O
Glass Lake
Andrews Creek
I CRK S D O
The Loch

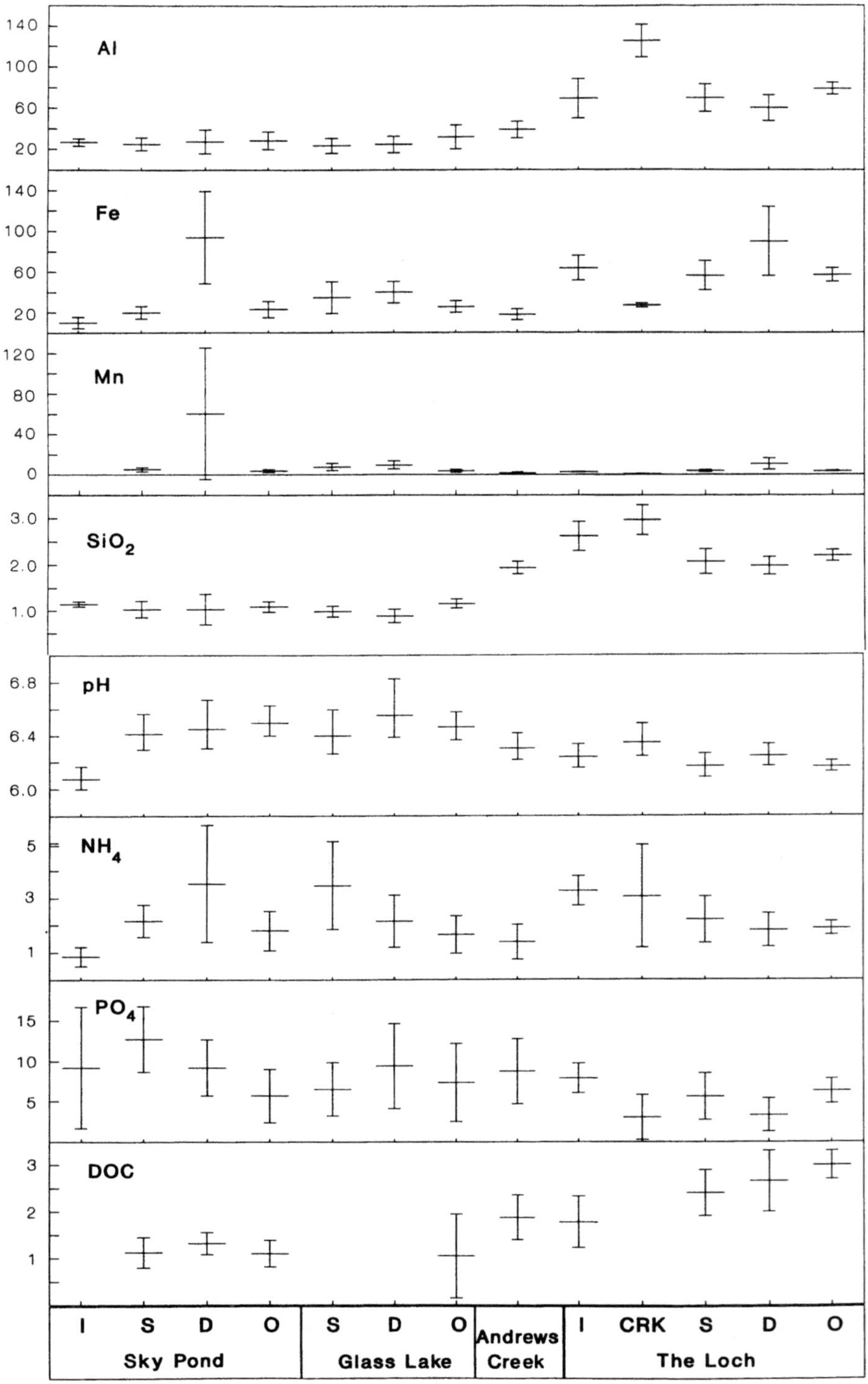

Figure 8.11 (*Continued*)

the end of the snowmelt period (Figures 8.7 and 8.11, samples sizes are shown in Table 8.7). Winter conditions persist in the bottom waters of most lakes into Phase II. This can be seen particularly in higher concentrations of base cations, acid-neutralizing capacity (ANC), and iron, and lower concentrations of nitrate in bottom waters of all three lakes (Figure 8.11). ANC and iron are significantly higher in Sky Pond bottom waters.

Spatial differences in solute concentrations become more pronounced than they were during the winter period. Solutes follow different patterns downstream from the inlet to Sky Pond, depending on their many biogeochemical controls and sources. Sky Pond inlet, at the very top of Icy Brook stream system, is the most dilute of all locations. All solutes measured are equal to or lower in concentration at Sky Pond inlet than anywhere else in LVWS, with the exception of NO_3, which is at its highest concentration. Phosphate has an average value of $9\,\mu g\,L^{-1}$ in the Sky Pond inlet, but is quite variable and can range above $15\,\mu g\,L^{-1}$. Sources of solutes to the Sky Pond inlet are precipitation, especially snow and snowmelt, the Taylor rock and ice glaciers, and bedrock weathering materials. Icy Brook at this elevation flows almost entirely under talus and till until it enters Sky Pond, so the influence from alpine soils is assumed to be minimal.

Chemical species that originate from processes occurring in the soils (as a result of root uptake of nutrients, mineralization of organic matter, production of organic anions, and mineral weathering) are expected to have higher concentrations in surface waters at the lower elevations of LVWS. The lower elevations are where soils are most abundant and nearest to surface waters. Species that follow this pattern are silica, aluminum, and dissolved organic carbon. Several authors have suggested that organic ions in soils may control weathering rates of primary minerals in the upper soil zone through complexation and through influencing soil solution pH (Graustein 1981; Antweiler and Drever 1982; Cronan 1984). Such a mechanism is postulated here to account for the concurrent increases in DOC and these metals, although Mast and Drever (1987) could find no evidence that the presence of oxalate increased the solution concentrations of Al or SiO_2 from the primary mineral oligoclase in controlled laboratory experiments. Because organic matter forms strong complexes with metals (Lindsay 1979; Cronan 1980; Cozzarelli et al. 1987), it may be that DOC becomes important to metal mobility after primary weathering has already taken place. Alkalinity and base cations should also have higher levels in surface waters below treeline as a result of soil activity (Cosby et al. 1985; Reuss and Johnson 1986). They do not, and I think this is because of the small proportion (6%) of LVWS that has had soil development. These species are also primary weathering products, which can come from anywhere within the basin area, making it difficult to distinguish a contribution solely from soils. Physical erosion from freeze-thaw cycles, debris flows, and avalanches, followed by snowmelt flushing, have been suggested by Mast et al. (1990) to be the major sources of cations, silica, and alkalinity to surface waters.

Ammonium concentrations are low throughout LVWS and are not significantly different from one site to another. Concentrations were highest in Sky Pond bottom waters, Glass Lake surface waters, and the inlet to The Loch. Sulfate and chloride do not vary significantly from one site to another downstream during Phase II, although both ions are significantly higher at The Loch outlet, the bottom of LVWS, than at the Sky Pond inlet, the top of the watershed. Sky Pond inlet and Andrews and Little Loch Creeks had the lowest SO_4 concentrations.

The variation in ion concentrations at The Loch outlet during snowmelt has been examined in detail by Denning et al. (in press) who distinguished between chemical inputs from directly melting snow and inputs that represent soil water pushed into surface waters as the soil water is replaced each spring by melting snow. The minimum yearly concentrations of pH, ANC, and silica each year correspond within 1 or 2 days with the time at which one lake volume has flowed through The Loch outlet. While this suggests simple replacement of one type of water with another, stream pH and ANC never approach the measured values in the snowpack, which suggests a watershed influence. The pH minimum is between 5.8 and 6.2 in streams versus snowpack pH values of 4.9 to 5.2. Stream ANC declines each spring to 25 $\mu eq\, L^{-1}$, while ANC in the snowpack is negative.

Chemical fluctuations used as additional signals of soil-derived water include DOC, Al, NO_3, NO_2, NH_4, PO_4 and K (Figure 8.12). Nutrients that are almost undetectable in surface waters through most of the year are present at their highest concentrations for a short period at the beginning of the spring melt period. Ammonium and PO_4, especially, are normally cycled between vegetation and microbial decomposition processes, but we postulate a break in the cycle occurs each winter when microbial decomposition processes continue but nutrients are not taken up by plants. This allows nutrients to accumulate in the soil water, and the presence of these species in surface waters is evidence that meltwater causes a pulse of soil water to flow into lakes and streams. The marked pulse of DOC early in the snowmelt period is accompanied by an almost identical concentration pattern ($r^2 = .93$) in total aluminum concentrations. As discussed previously and in Chapter 7 (this volume), Al solubility is highly related to its equilibrium relationship with organic carbon, and the concurrent presence of these two species in stream water is evidence of soil water flushing.

Contemporaneous fluctuations of SO_4 and NO_3 occur later in the snowmelt period over a 6-week period of time (Figure 8.12). These represent extended inputs from snow as the snowpack ripens and releases meltwater first in one part of LVWS then another (Denning et al. in press). A snowmelt "pulse" has been described by many authors (e.g., Johannessen and Henriksen 1978; Leivestad and Muniz 1979; Abrahamsen et al. 1990) as the fractionation of solutes with the earliest meltwater. Such a pulse is observed in snow meltwater in LVWS. By the time it actually appears in LVWS surface water

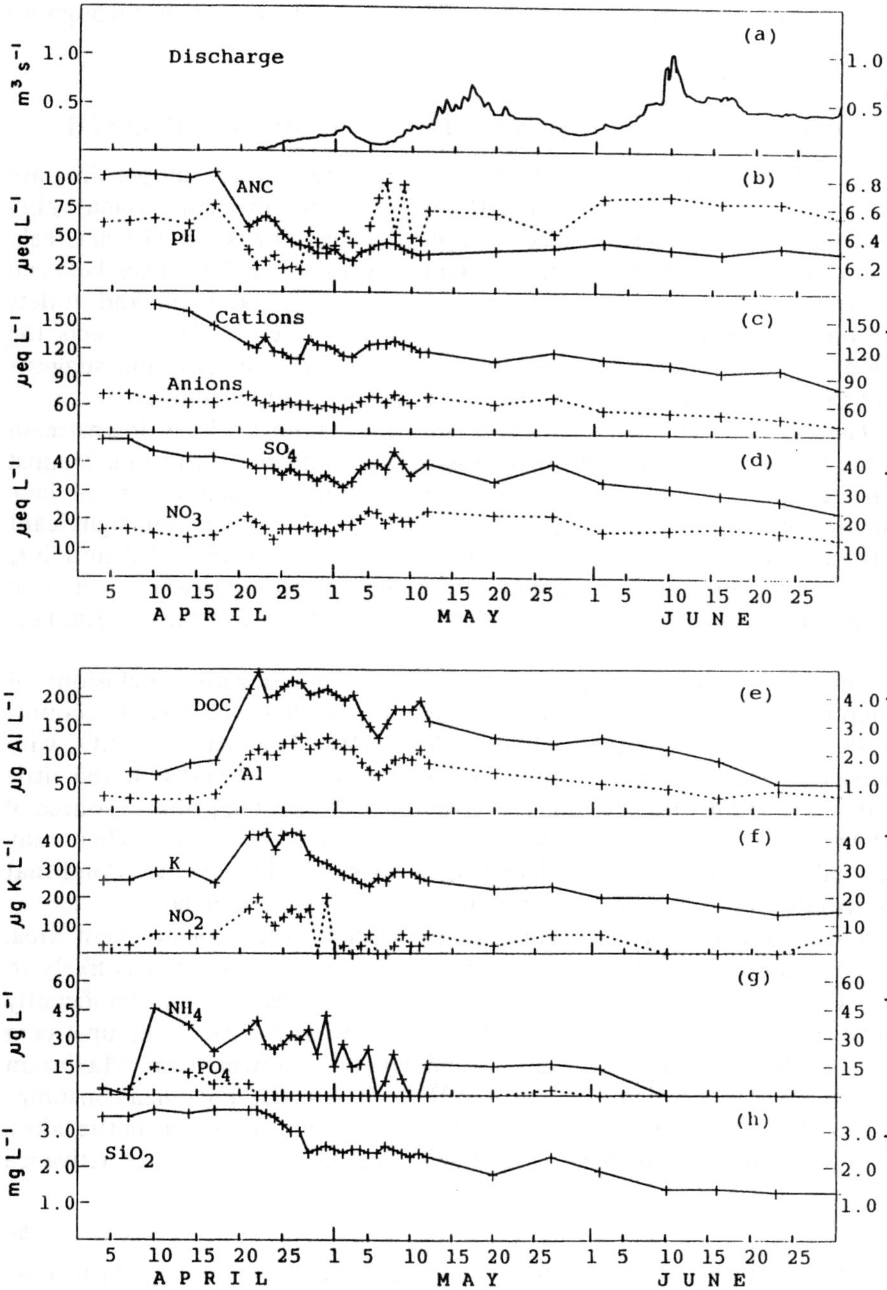

Figure 8.12. Stream discharge (a) and chemical parameters (b–h) at The Loch Outlet April through June 1987. Cations are Ca + Mg + Na + K; anions are $NO_3 + SO_4 + Cl$. (Reprinted by permission of Kluwer Academic Publishers.)

it is a snowmelt "smear" caused by differential melting of snow throughout the watershed.

Spatial Variability in Summer Lakes and Streams (Phase III)

Mean solute chemistry for each sampling site (Figure 8.13; sample sizes are shown in Table 8.7) during Phase III (July 15–September 30) is remarkably consistent across most watershed locations. An exception is Little Loch Creek. Wind mixing and flushing prevent significant ($p > .05$) differences between surface and deep samples in all three lakes. Further, lake inlets and outlets are not significantly different from the lakes themselves. This reflects the rapid turnover times for the lake water, even in late summer, and suggests that the lakes behave much as large stream pools in the summer.

There is very little increase in concentration for most solutes downstream in the watershed. Simple linear regressions of solute concentrations against subcatchment area (the area of the watershed above each stream or lake sample point) for all sites except Little Loch Creek show slight, but significant correlations ($p < .05$) for Ca ($r^2 = .81$), Na ($r^2 = .61$), ANC ($r^2 = .72$), and SO_4 ($r^2 = .66$). The increased amount of water that also accompanies the increased surface area may dilute, and thus dampen, the effect of increased drainage area.

The increase of Ca, Na, and ANC is in keeping with conclusions of Chapter 6 (this volume) and of Mast et al. (1990) that weathering accounts for the majority of observed solutes within LVWS. The increase in SO_4 with increased drainage area is not, however, because Mast (1989) found little evidence of sulfur-bearing minerals. Wet deposition is the primary source of sulfur for many lakes of the Front Range (Turk and Spahr 1989). There may be small amounts of sulfide minerals in the watershed below Sky Pond that contribute sulfur to lakes and streams of Loch Vale Watershed.

Silica does not show a significant correlation with subcatchment area, which was unexpected considering primary mineral weathering is likely to provide most SiO_2 to surface waters. Andrews Creek has a significantly higher mean SiO_2 concentration than Sky Pond and Glass Lake upstream and a slightly higher concentration than The Loch downstream. Maximum diatom densities occur in Phases I and II, not Phase III, thus silica consumption by diatoms is ruled out as a reason for lower concentrations in the lakes. Andrews Glacier, at the head of Andrews Creek shows evidence of active ice

Figure 8.13. Mean solute concentrations for Phase III, July 15 through September 30. Bars represent 2 standard errors of the mean. Data are presented for lake inlet (I); lake surface (S), sampled at 0.5 m below water surface; lake deep (D), sampled between 0.5 and 1.0 m above lake bottom; lake outlet (O); Andrews Creek; and Little Loch Creek (CRK). Sample sizes are found in Table 8.7. Values, μeq L^{-1}; except pH, pH units; Al, Fe, and Mn, μg L^{-1}; SiO_2 and DOC, mg L^{-1}.

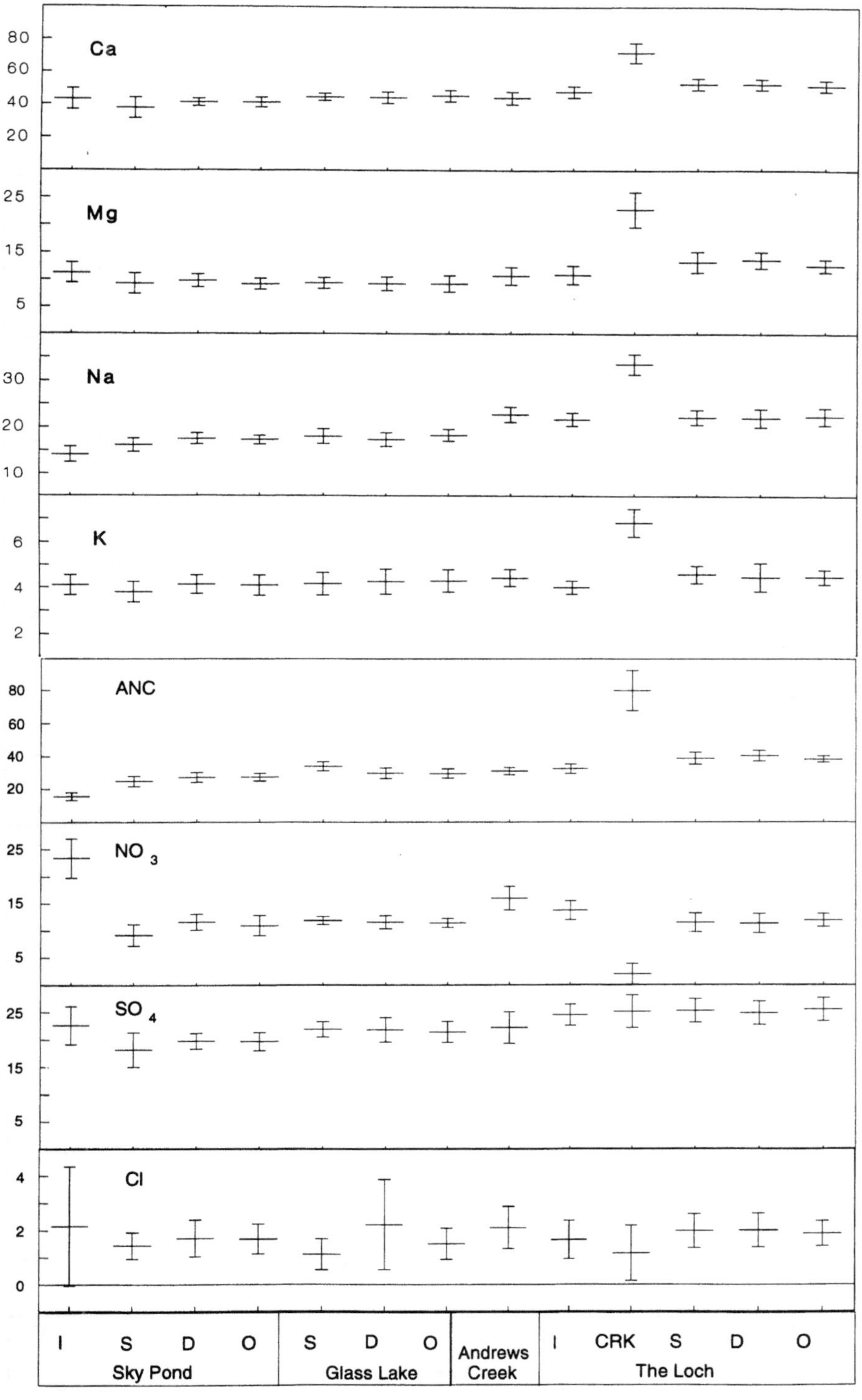

Ca
80
60
40
20
Mg
25
15
5
Na
30
20
10
K
6
4
2
ANC
80
60
40
20
NO 3
25
15
5
SO 4
25
15
5
Cl
4
2
0
I
S
D
O
Sky Pond
S
D
O
Glass Lake
Andrews Creek
I
CRK
S
D
O
The Loch

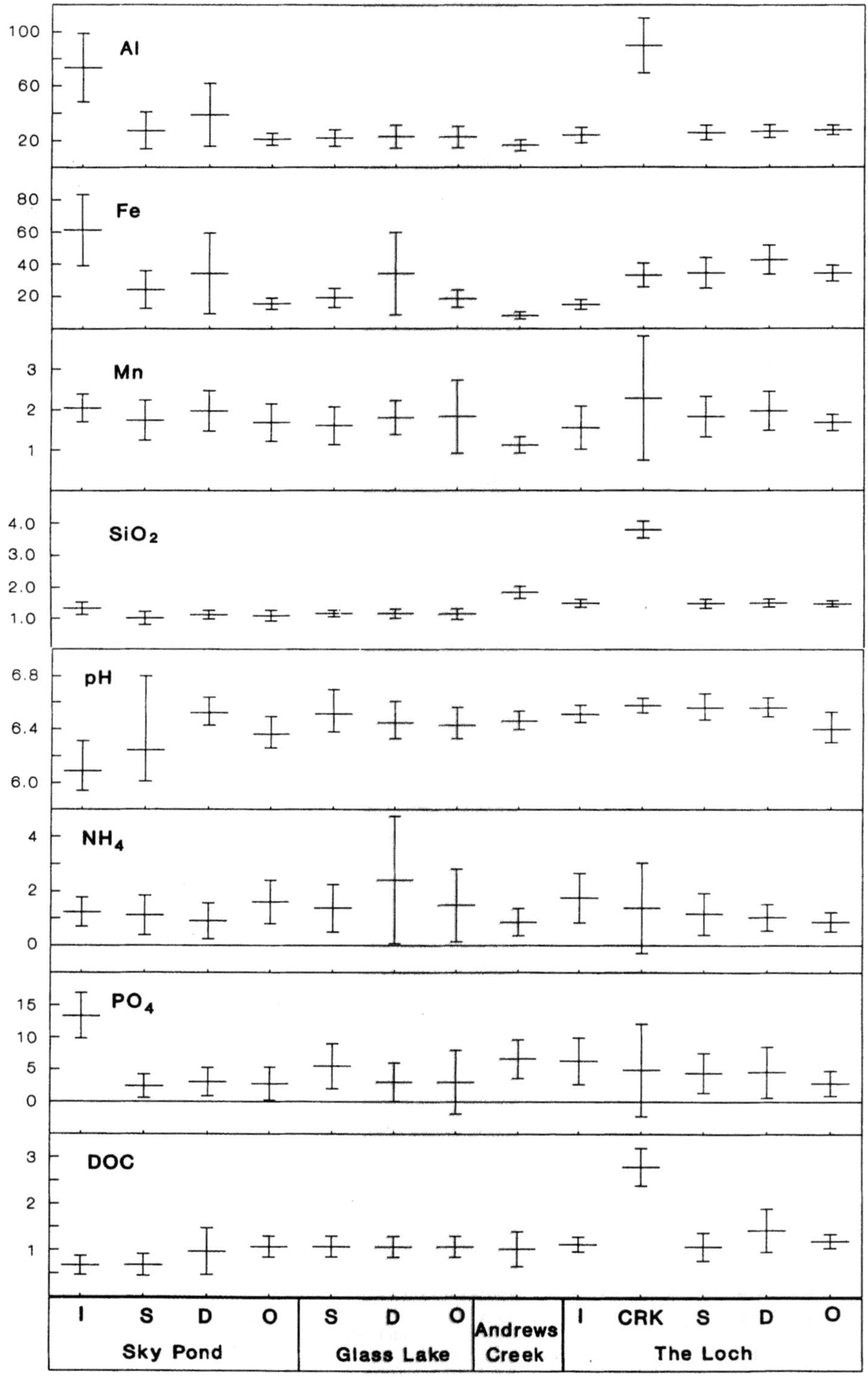

Figure 8.13 (*Continued*)

movements, so an alternative reason for higher silica in Andrews Creek is that glacial scouring enhances weathering. It is unknown why similar patterns are not displayed by other mineral weathering products.

Chloride concentrations are nearly constant at 2–3 μeq L^{-1} all locations in LVWS. Surface water Cl is nearly equal to the summer precipitation concentration of about 3 μeq L^{-1} (see Chapter 7, this volume) and suggests evapotranspiration is not very important between July and October.

Phosphate concentrations are significantly higher at the Sky Pond inlet during Phase III than anywhere else in the drainage. Below the inlet, PO_4 averages 2–6 μeq L^{-1}. At the inlet the concentration is 14 μeq L^{-1}. There should be little biological activity in Icy Brook above Sky Pond, and this may be why PO_4 is so much greater there. Phosphate in wet deposition is nearly always below detection limits for the National Atmospheric Deposition Program (these appear to be 20 μeq L^{-1}; NADP/NTN Data Base 1990), so a comparison of precipitation with stream concentrations is not possible.

Surface waters in Little Loch Creek show significantly elevated concentrations of base cations, ANC, SiO_2, Al, and DOC, and significantly decreased NO_3 concentrations. The concentrations are similar to those measured for soil solutions (see Chapter 7, this volume). It appears Little Loch Creek receives a much greater fraction of soil drainage water than other locations within LVWS. The south-facing slopes above Little Loch Creek are drier than the rest of LVWS, and snow melts off rapidly each summer. There is a lesser supply of yearly snowmelt, and no permanent snowfields, so water is much less available to Little Loch Creek than to other streams of LVWS.

Aluminum concentrations in Little Loch Creek are substantially higher than would be expected to result in an inorganic system from equilibrium with soil clay minerals (Drever 1988) at the pH found (mean pH, 6.6). These high levels of Al result from complexation with organic acids produced from the decomposition of soil organic matter (see Chapter 7, this volume; Walthall, 1985; Denning et al. in press). Low concentrations of NO_3 in this creek suggest greater biological uptake relative to the other sampling sites and are also consistent with soil solution samples. The higher concentrations of mineral weathering products may reflect longer contact time of water in the soil as opposed to channelized flow from snowfields and glacial ice.

Concentrations of NO_3 are very high throughout LVWS except in Little Loch Creek. The lakes have mean NO_3 values of 9 to 12 μeq L^{-1}, compared to summer precipitation of 15 μeq L^{-1}. Andrews Creek NO_3 averages 16 μeq L^{-1}. High surface water concentrations in eastern North America and in Europe have been attributed to HNO_3 in precipitation (Malanchuk and Nilsson 1989; Dillon and Molot 1990). Under natural conditions N is a limiting nutrient to plant growth, and historically watersheds show very low concentrations of both NO_3 and NH_4 (Goldman and Horne 1983; Hemond and Eshleman 1984; Likens 1985). Nitrate retention in lake waters has been related by Kelly et al. (1987) to lake depth and water residence time and by Dillon and Molot (1990) to catchment grade. These are all important to understanding high NO_3 in the subalpine systems of LVWS, which is characterized by steep catchment grades, a low percentage of forested area, shallow lakes, and rapid replacement times. Steep gradients enhance transport of precipitation directly to streams and then "pipe" it down and out of the watershed. The high percentage of rock and talus offers little opportunity for terrestrial biological uptake. Nitrate was rarely present in soil waters (Arthur 1990), so it does not appear that high NO_3 values in surface waters imply N saturation of the forested ecosystem, as has been observed by others (Henriksen and Brakke 1988). Shallow lakes and rapid flushing work to prevent contact of NO_3 and the lake bottom microbial communities primarily responsible for aquatic NO_3 consumption (Knowles and Lean 1987; Dillon and Molot 1990).

In contrast to NO_3, NH_4 concentrations were very low during the summer months. Ammonium is the preferred source of assimilation by organisms (Syrett 1981), so retention of most of the incoming NH_4 is expected. Ammonium did not differ significantly between any of the stream or lake sites within LVWS in Phase III.

The mean NO_3 concentration in Sky Pond inlet is 23 μeq L^{-1}, a significantly higher value than in precipitation. Evaporation of 30% of the incoming precipitation water can account for at most 5 μeq NO_3 L^{-1}, and this is not enough to account for all the observed NO_3. Sievering et al (1989) found that dry deposited N could account for 25%–50% of summertime wet NO_3 deposition under forest canopies at nearby Niwot Ridge, but there is no forest vegetation above Sky Pond. It is possible that lightning is a significant source of nitrate to high-elevation ecosystems. Lightning oxidizes atmospheric N_2, and the resultant nitrate can be deposited in wet deposition (because rain or hail often accompanies lightning events) or as dry deposited compounds. LeGrand and Delmas (1986) suggested that lightning at higher latitudes is the most likely source of nitrate in Antarctic snow. Most NO_x formed from lightning goes up, not down, into the upper troposphere where it can be carried long distances. It is possible that some amount of nitrate forms locally and serves as a source for wet deposited NO_3, although this is speculative. The mountains in and around Rocky Mountain National Park experience an extremely high frequency of lightning strikes (Hansen et al. 1978; Banta and Schaaf 1987).

Comparison to Other Western Lakes

The Loch was sampled in 1985 as a Special Interest Lake as part of the U.S. Environmental Protection Agency's Western Lake Survey (WLS; Eilers et al. 1986; Landers et al. 1986). The 719 lakes included in the WLS were selected by stratified random design, so that they represent the estimated 10,393 lakes in the western United States (Landers et al. 1986). The median elevation of lakes sampled in the southern Rockies was 3264 m, slightly higher than the elevation of The Loch, at 3048 m. The Loch is atypical of these Front Range lakes that were sampled (Table 8.8). Overall, The Loch has a much greater watershed area to lake area ratio than other lakes of the Front Range, the southern Rockies, and the western United States that were sampled by the WLS. It also has much more dilute waters. Mineral weathering contributes far less to solution chemistry of The Loch than to most other lakes within the Front Range, whereas Front Range lakes appear to be typical of other lakes within the southern Rockies. For example, the sum of the base cations for The Loch is 110.3 μeq L^{-1}, compared with median values of 352.5 μeq L^{-1} and 367.4 μeq L^{-1} for the Front Range and the southern Rockies, respectively. Acid-neutralizing capacity for The Loch is 48.9 μeq L^{-1}, compared with median values of 337.5 μeq L^{-1} and 317.0 μeq L^{-1} for the Front Range and the southern Rockies, and SiO_2 in The Loch is only 1.77 mg L^{-1} compared to 2.30 mg L^{-1} and 2.12 mg L^{-1} for the subpopulations. This suggests that LVWS bedrock is far more resistent to weathering than is typical of the rest of the southern Rockies and may be more vulnerable to increasing acidic atmospheric deposition.

Table 8.8. Comparison of Western Lake Survey (WLS) Population Statistics with The Loch[a]

	West ($n = 719$)	Southern Rockies ($n = 132$)	Front Range ($n = 51$)	The Loch ($n = 1$)
pH	7.16	7.60	—	6.55
Ca (μeq L^{-1})	92.4	233.1	222.6	68.3
Mg (μeq L^{-1})	26.4	91.5	91.6	16.9
Na (μeqL^{-1})	23.9	33.8	33.1	21.8
K (μeq L^{-1})	5.6	9.0	5.2	3.3
NH_4 (μeq L^{-1})	0.0	0.0	—	0.0
Cl (μeq L^{-1})	4.1	4.2	3.6	4.3
SO_4 (μeq L^{-1})	18.9	34.6	35.2	30.4
NO_3 (μeq L^{-1})	0.4	0.5	—	15.2
ANC (μeq L^{-1})	119.4	317.0	337.5	48.9
Total P (μg L^{-1})	4.7	8.1	—	9.5
SiO_2 (mg L^{-1})	2.21	2.12	2.3	1.77
Fe (μg L^{-1})	13.7	17.8	42.5	18.0
DOC (mg L^{-1})	1.21	1.48	1.51	0.87
Conductance (μS cm^{-1})	16.5	37.1	37.5	12.6
Site depth (m)	7.8	5.1	6.7	2.4
Lake area (ha)	4.6	3.5	3.7	5.0
WA:LA	23.7	23.5	23.8	132
Seechi disk transparency (m)	4.8	3.4	2.4	2.4

[a] All data are from Landers et al. (1986) except WA:LA for The Loch. The WLS reports a WA:LA of 650 for The Loch. The Loch was sampled as a special interest lake on 10/10/85. WA:LA denotes watershed area to lake area ratio.

Solution chemistry for The Loch is more characteristic of the West as a whole, and similar in composition to the very dilute lakes that are found in the Sierra Nevada (Landers et al. 1986). Dissolved organic carbon in The Loch is half that of the rest of the Rockies, probably because only 6% of LVWS supports evergreen forests, compared to 35.4% evergreen coverage for the southern Rockies as a whole. Sulfate and chloride concentrations in The Loch are comparable to concentrations found in the Front Range and in the southern Rockies, and this reflects their predominantly atmospheric origin (Turk and Spahr 1989). Sulfate concentrations in southern Rocky Mountain lakes are higher than the median SO_4 concentration for the West and may reflect regional emissions (see Chapter 4, this volume). Nitrate, as discussed previously, is much higher in The Loch than in other WLS lakes.

Conclusions

We conclude that lakes of Loch Vale Watershed are not representative of the population of western lakes. As such, their behavior cannot be generalized to other lakes. Loch Vale Watershed lakes, instead, represent the most sensitive water bodies of the western United States. This makes them good "indicators" of change, because their low base cation and ANC concentrations make them extremely sensitive to changes from environmental disturbance.

The most important differences between the surface water environment of LVWS and those in other regions result from the topography and the climate. The stream gradient of Icy Brook in LVWS is 15%. Hydrologic residence times are so short during Phases II and III that little or no difference in stream chemistry develops between alpine Sky Pond and the subalpine Loch. During Phases II and III, all surface water quantity and quality are much more hydrographically controlled than biogeochemically controlled. By contrast, Phase I is very different. Within-lake physical and biogeochemical processes including freeze-concentration, inverse thermal stratification, anoxia, and active biological consumption of nutrients, create very different lake conditions.

There are slight but distinct differences between the alpine and subalpine lakes of LVWS. Within-lake processes are important to the biogeochemistry of Sky Pond, particularly during Phase I. In contrast, the most important influence on The Loch is water from upstream. When the water is flowing, there is no difference between The Loch and Icy Brook. When the water stops, lake properties become apparent. For a short period each spring, soil water inputs are important to solution chemistry in The Loch. Little Loch Creek has very different behavior from the other water bodies of LVWS because of to its southerly aspect and small watershed.

References

Abrahamsen G, Seip HM, Semb A (1990) Long-term acidic precipitation studies in Norway. In: Adriano DC, Havas M, eds. Acidic Deposition and Aquatic Ecosystems: Case Studies, pp. 137–179. Springer-Verlag, New York.

Antweiler RC, Drever JI (1982) The weathering of a late Tertiary volcanic ash: importance of organic solutes. Geochim. Cosmochim. Acta 47:623–629.

Arthur MA (1990) The Effects of Vegetation on Watershed Biogeochemistry at Loch Vale Watershed, Rocky Mountain National Park, Colorado. Ph.D. Dissertation, Cornell University, Ithaca, New York.

Banta RM, Schaaf CB (1987) Thunderstrom genesis zones in the Colorado Rocky Mountains as determined by traceback of geosynchronous satellite images. Mon. Weather Rev. 115:463–476.

Barica J (1977) Effects of freeze-up on major ion and nutrient content of a prairie winterkill lake. J. Fish. Res. Bd. Can. 34:2210–2215.

Baron J, Bricker OP (1987) Hydrologic and chemical flux in Loch Vale Watershed, Rocky Mountain National Park. In: Averett RC, McKnight DM, eds. Chemical Quality of Water and the Hydrologic Cycle, Lewis, Chelsea, Michigan. pp. 141–157.

Baron J, Norton SA, Beeson DR, Herrmann R (1986) Sediment diatom and metal stratigraphy from Rocky Mountain lakes with special reference to atmospheric deposition. Can. J. Fish. Aquat. Sci. 43:1350–1362.

Bergmann MA, Welch HE (1985) Spring meltwater mixing in small arctic lakes. Can. J. Fish. Aquat. Sci. 42:1789–1798.

Canfield DE, Jr., Bachmann RW, Hoyer MV (1983) Freeze-out of salts in hardwater lakes. Limnol. Oceanogr. 28:970–977.

Carignan R, Tessier A (1988) The co-diagenesis of sulfur and iron in acid lake sediments of southwestern Quebec. Geochim. Cosmochim. Acta 52:1179–1188.

Cavari BZ, Phelps G (1977). Sensitive enzymatic assay for glucose determination in natural waters. Contribution 121 of the Israel Oceanography and Limnology Research, Ltd. Appl. Environ. Microbiol, 33:1237–1243.

Cook RB, Kelly CA, Schindler DW, Turner MA (1986) Mechanisms of hydrogen ion neutralization in an experimentally acidified lake. Limnol. Oceanogr. 31:134–148.

Cosby BJ, Hornberger GM, Galloway JN, Wright RF (1985) Modeling the effects of acid deposition: assessment of a lumped parameter model of soil water and streamwater chemistry. Water Resour. Res. 21:51–63.

Cozzarelli IM, Herman JS, Parnell RA, Jr. (1987) The mobilization of aluminium in a natural soil system: effects of hydrologic pathways. Water Resour. Res. 23:859–874.

Cronan CS (1980) Solution chemistry of a New Hampshire subalpine ecosystem: a biogeochemical analysis. Oikos 34:272–281.

Cronan CS (1984) Chemical weathering and solution chemistry in acid forest soils: differential influence of soil type, biotic processes and H^+ deposition. In: Drever JI, ed. The Chemistry of Weathering, pp. 55–74. D. Reidel, Dordrecht, The Netherlands.

Degens ET (1969) Biogeochemistry of stable carbon isotopes. In: Eglington G, Murphy MTJ, eds. Organic Geochemistry, Methods and Results, pp. 304–329. Springer-Verlag, New York.

Denning AS, Baron J, Mast MA, Arthur MA. Hydrologic pathways and chemical composition of runoff during snowmelt in Loch Vale Watershed, Rocky Mountain National Park. Water Air Soil Pollut. 55 (in press).

Dickson W (1978) Effects of acidification of Swedish lakes. Verh. Int. Ver. Limnol. 20:851–856.
Dillon PJ, Molot LA (1990) The role of ammonium and nitrate retention in the acidification of lakes and forested catchments. Biogeochemistry 11:23–43.
Drever JI (1988) The Geochemistry of Natural Waters, 2d Ed. Prentice-Hall, Englewood Cliffs.
Driscoll CT, Baker JP, Bisogni JJ, Schofield CL (1984) Aluminum speciation and equilibria in dilute acidic surface waters of the Adirondack region of New York State. In: Bricker OP, ed. Geological Aspects of Acid Rain., pp. 55–75. Ann Arbor Science, Ann Arbor, Michigan.
Eilers JM, Kanciruk P, McCord RA, Overton WS, Hook L, Blick DJ, Brakke DF, Kellar P, Silverstein ME, Landers DH (1986) Characteristics of Lakes in the Western United States: Vol. II Data Compendium for Selected Physical and Chemical Variables. EPA-600/3-86/054B, U.S. Environment Protection Agency, Washington, DC.
Gibson JH, Galloway JN, Schofield CL, McFee W, Johnson R, McCarley S, Dise N, Herzog D (1983) Rocky Mountain Acidification Study. FWS/OBS-80/40.17, Division of Biological Service, Eastern Energy and Land Use Team, U.S. Fish and Wildlife Service, Washington, D.C.
Giblin AE, Likens GE, White D, Howarth RW (1990) Sulfur storage and alkalinity generation in New England lake sediments. Limnol. Oceanogr. 35:852–869.
Goldman CR, Horne AJ (1983) Limnology. McGraw-Hill, New York.
Graustein WC (1981) The Effects of Forest Vegetation on Solute Acquistion and Chemical Weathering: A Study of the Tesuque Watersheds near Santa Fe, New Mexico. Ph.D. Dissertation, Yale University, New Haven, Connecticut.
Hansen WR, Chronic J, Matelock J (1978) Climatography of the Front Range Urban Corridor and Vicinity, Colorade. U.S. Geol. Surv. Prof. Pap. No. 1019, U.S. Government Printing Office, Washington, DC.
Hemond HF, Eshleman KN (1984) Neutralization of acid deposition by nitrate retention at Bickford Watershed, Massachusetts. Water Resour. Res. 20:1718–1724
Henriksen A, Brakke DF (1988) Increasing contributions of nitrogen to the acidity of surface waters in Norway. Water Air Soil Pollut. 42:183–201.
Hutchinson GE (1975) A Treatise on Limnology, Vol. I, Part 1: Geography and Physics of Lakes. Wiley, New York.
Johannessen M, Henriksen A (1978) Chemistry of snow meltwater:changes in concentration during melting. Water Resour. Res. 14:615–619.
Johnson NM, Driscoll CT, Eaton JS, Likens GE, McDowell WH (1981) Acid rain, dissolved aluminum and chemical weathering at the Hubbard Brook Experimental Forest, New Hampshire. Geochim. Cosmochim. Acta 45:1421–1437.
Kaufmann MR (1981) Automatic determination of conductance, transpiration, and environmental conditions of forest trees. For. Sci. 27:817–827.
Kelly CA, Rudd JWM, Hesslein RH, Schindler DW, Dillon PJ, Driscoll CT, Gherini SA, Hecky RE (1987) Prediction of biological acid neutralization in acid-sensitive lakes. Biogeochemistry 3:129–140.
Knight CA (1987) Slush on lakes. In: Loper DE, ed. Structure and Dynamics of Partially Solidified Systems, pp. 455–465. Martinus Nijhoff, Dordrecht, The Netherlands.
Knight CA (1988) Formation of slush on floating ice. Cold Reg. Sci. Technol. 15:33–38.
Knowles R, Lean DRS (1987) Nitrification: a significant cause of oxygen depletion under winter ice. Can. J. Fish. Aquat. Sci. 44:743–749.
Landers DH, Eilers JM, Brakke DF, Overton WS, Schonbrod RD, Crowe RT, Linthurst RA, Omernik JA, Teague SA, Meier EP (1986) Characteristcs of Lakes

in the Western United States, Vol. I: Population Descriptions and Physicochemical Relationships. EPA-600/3-86/054a, U.S. Environmental Protection Agency. Washington D.C.

LaZerte BD (1983) Stable carbon isotope ratios: implications for the source of sediment carbon and for phytoplankton carbon assimilation in Lake Memphramagog, Quebec. Can. Jour. Fish. Aquat. Sci. 40:1658–1666.

LeGrand MR, Delmas RJ (1986) Relative contributions of tropospheric and stratospheric sources to nitrate in Antarctic snow. Tellus 38B:236–249.

Leivestad H, Muniz IP (1979) Fishkill at low pH in a Norwegian river. Nature (London), 259:391–392.

Likens GE (1985) An Ecosystem Approach to Aquatic Ecology: Mirror Lake and Its Environment. Springer-Verlag, New York.

Likens GE, Bormann FH, Pierce RS, Eaton JS, Johnson NM (1977) Biogeochemistry of a Forested Ecosystem. Springer-Verlag, New York.

Lind OT (1979) Handbook of Common Methods in Limnology, 2d Ed. CV Mosby, St. Louis, Missouri.

Lindsay WL (1979) Chemical Equilibrium in Soils. Wiley, New York.

Malunchuk JL, Nilsson J, eds (1989) The Role of Nitrogen in the Acidification of Soils and Surface Waters. Miljørapport 1989:10, Nordic Council of Ministers, Denmark.

Mast MA (1989) A Laboratory and Field Study of Chemical Weathering with Special Reference to Acid Deposition. Ph.D. Dissertation, University of Wyoming, Laramie.

Mast MA, Drever JI (1987) The effect of oxalate on the dissolution rates of oligoclase and tremolite. Geochim. Cosmochim. Acta 51:2559–2568.

Mast MA, Drever JI, Baron J (1990) Chemical weathering in the Loch Vale Watershed, Rocky Mountain National Park, Colorado. Water Resour. Res. 26:2971–2978.

McKnight DM, Brenner M, Smith RL, Baron J, Spaulding SA (1986) Seasonal Changes in Phytoplankton Populations and Related Chemical and Physical Characteristics in Lakes in Loch Vale, Rocky Mountain National Park, Colorado. Water Resources Investigations Report 86-4101, U.S. Geological Survey, Denver, Colorado.

McKnight DM, Miller C, Smith RL, Baron J, Spaulding S (1988) Phytoplankton Populations in Lakes in Loch Vale, Rocky Mountain National Park, Colorado. Water-Resources Investigations Report 88-4115, U.S. Geological Survey, Denver, Colorado.

NADP/NTN (1990) National Atmospheric Deposition Program. Tape of weekly data. National Atmospheric Deposition Program (IR-7)/National Trends Network, July 1978–January 1989. Magnetic tape, 9 track, 1600 cpi, ASCII. NADP/NTN Coordination Office, Natural Resource Ecology Laboratory, Colorado State University, Fort Collins, Colorado.

Nilsson SI (1985) Budgets of aluminum species, iron, and manganese in the Lake Gårdsjön catchment in SW Sweden. Ecol. Bull. 37:120–132 (Stockholm, Sweden).

Nriagu JO, Soon YK (1985) Distribution and isotropic composition of sulfur in lake sediments of northern Ontario. Geochim. Cosmochim. Acta 49:823–834.

Pennak RW (1963) Rocky Mountain states. In: Frey DG (ed) Limnology in North America, pp. 349–370. University of Wisconsin Press, Madison.

Psenner R (1988) Alkalinity generation in a soft-water lake: watershed and in-lake processes. Limnol. Oceanogr. 33:1463–1475.

Rau G (1980) Carbon-13/carbon-12 variation in subalpine lake aquatic insects: food source implications. Can. J. Fish. Aquat. Sci. 37:742–746.

Reuss JO, Johnson DW (1986) Acid Deposition and the Acidification of Soils and Waters. Ecological Studies 59, Springer-Verlag, New York.

Rudd JWM, Kelly CA, Furutani A (1986) The role of sulfate reduction in long-term accumulation of organic and inorganic sulfur in lake sediments. Limnol. Oceanogr. 31:1281–1292.

Rudd JWM, Kelly CA, Schindler DW, Turner MA (1990) A comparison of the acidification efficiencies of nitric and sulfuric acids by two whole-lake acidification experiments. Limnol. Oceanogr. 33:663–679.

Schafran GC, Driscoll CT (1987) Comparison of terrestrial and hypolimnetic sediment generation of acid neutralizing capacity for an acidic Adirondack lake. Environ. Sci. Technol. 21:988–993.

Schindler DW (1986) The significance of in-lake production of alkalinity. Water Air Soil Pollut. 30:931–944.

Schindler DW, Hesslein RH, Wageman R, Broecker WS (1980) Effects of acidification on mobilization of heavy metals and radionuclides from the sediments of a freshwater lake. Can. J. Fish. Aquat. Sci. 37:373–377.

Schofield CL (1976) Acid precipitation: effects on fish. Ambio 5:228–230.

Sievering H, Braus J, Caine J (1989) Dry deposition of nitrate and sulfate to coniferous canopies in the Rocky Mountains. In: Olson RK, LeFohn AS, eds. Effects of Air Pollution on Western Forests, pp. 171–176. APCA Transactions Series, Air and Waste Management Association, Pittsburgh, Pennsylvania.

Spaulding SA (1991) Phytoplankton Community Dynamics under Ice-Cover in The Loch, a Lake in Rocky Mountain National Park. M.S. Thesis, Colorado State University, Fort Collins.

Stoddard JL (1987) Alkalinity dynamics in an unacidified alpine lake, Sierra Nevada, California. Limnol. Oceanogr. 32:825–839.

Stumm WA, Morgan JJ (1981) Aquatic Chemistry: An Introduction Emphasizing Chemical Equilibria in Natural Waters, 2d Ed. Wiley, New York.

Syrett PJ (1981) Nitrogen metabolism of microalgae. In: Platt T, ed. Physiological Bases of Phytoplankton Ecology. Can. Bull. Fish. Aquat. Sci. 210:182–210.

Turk JT, Adams, DB (1983) Sensitivity to acidification of lakes in the Flat Tops Wilderness Area, Colorado. Water Resour. Res. 19:346–350.

Turk JT, Campbell DH (1987) Estimates of acidification of lakes in the Mt. Zirkel Wilderness Area, Colorado. Water Resour. Res. 23:1757–1761.

Turk JT, Spahr NE (1989) Rocky Mountains: controls on lake chemistry. In: Adriano DC, Havas M, eds. Acidic Deposition and Aquatic Ecosystems: Case Studies, pp. 181–208. Springer-Verlag, New York.

Walthall PM (1985) Acidic Deposition and the Soil Environment of Loch Vale Watershed in Rocky Mountain National Park. Ph.D. Dissertation, Colorado State University, Fort Collins.

Wetzel RG (1983) Limnology. Saunders, Philadelphia.

Whipple GC (1898) Classifications of lakes according to temperature. Am. Nat. 32:24–33.

White JR, Driscoll CT (1987) Manganese cycling in an acidic Adirondack lake. Biogeochemistry 3:87–104.

Wright RF, Henriksen A, Morrison B, Caines LA (1980) Acid lakes and streams in the Galloway area, southwestern Scotland. In: Dabløs D, Tollan A, eds. Ecological Impact of Acid Precipitation, pp. 248–249: Proceedings of an International Conference, Sandefjord, Norway, March 11–14, 1980 SNSF Project, Oslo-Ås.

Zlotnik I, Dubinsky Z (1989) The effect of light and temperature on DOC excretion by phytoplankton. Limnol. Oceanogr. 34:831–839.

9. Aquatic Biota

Sarah A. Spaulding, Mitchell A. Harris, Diane M. McKnight, and Bruce D. Rosenlund

From its headwaters at Taylor Glacier near the Continental Divide to The Loch, Icy Brook drops more than 400 m in just over 2 km. During its course, Icy Brook traverses two major vegetational zones, the alpine and subalpine (Marr 1961). Given the high elevation, relatively inert and insoluble granitic bedrock, and sparse vegetational cover, waters of these zones are characterized by low water temperatures, low primary productivity, and low allochthonous nutrient inputs. The lakes and streams of Loch Vale Watershed (LVWS) are relatively unaltered ecosystems. Therefore, it was a natural extension of the LVWS project to study the organisms found there. This chapter describes the phytoplankton, macroinvertebrate, and fish communities of Icy Brook, Sky Pond, Glass Lake, and The Loch.

Seasonality of Phytoplankton in Loch Vale Lakes

Characterization of Alpine Lakes

Subalpine and alpine lakes are generally low in phytoplankton abundance (Table 9.1). Lakes in the Front Range of the Rocky Mountains are no exception (Pennak 1968; Keefer and Pennak 1977; Herrmann 1978). Low nutrient levels and cold temperatures result in low productivity and biomass. Although a vast colonization pool of algae are present in the form of wind-deposited spores, only those physiologically able to withstand the low

Table 9.1. Algal Taxa Collected from Lakes in Loch Vale Watershed (Sky Pond and The Loch) in Rocky Mountain National Park, 1984–1989[a]

Taxa	Winter	Spring	Summer	Fall
BACILLARIOPHYTA				
Achnanthes affinis Grun.		+	+	+
Achnanthes austriaca var. *helvetica* Hust.			+	
Achnanthes clevei Grun.			+	
Achnanthes detha Hohn & Hellr.		+	+	+
Achnanthes lanceolata (Bréb) Grun.		+	+	
Achnanthes levanderi Hust.			+	+
Achnanthes linearis (Wm. Smith) Grun.	+	+		+
Achnanthes linearis var. *curta* H. L. Smith			+	
Achnanthes marginulata Grun.		+	+	
Achnanthes microcephala (Kütz.) Grun.		+	+	
Achnanthes minutissima Kütz.		+	+	
Anomoeneis serians var. *brachysira* (Bre̋b ex. Kütz.) Hust.				+
Anomoeneis serians var. *famliaris*	+		+	
Asterionella formosa var. *formosa* Hass.	++	++	++	++
Caloneis bacillum (Grun.) Cleve			+	+
Caloneis vetricosa var. *alpina* (Ehrenb.) Meist			+	
Caloneis sp.			+	
Cyclotella sp.		+		
Cyclotella stelligera Cleve & Grun.	+	+	+	+
Cymbella lunata Wm. Smith			+	
Cymbella minuta for. *latens* (Krasske) Reim.			+	
Cymbella minuta Hilse		+	+	
Cymbella minuta var. *silesica* (Busch. ex Rabh.) Reim.			+	
Diatoma anceps (Ehrenb.) V.H.		+	+	
Diatoma hiemale var. *mesodon* (Ehrenb.) Grun.	+	+	+	+
Eunotia bigibba var. *pumila* Grun.			+	
Eunotia incisa Wm. Smith ex Greg.		+	+	
Eunotia pectinalis var. *minor* (Kütz.) Rabh.			+	
Eunotia pectinalis (O.F. Müll.) Rabh.			+	
Eunotia sp.	+			
Fragilaria capucina Desm.			+	
Fragilaria construens var. *venter* (Ehrenb.) Grun.			+	
Fragilaria crotonensis Kitton		+	+	+

Table 9.1. (*Continued*)

Taxa	Winter	Spring	Summer	Fall
Fragilaria leptostauron (Ehrenb.) Hust.	+			
Fragilaria leptostauron var. *dubia* (Grun.) Hust.		+	+	+
Fragilaria pinnata (Ehrenb.)	+	+	+	+
Fragilaria vaucheriae (Kütz.) Peters	+		+	
Fragilaria virescens Ralfs		+	+	+
Frustulia rhomboides (Ehr.) De Toni			+	+
Gomphonema angustatum (Kütz.) Rabh.		+	+	
Hannaea arcus Patr.	+	+	+	
Hannaea arcus var. *amphioxys*			+	
Hantzschia amphioxys (Ehrenb.) Grun.			+	
Melosira lirata (Ehrenb). Kütz.	+	+	+	+
Meridion circulare (Grev.) Ag.	+	+	+	
Meridion circulare var. *constrictum* (Ralfs) V.H.			+	
Navicula arvensis Hust.			+	
Navicula contenta for. *biceps* (Arn.) Grun.			+	+
Navicula cryptocephala Kütz.			+	
Navicula elginensis var. *lata*		+		+
Navicula luzonensis Hust.		+	+	
Navicula minima Greve.	+	+	+	+
Navicula minuscula Grun.				
Navicula notha Wallace				
Navicula pseudoscutiformis Hust.		+	+	+
Navicula pupula Kütz.	+		+	
Navicula radiosa Kütz.	+		+	
Navicula schmassmannii Hust.			+	+
Navicula subbacillum Hust.			+	
Navicula sp.	+	+	+	
Navicula viridula (Kütz.) Kütz.		+		
Navicula viridula var. *avenacea* (Bréb. ex Grun.) V.H.	+	+		
Nitzschia dissipata (Kütz.) Grun.	+			
Nitzschia frustulum (Kütz.) Grun.			+	+
Nitzschia linearis Wm. Smith			+	
Nitzschia microcephala Grun.		+	+	
Nitzschia palea var. *palea* (Kütz.) W. Smith	+	+	+	
Nitzschia sp.				+
Pinnularia abaujensis var. *rostrata* (Patr.) Patr.			+	
Pinnularia borealis Ehrenb.	+	+	+	
Pinnularia sp.		+		

(*Continued*)

Table 9.1. (*Continued*)

Taxa	Winter	Spring	Summer	Fall
Rhopalodia gibba (Ehrenb.) O.F. Müll.		+		
Stauroneis phoenicenteron (Nitz.) Ehrenb.		+		
Stauroneis smithii Grun.		+	+	
Surirella sp.			+	
Synedra parasitica (Wm. Smith) Hust.		+		
Synedra radians Kütz.	+	+	+	+
Synedra rumpens Kütz.	+	+	+	+
Synedra rumpens var. *familiaris* (Kütz.) Hust.	+	+	+	+
Synedra sp.	+		+	
Tabellaria flocculosa (Roth) Kütz.	+			+
CHLOROPHYTA				
Actinotaenium sp.	+			
Ankistrodesmus convolutus Corda		+		
Ankistrodesmus falcatus var. *acicularis* (A. Brown) G.S. West	+	++		+
Ankistrodesmus nannoselene Skuja		+		
Ankistrodesmus sp.	+			
Ankyra judayi (G.M. Smith) Fott	+		+	
Chlamydomonas dinobryonii G.M. Smith			+	
Chlamydomonas spp.	+	++	++	+
Chlorella ellipsoidea Gerneck	+	++	++	+
Chlorella vulgaris Beyerinck	+			
Chlorella sp.	+	+	+	+
Chlorococcum humicola (Naeg.) Rabh.	+			
Chlorococcum infusionum (Schrank) Meneghini	+	++	++	+
Chlorogonium sp.	+	+	+	
Chlorococcum sp.	+	++		+
Closterium sp.		+		
Coccomyxa dispar Schmidle	+			
Coccomyxa minor Schmidle	+			
Corone sp.		+	+	
Cosmarium sp.			+	
Crucigenia sp.		+		
Desmidium sp.	+			
Dictyosphaerium ehrenbergianum (Naeg.)			+	
Dictyosphaerium pulchellum Wood	+	+		+
Dictyosphaerium sp.		++	++	+
Elakotothrix viridis (Snow) Printz				+
Eudorina elegans Ehrenb.		++	++	+
Eutetrasporis sp.		++	++	

Table 9.1. (*Continued*)

Taxa	Winter	Spring	Summer	Fall
Gloeococcus sp.		+	+	
Gloeocystis spp.		++	+	+
Golenkinia radiata (Chod.) Willie			+	
Gonatozygon sp.			+	
Gonium sociale (Duj.) Warming	+		+	
Gonium sp.		++	++	+
Hyalotheca sp.		++		
Kirchneriella lunaris (Kirch.) Moebius	+	++		
Kirchneriella sp.	+	+		+
Mesotaenium sp.		+		
Micratinium sp.		++		
Microspora sp.		+		
Nephrocytium limneticum (G.M. Smith) G.M. Smith		++	+	
Nephrocytium sp.	+	+		
Oocystis sp.		+	+	
Pandorina morum (Müll.) Bory		+	+	
Pandorina sp.		+		
Pleodorina californica Shaw		++	+	
Pteromonas sp.			+	
Scenedesmus abundans (Kirch.) Chod.			+	
Scenedesmus quadricauda (Turp) Bréb.		+		
Scenedesmus quadricauda var. *maximus* West & West			+	
Scenedesmus serratus (Corda) Bohlin	+	++	+	+
Scenedesmus sp.	+	+		+
Schroederia setigera (Schroed.) Lemm.		+		
Selenastrum minutum (Naeg.) Collins			+	
Selenastrum sp.	+			
Sphaerozosma sp.			+	
Sphondylosium planum (Wolle) West & West		+		
Spirogyra sp.			+	
Staurastrum sp.			+	
Treubaria sp.			+	
Ulothrix sp.	+	+	+	
CHRYSOPHYTA				
Chrysosphaera sp.				+
Dinobryon cylindricum var. *alpinum* (Imhof) Bachmann	+			
Dinobryon cylindricum Imhof		+		
Dinobryon divergens Imhof	+	++	+	+

(*Continued*)

Table 9.1. (*Continued*)

Taxa	Winter	Spring	Summer	Fall
Dinobryon pediforme (Lemm.) Steinecke			+	
Dinobryon sociale Ehrenb.				
Dinobryon sertularia Ehrenb.	+	+		+
Dinobryon sp.	++			
Epichrysis sp.	+	+		+
Kephryion sp.	+	+		
Kephryion spirale (Lack) Conrad		+		
Mallomonas alpina Pascher & Ruttner	+			
Mallomonas akrokomas Ruttner				+
Mallomonas akrokomas var. *parvula* Conrad		+		
Mallomonas sp.	+	+		+
Ochromonas (?) sp.	+			
Small chrysophyte flagellate	+	+	+	+
CRYPTOPHYTA				
Chilomonas sp.			+	
Chroomonas sp.		++	++	
Cryptomonas alpina Chod.			+	
Cryptomonas erosa Ehrenb.	+	+	+	+
Cryptomonas marsonii Skuja.	+	++	+	+
Cryptomonas ovata Ehrenb.		+		
Cryptomonas rostrata Troitzkaja		+		
Cryptomonas sp.	+	+		+
Rhodomonas minuta Skuja.	+	++	++	+
CYANOPHYTA				
Anabaena sp.			+	+
Aphanocapsa delicattissima West & West		+	+	+
Aphanocapsa elachista West & West				+
Aphanothece sp.	+	++		+
Arthrospira sp.				
Chroococcus dispersus (Keissl.) Lemm.	+	++	+	+
Chroococcus limneticus Lemm.	+	+	+	
Chroococcus minimus (Keissl.) Lemm.	+			
Chroococcus turgidus (Kütz.) Naeg.			+	
Chroococcus varius A. Braun			+	
Chroococcus sp.		+		
Dactylococcoposis acicularis Lemm.	+	+		+
Dactylococcopsis fasciculatus Lemm.		+	+	
Dactylococcopsis sp.		+		
Gloeothece sp.				

(*Continued*)

Table 9.1. (*Continued*)

Taxa	Winter	Spring	Summer	Fall
Lyngbya limnetica Lemm.		+	+	
Lyngbya nana Tilden	+	+	+	+
Merismopedia sp.				
Microcystis sp.		+		
Nostoc paludosum Kutz.			+	
Oscillatoria angusta Koppe		++		
Oscillatoria angustissima West & West		+		
Oscillatoria limnetica Lemm.	+	++	++	++
Oscillatoria sp.	+	+	+	+
Oscillatoria subbreviss var. *minor* Schmidle			++	
Phormidium sp.	+	+	+	
Rhabdoderma elongatus Naeg.	+	++		
Rhabdoderma iregulare (Naumann) Geitler		+		
Rhabdoderma lineare Schmidle & Laut.		++	+	
Rhabdoderma sp.	+			
Rhabdogloea sp.	+			
Symplocastrum sp.		+		
Synechococcus sp.	+	+		
EUGLENOPHYTA				
Phacus sp.				
Trachelomonas sp.		+		
PYRROPHYTA				
Peridinium bipes var. *travectum* Stein	+			
Peridinium cinctum (Müll.) Ehrenb.	+			
Peridinium inconspicuum Lemm.	+	+		
Peridinium sp.	+	+		

[a]Relative abundances based on maximum values from the two lakes and indicated by season: "+", occurred at densities < 1000 cells ml^{-1}; "+ +", occurred at densities > 1000 cells ml^{-1} in at least one collection. Identification made by R. G. Dufford, D. R. Beeson, and S. A. Spaulding.

temperatures and extreme seasonality of light and hydrologic regimes are present in significant numbers. These and other physical factors exert a strong control on the composition and biomass of the phytoplankton community. In addition, ecological factors such as algal parasitism, nutrient competition, and food web interactions may influence phytoplankton populations (Canter and Lund 1968; Kalff and Knoechel 1978; Bird and Kalff 1986; Carpenter et al. 1985; Carpenter and Kitchell 1988). The most common phytoplankton in LVWS include *Asterionella formosa* Hass. in the spring and *Oscillatoria limnetica* Lemm. in the autumn.

Extreme variation of the physical environment occurs during the course of a year. Snowmelt results in rapid flushing of the lakes. The melting of snow continues into the summer and causes water temperatures to remain low for most of the year (see Chapter 8, this volume). Lakes above 3200 m in the Front Range have complete ice cover for more than 230 days of the year (Keefer and Pennak 1977), and some small lakes above 3800 m are covered by ice for more than 300 days of the year.

Spring

The annual hydrologic regime has a dominant influence on phytoplankton populations, and at Loch Vale snowmelt dominates the hydrology. Peak runoff during spring snowmelt is characteristic of mountain drainages with winter snow cover. The size of the annual snowpack combined with spring temperatures determines intensity and duration of snowmelt, which affects algal growth. Several investigators (Brook and Woodward 1956; Dickman 1969; Larsson 1972; O'Connell and Andrews 1987) have documented the inverse relationship between lake replacement time and both phytoplankton and zooplankton abundance. During the period of ice cover, flushing rates are near zero. However, in spring and summer The Loch has very rapid flushing rates of as much as 0.71 lake replacement day^{-1} (McKnight et al. 1988). For phytoplankton to be present in lakes with rapid replacement, they must have rapid growth rates or an upstream source. Both of these were observed in The Loch in 1984 and 1985.

A spring increase in the diatom *Asterionella formosa* Hass. occurred in The Loch in which cell abundances reached a maximum of up to 9000 cells ml^{-1} (Figure 9.1). Growth rates were calculated for *A. formosa* based on lake flushing rate, cell numbers in the lake, and cell numbers at the inlet and outlet of the lake (McKnight et al. 1990). *Asterionella formosa* collected at The Loch inlet were assumed to have flushed downstream from Sky Pond and Glass Lake. The calculated growth rates in The Loch ranged from 0.15 to 0.49 cell day^{-1} (Table 9.2). These values are lower but comparable to growth rates in algal cultures grown in optimal conditions (Reynolds 1984).

Primary productivity was measured in situ using ^{14}C on three dates in 1985 (McKnight et al. 1990). Carbon fixation rates were greatest on June 3, during the *A. formosa* bloom (Table 9.3). The photosynthetic rate at the surface was 46.0 $\mu g\,C\,L^{-1}\,hr^{-1}$, nearly 10-fold higher than photosynthetic rate later in the year. Inhibition of algal growth may occur at the surface of lakes with high light intensities (Reynolds 1984). On the basis of the three sample dates, there was no evidence that photoinhibition occurs in The Loch.

Sky Pond has a larger volume and higher abundance of phytoplankton than The Loch. In 1984, when the peak cell abundance of *A. formosa* was 9000 cells ml^{-1} in The Loch, Sky Pond had concentrations of greater than

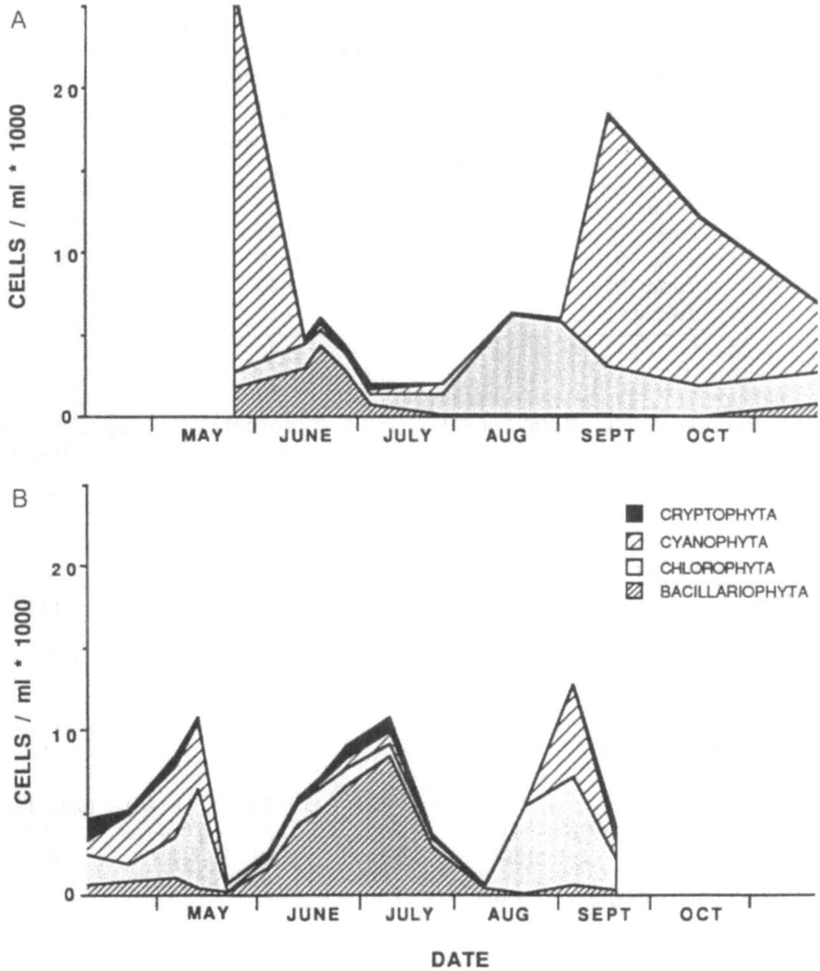

Figure 9.1. Phytoplankton abundance in cells ml^{-1} in The Loch during period of open water: (A) 1984; (B) 1985.

30,000 cells ml^{-1} (Figure 9.2). The higher abundances were reflected in the concentration of chlorophyll *a*. At most times of the year, the concentration of chlorophyll *a* in Sky Pond was greater than in The Loch but changes in chlorophyll *a* were not always coincident with changes in cell numbers (Figure 9.3). The concentration of chlorophyll *a* is an estimate of biomass, but cell numbers may not necessarily represent biomass because chlorophyll content of different algal species varies widely.

Table 9.2. Relationship Between Flushing Rate of Lake and Estimated Net Growth Rate of *Asterionella formosa* in The Loch During 1985 Snowmelt Period (from McKnight et al. 1988)

Date	Flushing Rate (day^{-1})	Estimated Rate of Growth (day^{-1})
June 3	0.56	0.35
June 12	0.65	0.35
June 19	0.71	0.19
June 10	0.56	0.49
July 23	0.55	0.15

Table 9.3. Photosynthetic Rate ($\mu g\, C\, L^{-1}\, hr^{-1}$), Chlorophyll *a* ($\mu g\, L^{-1}$), and Photosynthetically Active Radiation (PAR) ($\mu E\, m^{-2}\, s^{-1}$) as Function of Depth in The Loch During 1985 Open Water Period (from McKnight et al. 1990)

		Depth (m)				
Date	Parameter	Surface	1	2	3	4
Jun 3	Mean photosynthetic rate	46.0	61.0	9.0	10.0	19.0
	(range)	(40–52)	(60.6–61.3)	(7.6–10.4)	(7.4–13.1)	(17–20)
	Chlorophyll *a*	1.2	—	1.3	—	1.4
	PAR (2:30 P.M.)	750	—	170	—	—
Aug 8	Mean photosynthetic rate	3.7	3.8	3.7	3.7	1.0
	(range)	(3.1–4.3)	(3.5–4.3)	(3.0–4.4)	(3.1–4.3)	(0.95–1.05)
	Chlorophyll *a*	1.0	—	1.3	—	1.5
	PAR (10:45 A.M.)	1250	750	650	300	400
Sep 17	Mean photosynthetic rate	8.0	7.4	—	—	—
	(range)	(7.3–8.3)	(6.9–7.9)	—	—	—
	Chlorophyll *a*	1.4	—	1.5	—	1.0
	PAR (11:40 A.M.)	1900	950	—	—	—

Autumn

Large numbers of cells of *Oscillatoria limnetica* Lemm. occurred in both lakes during the autumn of 1984 and 1985 (see Figures 9.1 and 9.2). In Sky Pond, cell numbers were again much greater than in The Loch and reached peak abundances of 400,000 cells ml^{-1}. Although *O. limnetica* cells are small compared to many chlorophyte and diatom cells, the large numbers represent a substantial increase in biomass. Primary productivity in The Loch on September 17, 1985, ranged between 6.9 and 8.3 $\mu g\, C\, L^{-1}\, hr^{-1}$. These rates were measured during the latter period of the *O. limnetica* bloom, and may

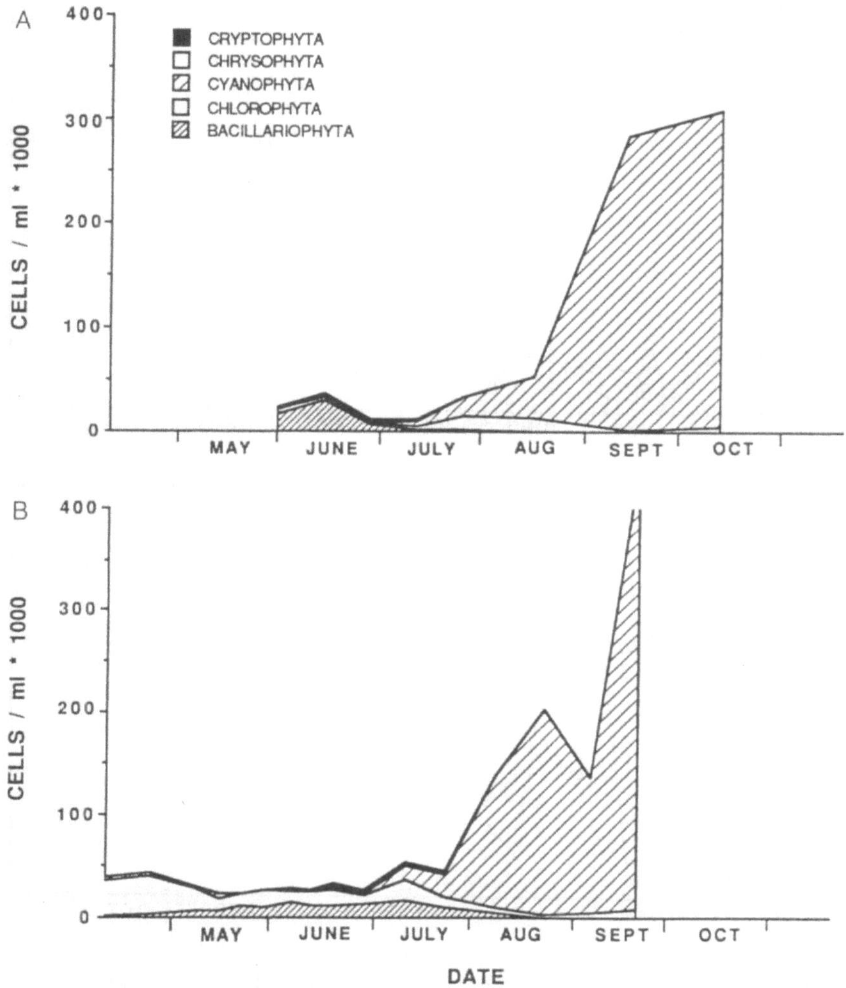

Figure 9.2. Phytoplankton abundance in cells ml^{-1} in Sky Pond during period of open water: (A) 1984; (B) 1985.

not be representative of the actively growing population (McKnight et al. 1990). Late summer cyanophyte blooms are common in many temperate lakes (Wetzel 1983). In alpine lakes, however, there has been little documentation of such blue-green blooms. McKnight (unpublished data) found that Bierstadt Lake in Rocky Mountain National Park contained several cyanophyte species in high abundance (100,000 cells ml^{-1}) in September 1987. In three other nearby lakes, however, the phytoplankton were dominated by chrysophytes, cryptophytes (Bear and Nymph lakes), or chlorophytes (Sprague Lake).

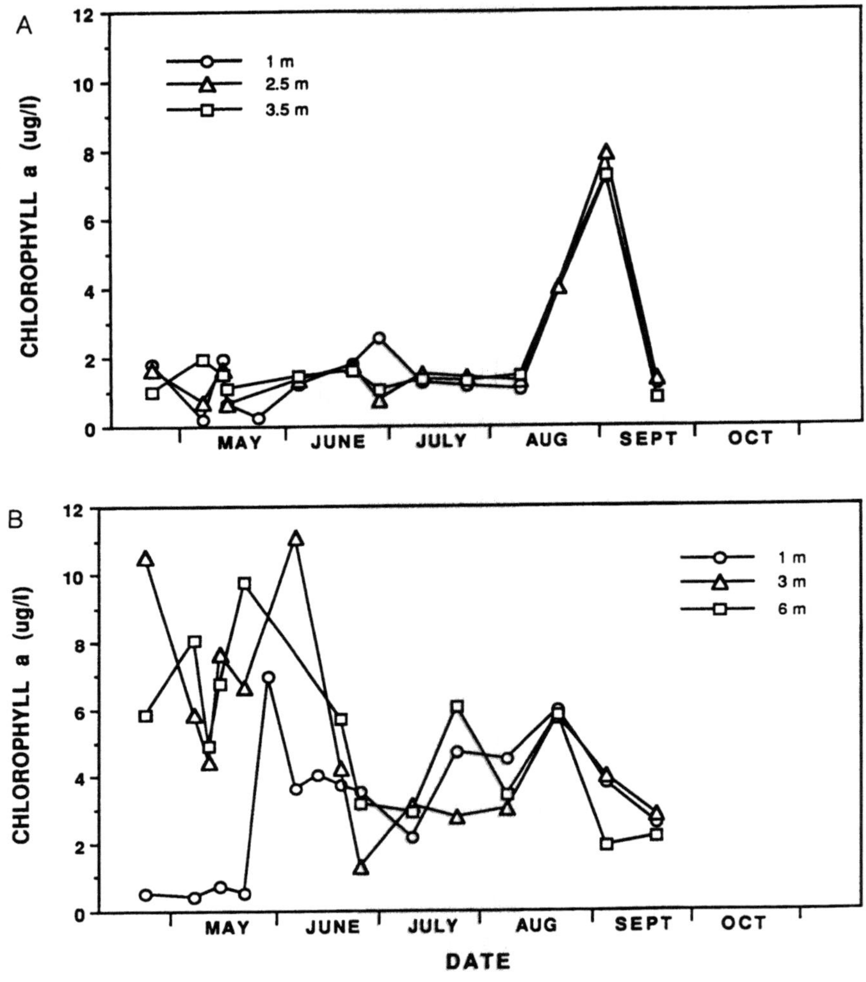

Figure 9.3. Concentration of chlorophyll *a* during period of open water in 1985: (A) The Loch; (B) Sky Pond.

Winter

Many Front Range lakes are subject to high winds, and snow is blown from the ice surface. Therefore, snow often does not accumulate on Front Range lakes, as it may on other temperature lakes. Light may penetrate the ice and result in the growth of winter algae populations.

In The Loch, peak cell numbers of 10,000–15,000 cells ml^{-1} in winter (Figure 9.4) are comparable to peak summer concentrations (McKnight et al. 1988). The diatom *A. formosa* is an important component of the winter phytoplankton in The Loch. Several species of chlorophytes, *Coccomyxa dispar* Schmidle, *Chlamydomonas* sp., *Ankistrodesmus falcatus* var. *acicularis*

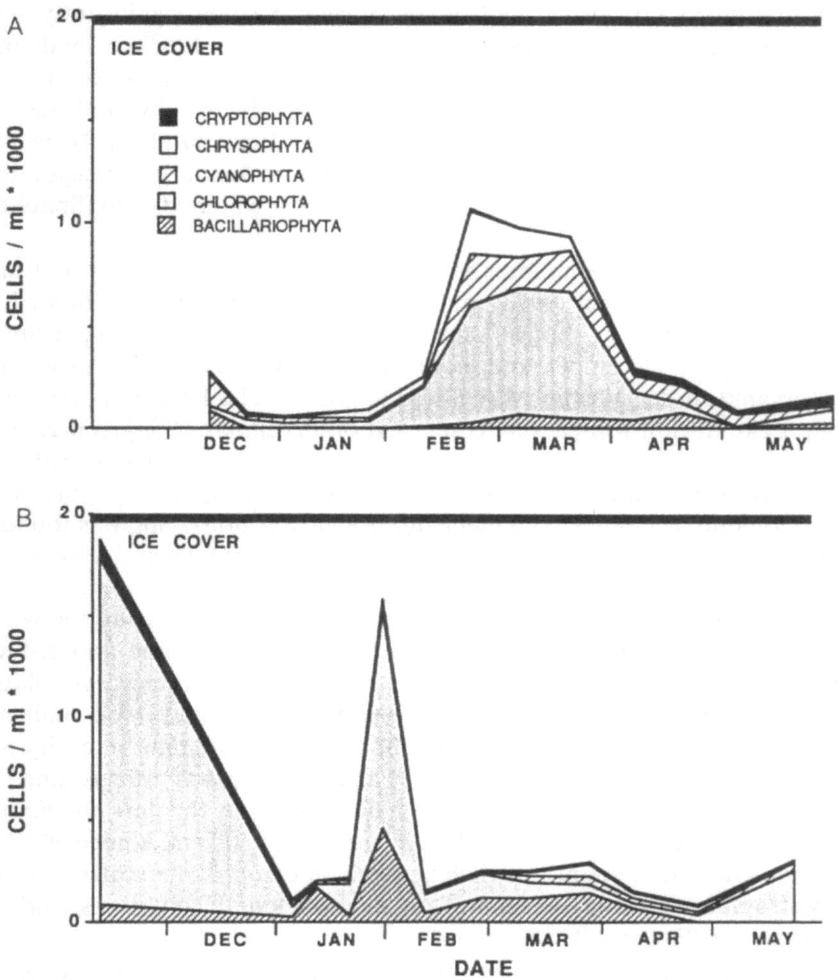

Figure 9.4. Phytoplankton abundance in cells ml^{-1} in The Loch during period of ice cover: (A) 1987–1988; (B) 1988–1989.

G. S. West, and Chlorococcales are also present (see Table 9.1). Although the same dominant (and most of the rare) taxa were present in the winters of 1987–1988 and 1988–1989, they varied in time of occurrence and abundance (Spaulding 1991). There was a bloom of the chrysophyte *Dinobryon* sp. in February 1988 and again in March 1989. Although it was not the most numerous alga during that time, *Dinobryon* sp. has relatively large cells and made the greatest contribution to total biomass.

There is an early winter maximum of phytoplankton abundance in winter (Figure 9.4). This peak may be attributed to increased concentration of solutes in the "freezing-out" process. Bodies of water with a large surface area-to-volume ratio, such as The Loch, may become significantly more concentrated when the water freezes (Canfield et al. 1983). A midwinter minimum in phytoplankton abundance follows the initial peak. Solar radiation is lowest in winter; the sun is at a low angle in the sky, and the high ridges of Loch Vale shade the lake surface from direct sun from November until February. An increase in algal abundance in January corresponds to increasing light levels on the ice surface and in the water column (Spaulding 1991).

High winter algal abundances have been documented in other Front Range alpine lakes. Herrmann (1978) and Keefer and Pennak (1978) found high densities of algae in Long Lake. Keefer (1969) found peak primary productivity in December and January in Long Lake, with the greatest phytoplankton densities in May under ice cover. The large populations in Long Lake during winter were composed of unicellular blue-green algae of uncertain taxonomic affinity that ranged in size from 2 to 6 μm. Although they were present throughout all seasons, they made up a large part of the winter bloom. The blue-green filamentous alga, *Shizothrix* sp., was found to reach a peak of 12 million cells liter^{-1}, and disappeared when the ice melted (Keefer and Pennak 1977; Herrmann 1978).

Ellsworth (1983) found that algae reach high numbers during the winter in Front Range ponds because of long water residence time and reduced numbers of zooplankton grazers. The absence of wind-induced circulation and the resulting exchange of atmospheric gases may also influence phytoplankton beneath ice cover (Wright 1964). At times light intensity may become too low for photosynthesis, yet algal populations are sustained. In some cases, facultative heterotrophs are able to live in the dark indefinitely on dissolved organic substances (Hutchinson 1967; Kalff and Knoechel 1978). Chrysophytes are thought to switch between autotrophic, heterotrophic, and phagotrophic nutrition depending on environmental conditions and cell requirements (Sangren 1988).

The concentrations of chlorophyll *a* during the period of ice cover are comparable to summer values in The Loch (Figure 9.5) and correspond well to the graphs of cell numbers, except for the chlorophyte peak in February of 1989. This peak does not appear in the chlorophyll *a* concentrations and remains unexplained.

During the winter of 1987–1988, infestations of the fungal chytrid parasite were observed on *A. formosa* and occasionally on the chlorophyte *A. falcatus* var. *acicularis*. A decline in *A. formosa* was observed following heavy chytrid infestations. Fewer chytrids were present in 1988–1989, and *A. formosa* was found in greater numbers. Koob (1966) found chytrid parasites to be numerous in lakes in the Rawah Mountains of Colorado, but they were not related to the decline of *A. formosa*.

Distinct seasonal cycles in phytoplankton biomass and species composition occur in Loch Vale lakes. The diatom *A. formosa* may be abundant during the period of spring snowmelt and grows at rates high enough to

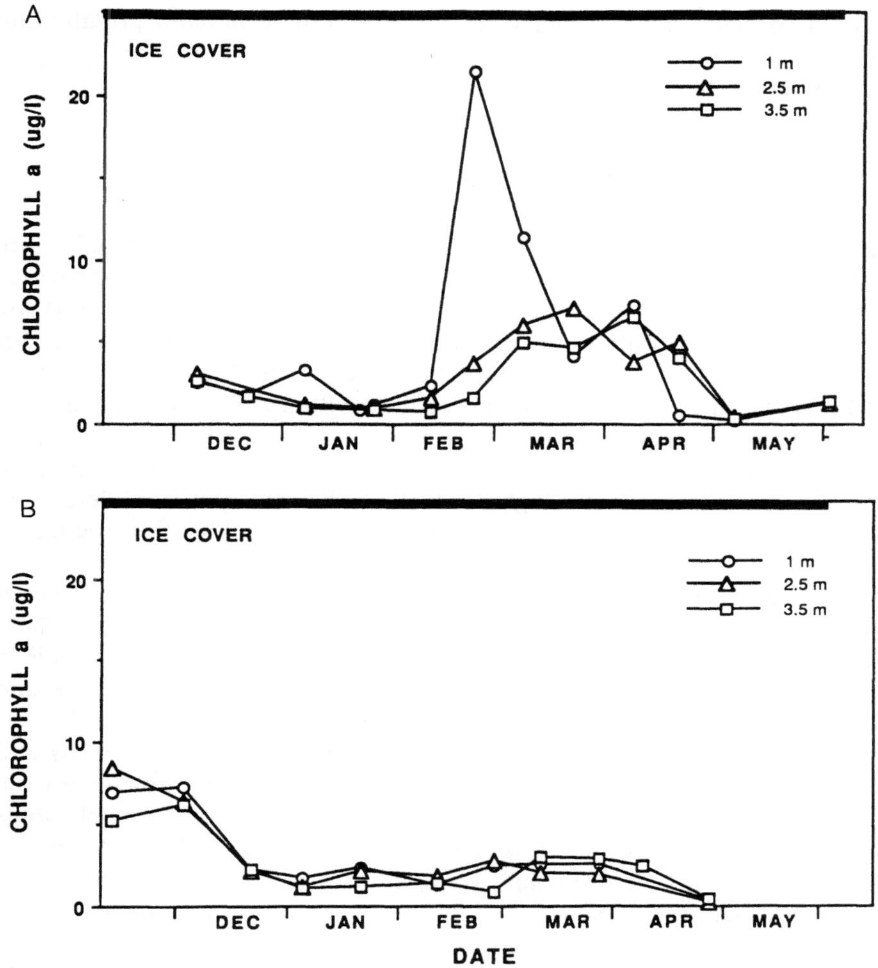

Figure 9.5. Concentration of chlorophyll *a* during period of ice cover in The Loch: (A) 1987–1988; (B) 1988–1989.

maintain substantial populations during rapid lake water replacement. Summer populations of green algae are at a minimum followed by an autumn bloom of *O. limnetica.* During the winter, water renewal is at a minimum and algae appear to be limited by low light intensity. Because of low incident sun angles, a "bloom" occurs under the ice when the lakes are no longer shaded. Despite generalizations of the successional sequence, there are differences among Loch Vale lakes and in the same lake between years. Lake flushing, daylength, light penetration, lake mixing, temperature, and nutrient loading all change on a seasonal basis as well as interacting to form the planktonic environment. Beyond physiological limitations, biological interactions influence community composition during some times of the year. Fungal chytrid parasites appear related to the decline of *A. formosa* populations, and zooplankton grazing and fish predation probably also affect alpine phytoplankton communities.

Aquatic Macroinvertebrates

Substrata in the streams consist predominately of cobble and pebble, with little organic matter. Alpine and subalpine streams are often frozen or covered with snow between November and June of each year. When streams are open, the turbulent waters are saturated with oxygen. Rooted vascular plants are absent. Because of the steep topographic gradient, debris dams are rare, and this may limit the retention of nutrients within the stream system.

More than 60 species of macroinvertebrates, primarily aquatic insects, occur in the riffles of Icy Brook (Table 9.4). Many of these species are geographically widespread throughout the North American cordillera, and comparable species composition has been reported from nearby streams by Elgmork and Sæther (1970) and Ward (1986).

Two riffle sites on Icy Brook were intensively sampled during the summer and autumn of 1988. A modified Hess bottom sampler (Merritt et al. 1984b) with a 280-μm mesh net and 17-cm bottom opening was used for sampling. Additionally, qualitative aerial collecting for terrestrial adult stages and observation was performed around The Loch (Harris, unpublished data).

One riffle site is in an alluvial meadow, below late-lying snowbeds and seeps, as the stream enters the subalpine zone and directly upstream from the confluence of Andrews Creek. The stream substrate is composed of cobble with some coarse sand, which is less than 10 cm above bedrock. Common meadow herbs include *Cardamine cordifolia, Senecio triangularis, Mertensia ciliata, Caltha leptasepala*, and the graminoids *Carex aquatilis* and *Calmagrostis canadensis.* The elevation is approximately 3160 m.

The other riffle site is forested with a 69% canopy cover of subalpine fir [*Abies lasiocarpa* (Hook.) Nutt.] and Engelman spruce [*Picea engelmanni* (Parry) Engelm.]. The site is about 30 m upstream from The Loch, at an elevation of approximately 3115 m. The stream substrate is dominated by cobble and small boulders. The moss *Fontinalis* sp. is abundant on many boulders.

Macroinvertebrate Community Structure an Function

Aquatic insects in the orders Ephemeroptera (mayflies), Plecoptera (stoneflies), Trichoptera (caddisflies), and Diptera (true flies) comprise more than 80% of the fauna of Icy Brook (Table 9.5). Non-insect groups such as the oligochaeta (aquatic worms), Nematoda (roundworms), Acarina (mites), and Tricladida (flatworms) are also numerically important. The dipteran midge family, Chironomidae, comprises more than 48% of the individuals in the

Table 9.4. Macroinvertebrate Taxa Collected from Icy Brook in Rocky Mountain National Park Elevation Approximately 3115–3160 m[a]

Taxa	Relative Abundance	
	Meadow	Canopy
EPHEMEROPTERA		
Baetidae		
Acentrella carolina (Banks)	U	
Baetis bicaudatus Dodds	A	A
Ephemerellidae		
Drunella coloradensis (Dodds)	C	U
Drunella doddsi Needham		U
Heptageniidae		
Cinygmula ramaleyi (Dodds)	A	A
Epeorus deceptivus (McDunnough)		C
Epeorus longimanus (Eaton)	U	
Rhithrogena hageni Eaton	*	*
Rhithrogena robusta Dodds	U	U
Siphlonuridae		
Ameletus velox Dodds	C	C
PLECOPTERA		
Capniidae		
Capnia gracilaria Claassen	U	U
Chloroperlidae		
Alloperla pilosa Needham and Claassen	U	U
Sweltsa borealis (Banks)	*	*
Sweltsa coloradensis (Banks)	*	*
Sweltsa lamba (Needham and Claassen)	C	C
Leuctridae		
Paraleuctra vershina Gaufin and Ricker	*	*
Nemouridae		
Zapada cinctipes (Banks)	#	#
Zapada haysi (Ricker)	#	#
Zapada spp.	A	A
Perlodidae		
Isoperla fulva Claassen		U
Isoperla sobria (Hagen)		U
Kogotus modestus (Banks)	C	C
Megarcys signata (Hagen)	C	C
Pictetiella expansa (Banks)		U
Taeniopterygidae		
Taenionema nigripenne (Banks)		U
TRICHOPTERA		
Glossosomatidae		
Glossosoma sp.		U
Limnephilidae		
Asynarchus nigriculus (Banks)	*	*

(*Continued*)

Table 9.4. (*Continued*)

Taxa	Relative Abundance	
	Meadow	Canopy
Chyrandra centralis (Banks)	*	*
Discosmoecus atripes (Hagen)	C	
Ecclisomyia conspersa (Banks)	*	*
Ecclisomyia maculosa Banks	U	U
Psychoglypha subborealis (Banks)	*	*
Psychoronia costalis (Banks)	U	
Rhyacophilidae		
Rhyacophila alberta Banks	*	*
Rhyacophila angelita Banks		U
Rhyacophila brunnea Banks	U	U
Rhyacophila coloradensis Banks		C
Rhyacophila hyalinata Banks		C
Rhyacophila verrula Milne	U	C
DIPTERA		
Blephariceridae		
Agathon elegantulus Röder	C	
Ceratopogonidae		
Bezzia sp.	U	U
Chironomidae		
Tanypodinae		C
Ablabesmyia sp.	#	#
Chironominae		
Chironomini	U	C
Polypedilum spp.	#	#
Tanytarsini	A	A
Microspectra spp.	#	#
Diamesinae	U	U
Diamesa spp.	#	#
Pagastia partica (Roback)	#	#
Pseudodiamesa sp.	#	#
Orthocladiinae	A	A
Crictopus spp.	#	#
Eukiefferiella group	#	#
Orthocladius spp.	#	#
Parametriocnemus sp.	#	#
Parorthocladius sp.	#	#
Rheocrictopus spp.	#	#
Tvetenia sp.	#	#
Empididae		
Clinocera sp.	C	C
Simuliidae		
Prosimulium onychodactylum Dyar and Shannon	U	U
Prosimulium travisi Stone	C	A
Simulium tuberosum (Lundstrom) [complex]	*	*

Table 9.4. (*Continued*)

Taxa	Relative Abundance	
	Meadow	Canopy
Tipulidae		
Antocha sp.		U
Dicranota spp.	C	A
Hexatoma sp.		U
ACARINA	A	A
NEMATODA	A	A
OLIGOCHAETA	A	A
TRICLADIDA		
Polycelis coronata (Girard)	C	C

[a]Voucher specimens are deposited in Colorado State University Insect Collection, Fort Collins, Colorado. Relative abundance by site based on immatures collected in benthic samples: U (uncommon), < 10; C (common), 10–100; A (Abundant), > 100; *, collected as adult only; #, not identified to this taxonomic level.

Table 9.5. Percentage of Macroinvertebrate Individuals Collected by Taxonomic Group for Each Site and Icy Brook Sites Combined

	Site		
	Meadow (%)	Canopied (%)	Icy Brook (%)
Ephemeroptera	19.4	3.9	9.8
Plecoptera	2.3	7.7	5.6
Trichoptera	0.4	0.7	0.6
Diptera	48.7	77.1	66.2
Non-insect	29.3	10.6	17.8
Total	100.1	100.0	100.0

meadow site. Next in order of abundance at the meadow riffle are oligochaete worms (14%), Nematoda (13%), baetid mayflies (10%, almost exclusively *Baetis bicaudatus* Dodds), and the heptageniid mayflies [8%, primarily *Cinygmula ramaleyi* (Dodds)]. At the canopied riffle, chironomids represent 74% of the individuals, the nemourid stoneflies, 7% [primarily *Zapada cinctipes* (Banks) and *Z. haysi* (Ricker)], and Oligochaeta, 5%.

Macroinvertebrate density averages almost 16,700 individuals m^{-2} at the meadow site and approximately 26,900 individuals m^{-2} at the canopied

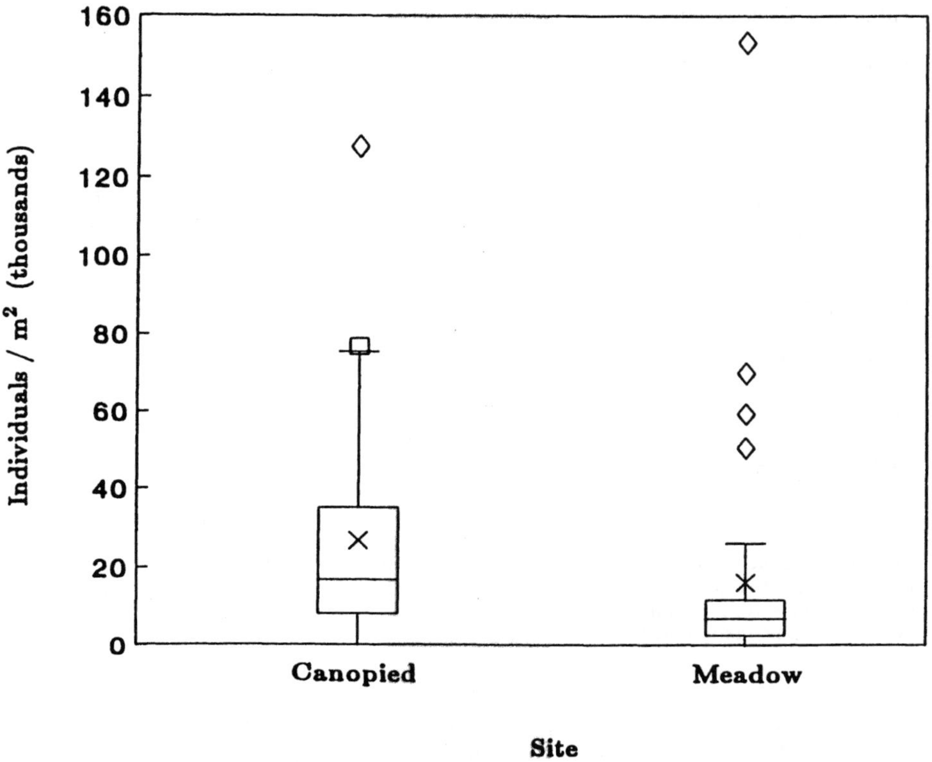

Figure 9.6. Box-and-whisker plots (Tukey 1977, SAS Institute Inc. 1988) of individuals m^{-2} at meadow and canopied sites on Icy Brook; sampling dates July–October 1988. Bottom, middle, and top lines of box represent 25, 50, and 75 percentiles, respectively; x, sample mean; $n = 30$ for each site. Vertical lines (whiskers) extend to 1.5 box lengths. Extreme values are marked with square (within 3 box lengths) or diamond (more extreme).

location (Figure 9.6). Macroinvertebrate density estimates for Icy Brook are much higher than have been reported from other nearby streams at similar altitudes. Short and Ward (1980) reported 1467 and 774 individuals m^{-2} in consecutive years for Joe Wright Creek at an elevation of 3045 m. At a 3109 m elevation site on the Middle St. Vrain Creek, Ward (1986) reported less than 2000 individuals m^{-2}.

Aquatic macroinvertebrates play a major role in organic matter processing in streams and lakes. They influence the cycling of organic material through feeding and physical interaction with the sediment (see review in Merrit et al. 1984a). Morphological-behavioral feeding mechanisms are used to classify species into functional feeding groups. Functional group abundance reflects trophic organization of the community and food availability in a stream (Hawkins and Sedell 1981).

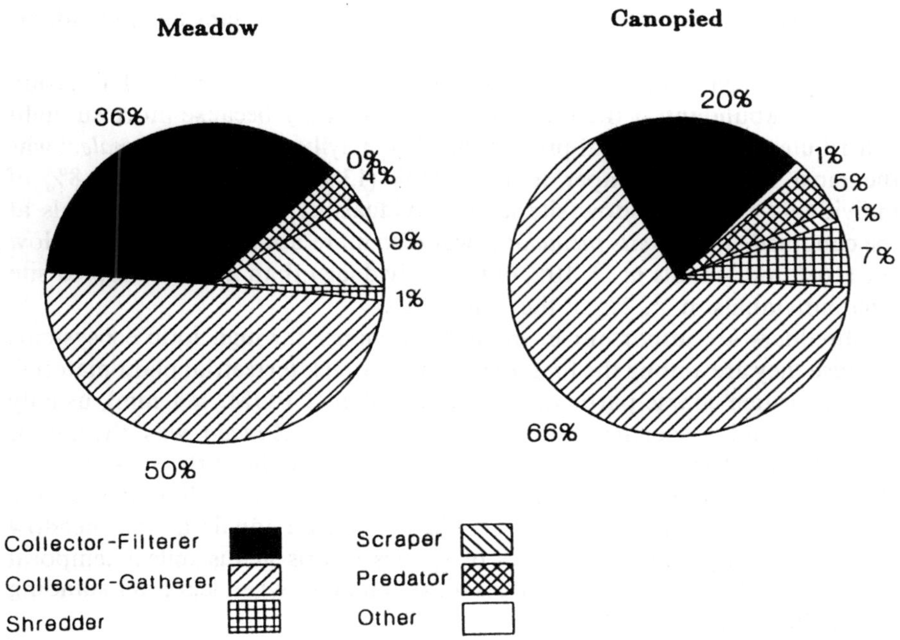

Figure 9.7. Pie diagram of functional groups at meadow and canopied sites on Icy Brook; sampling dates July–October 1988.

A functional group analysis was performed on the invertebrates of Icy Brook (Figure 9.7). Insects were classified according to their primary functional group as listed by Merritt and Cummins (1984). Non-insect classification was based on the feeding habits summarized in Pennak (1978). The benthic macroinvertebrate fauna of Icy Brook was clearly dominated by species considered to be collector-gatherers and collector-filterers.

Collector-gatherers feed on fine particulate organic matter (FPOM, < 1 mm in diameter) that has been deposited on the substrate, whereas the collector-filterers utilize specialized mouthparts to filter FPOM from the water column. The collector-gatherers were represented primarily by the chironomid subfamily Orthocladiinae, the oligochaete worms, and the mayfly *Baetis bicaudatus*. Considered a collector-gatherer for this study, *Baetis bicaudatus* is sometimes classified as a generalized collector-gatherer/scraper or a scraper. Radar and Ward (1989) found in a study in the upper Colorado River that diatoms formed a major part of the diet of several *Baetis* species.

The midge tribe Tanytarsini (*Microspectra* spp.) and the black flies (Simuliidae) were the major collector-filterers present. The Tanytarsini reached their greatest abundance, an average of 14,240 individuals m^{-2} for both sites, in September and October. The blackfly *Prosimulium travisi* Stone, while not as numerically abundant as the midges, was one of the most abundant organisms at the canopied site through mid-August, averaging 890 individuals m^{-2}.

Scrapers shear attached algae from submerged solid substrates. This group was more abundant at the meadow site, presumably because more sunlight contributes to greater algal production. The mayfly *Cinygmula ramalevi* was the most abundant scraper in Icy Brook (1380 individuals m^{-2}, 8% of individuals at the meadow site; 240 individuals m^{-2}, 1% of individuals at the canopied site). *Baetis bicaudatus* was also more abundant at the meadow site (1680 individuals m^{-2}, 10% of the individuals) than at the canopied site (660 individuals m^{-2}, 2% of the individuals).

Shredders are important in the flow of energy through lotic ecosystems because they are among the first organisms to attack terrestrial plant material. They consume coarse particulate organic matter (CPOM, >1 mm), usually leaf litter and associated microbes. The stonefly *Zapada* was the major shredder at both sites and was more abundant throughout the season in the closed canopy site (1790 individuals m^{-2}, 7% of the individuals at the canopied site; 230 individuals m^{-2}, 1% of the individuals at the meadow site). Reportedly, the life histories of species in this genus reflect temporal patterns of CPOM inputs; *Zapada* is most abundant when leaf litter standing crop is high (Short and Ward 1981).

Several predators that were commonly encountered included perlodid and chloroperlid stoneflies, the tipulid (crane fly) larva *Dicranota,* and aquatic mites. Predators maintained constant relative abundances through the sampling period and between sites. Predators averaged 3%–6% of the individuals at the meadow site and 3%–8% at the canopied site.

The river continuum concept (Vannote et al. 1980) provides a model for functional group organization relative to gradients of food resources along the longitudinal (downstream) profile of a stream. In shaded headwater streams, more shredders are expected where there are large allochthonous inputs (CPOM) from riparian vegetation. The maximum density of scrapers occurs where incident solar radiation is high because of the greater algal production in sunnier areas (Hawkins et al. 1982). The large percentage of detritivore collectors instead of shredders in Icy Brook indicates that a significant portion of the food base consists of FPOM. Where the canopy is absent, the importance of autochthonous production increases. Icy Brook differs from eastern North American streams where the river continuum concept originated; there, headwater streams receive substantial amounts of CPOM from nearby riparian vegetation.

Aquatic Invertebrates of High-Elevation Rocky Mountain Streams

Mountain benthic macroinvertebrate assemblages are typically classified relative to altitudinal zonation patterns (Dodds and Hisaw 1925; Allan 1975; Ward 1986). Hynes (1970) cautioned that zonation is useful for general description but that its precise ecological value is uncertain. Ward (1986) presented the St. Vrain River system as a gradient of changing environmental conditions from the headwaters to the plains of classified the benthic invertebrates according to their altitudinal distributional pattern. Other studies that have looked at altitudinal zonation in the Rocky Mountains include the pioneering study of Dodds and Hisaw (1925), Knight and Gaufin's (1966) study of the Plecoptera in a western Colorado drainage, Elgmork and Sæther's (1970) 5-day study in the alpine of North Boulder Creek, Mecom's (1972) study of the Trichoptera of the St. Vrain, and Allan's (1975) distributional study of Cement Creek. Typically, the aquatic insect species occurring in Icy Brook are restricted to mountain streams, often preferring higher gradient water courses; few [e.g., the stonefly *Sweltsa borealis* (Banks)] are found only in high elevation headwater reaches.

The cold-adapted species of Icy Brook have life histories and behavioral mechanisms that reflect the extreme conditions. Temperature is the factor most often cited as regulating the altitudinal distribution of the lotic fauna (Dodds and Hisaw 1925; Knight and Gaufin 1966; Ward and Stanford 1982; Ward 1986). These species respond to main temperature effects including diel fluctuations, seasonal extremes, and duration of the ice-free season (Dodds and Hisaw 1925; Ward and Stanford 1982). Food availability and the absence of vascular plants are other factors that vary with elevation and contribute to the composition of the high-altitude benthos (Ward 1986). Some factors that cross altitudinal zones, such as water velocity and substrate composition, may play a role in local distributions of macroinvertebrates but do not influence longitudinal distributions (Allan 1975). Predation pressure and competition may have an altitudinal component, but these factors have not been studied (Ward 1982).

Life Histories of Selected Species of Icy Brook

Ephemeroptera

Baetis bicaudatus and *Cinygmula ramaleyi* accounted for 92% of the individuals of the 10 mayfly species collected in Icy Brook. *Baetis bicaudatus*, originally described from Colorado (Dodds 1923), inhibits a wide variety of streams and rivers above 1300m. It is often the only mayfly occurring in small, cold subalpine and alpine streams (Jensen 1966). The nymph's fusiform, streamlined body shape allows in to occupy the tops and sides of rocks even in swift water. Morihara and McCafferty (1979) indicated that this species overwinters in the egg stage. This species has an extended emergence pattern, with adults present through much of the summer.

Cinygmula ramaleyi, also described from Colorado (Dodds 1923), is a common species of boreal western North America occurring in medium to small streams between 2438 and 3353m (foothills to alpine zone). The flattened nymphs can be found clinging to the tops and sides of exposed rocks in swift water. This species has a univoltine life cycle with adults emerging from August to September. Short and Ward (1980) also reported *Cinygmula* to be one of the most abundant mayflies in their study of Joe Wright Creek, a tributary of the Poudre River in north-central Colorado.

Two other mayflies, *Drunella coloradensis* (Dodds) and *Ameletus velox* Dodds, were often common in benthic collections. Both of these species were also described from Colorado (Dodds 1923). *Drunella coloradensis* is common in many of the coldwater streams of Rocky Mountain National Park at elevations above 2000 m. *Ameletus velox,* a powerful swimmer, can be found in the slower moving reaches and margins of Icy Brook. Adults are present from July to early August.

Plecoptera

This order of insects is strongly associated with cool running waters. In a study of altitudinal distribution of Plecoptera in the Gunnison River drainage of western Colorado, Knight and Gaufin (1966) found that the greatest number of stonefly species occurred between 2100 and 2700 m. Above 2700 m they found a decreasing number of species but a greater percentage of predatory species. Of the 14 species found in Icy Brook, 9 are considered predators and 5 shredders.

The widespread and primarily western genus *Zapada* (*Z. cinctipes*, *Z. haysi*) comprised 60% of the stonefly individuals at the meadow site and 87% of the stoneflies collected in the canopied site. *Zapada* species are found in small to large, cold, stony rivers from the foothills to the alpine. This stonefly is considered a shredder, and the life histories of several species are relatively well known (Mutch and Pritchard 1982, 1984; Cather and Gaufin 1976). Short and Ward (1981) reported a univoltine life cycle for *Z. cinctipes* in Little Beaver Creek, a nearby stream.

Several genera of Perlodidae and Chloroperlidae were commonly collected throughout the study. *Megarcys signata* (Hagen), originally described from Colorado, is the most common large stonefly in small streams of the higher elevations of the southern Rocky Mountains. Nymphs are found among cobble substrate or debris and are omnivorous, feeding on diatoms, chironomids, and small mayflies (Richardson and Gaufin 1971; Cather and Gaufin 1975; Allan 1982). Early instars were present in samples in August and September. The life cycle of this species is univoltine, and growth is most rapid in early summer (Allan 1982). Adults are present from late June to August.

Sweltsa lamba (Needham and Claassen) and *S. borealis* are two stoneflies often abundant in Rocky Mountain streams. Although little biological information is available, it appears *S. lamba* may be univoltine and *S. borealis* semivoltine in Icy Brook. Adults of *S. lamba* and *S. borealis* were collected from late June to late August.

Trichoptera

The most abundant caddisfly larvae in Icy Brook were species of the free-living rheophilic genus *Rhyacophila*. These larvae are large, up to 23 mm long (Wiggins 1977), and the genus is easily recognized. At least 15 species are known from Colorado (Herrmann et al. 1986). Species are confined primarily to cold mountain streams (Ross 1956). Congenerics often occur in the same riffle. There are at least six *Rhyacophila* species in Icy Brook (see Table 9.4). Most are predaceous on other aquatic insects, although some feed on plant material. *Rhyacophila verrula* Milne, a common late-emerging species (September to October) at both study sites, feeds on algae and other plant material (Smith 1968). The other common species found in this study was *R. brunnea* Banks, a predator, which overwinters as a third or fourth instar and pupates from May to June. Adults are active from June to August.

Among the largest of the Trichoptera is *Dicosmoecus atripes* (Hagen). Its large ballast case, up to 41 mm long (Wiggins 1977), is often attached to the underside of rocks in Icy Brook. *Dicosmoecus atripes* is apparently semivoltine, with larvae reaching their fifth and final instar by August or early September of the first season. After overwintering, the larvae become active from late May or early June of the second year, after which they then initiate a prepupal resting phase (Wiggins and Richardson 1982). Adults then emerge from September to October. Larvae are considered generalized predator-shredders (Wiggins and Richardson 1982).

Diptera

There are several dipteran families that are wholly or primarily aquatic in their immature stages and are important in Icy Brook. They include Chironomidae, Empididae, Simuliidae, and Tipulidae. Members of the midge family Chironomidae are found in great density and diversity in Icy Brook, as they are in freshwater ecosystems in general. Chironomids exhibit a wide range of feeding strategies and life cycles. Few species appear to be restricted to one feeding method (Pinder 1986). They are often found in great numbers and account for a substantial proportion of the biomass of some lotic systems.

Oligochaeta

The aquatic worms (Oligochaeta) represented 14% of the individuals at the meadow site and 5% at the canopied site. Ward (1986) found the tubificid *Limnodrilus hoffmeisteri* at its maximum density at comparable altitudes in the St. Vrain River. Elgmork and Sæther's (1970) 'Vermes', apparently a combination of oligochaetes and nematodes (Bushnell et al. 1982), were the most abundant group in their study of an alpine stream. Most aquatic oligochaetes ingest substrate to 2 to 3 cm below the water-substrate interface, thereby mixing the bottom sediment (Pennak 1978).

Macroinvertebrate Fauna of The Loch

Little information is currently available on benthic macroinvertebrates of alpine and subalpine lakes of the southern Rocky Mountains. Bushnell et al. (1982) reviewed studies concerning the chemical, physical, and biological attributes of alpine lakes. A species list of macroinvertebrates inhabiting alpine lakes in the Green Lakes Valley, west of Boulder, Colorado, is provided by Bushnell et al. (1987). Community structure and function are largely unknown in Colorado alpine lakes. Typically, the littoral benthic zones of these lakes are dominated by Trichoptera and Diptera. Other groups, such as Ephemeroptera, Coleoptera, and Mollusca, are less abundant and often are absent.

Table 9.6. Macroinvertebrate Fauna of The Loch

EPHEMEROPTERA
Baetidae
Callibaetis sp.
Siphlonuridae
Ameletus velox Dodds
Siphlonurus occidentalis (Banks)
PLECOPTERA
Nemouridae
Amphinemura banksi Baumann and Gaufin
TRICHOPTERA
Limnephilidae
Limnephilus abbreviatus Banks
Limnephilus coloradensis (Banks)
Limnephilus gioia Denning
Limnephilus picturatus McLachlan
COLEOPTERA
Dytiscidae
Agabus tristis Aube
DIPTERA
Chaoboridae
Eucorethra underwoodi Underwood
Chironomidae
Orthocladiinae
Simuliidae
Simulium tuberosum (Lundstrom) [complex]
MOLLUSCA
Sphaeriidae
Pisidium sp.

The macroinvertebrate fauna of The Loch has not been extensively studied. Qualitative sampling indicated that Trichoptera, almost exclusively *Limnephilus* spp. and chironomid midges, primarily Orthocladiinae, were numerically abundant in the littoral benthic zone (Table 9.6). Larval cases of *L. coloradensis*, and to a lesser extent *L. picturatus*, were often conspicuous in the shallower zones of The Loch. Other macroinvertebrate groups including the mayflies *Ameletus velox*, *Siphlonurus occidentalis* (Banks), and *Callibaetis* sp., the dytiscid beetle *Agabus tristis* Aube, and the fingernail clam *Pisidium* sp. appear to be relatively uncommon. Two additional species occur only near the inlet and outlet of The Loch: the blackfly *Simulium tuberosum* (Lundstrom) complex and the nemourid stonefly *Amphinemura banksi* Baumann and Gaufin. Larvae of the phantom midge species *Eucorethra underwoodi* Underwood were also found sporadically in shallow areas of The Loch. Immatures of this midge are typically found in small snowmelt pools.

The species composition and diversity of Icy Brook is comparable to stream systems reported from other sites in the southern Rocky Mountains. Although the fauna of Icy Brook is dominated numerically by only a few taxa, most aquatic groups of macroinvertebrates are represented. Chironomid midges comprise 66% of the individuals, mayflies 10%, and oligochaetes 8%. Many of the aquatic insect species occurring in Icy Brook are restricted to cold mountain streams. There are differences in the structure and function of riffle communities between open and closed sites. Shredders are more common in the canopied site and scrapers at the meadow site. The predominance of collector taxa indicates the importance of FPOM as energy in the stream.

Fish

The high elevation and natural barriers to fish migration suggest that LVWS waters were originally fishless (Rosenund and Stevens 1990). A fish hatchery established in Estes Park by Lord Dunraven in the late 1890s was the most likely source for the initial cutthroat trout (*Salmo clarki stomias*) introductions to The Loch and Glass Lake. These trout are native to the Arkansas River and Platte River drainages. From the 1920s to 1968, Colorado River cutthroat trout (*Salmo clarki pleuriticus*), rainbow trout (*Salmo gairdneri*), and brook trout (*Salvelina fontinalis*) were obtained from the Leadville fish hatchery (Rosenlund and Stevens 1990). All three lakes have naturally reproducing fish populations today. These fish survive on a diet of larval insects, including mayflies and chironomids.

According to Rosenlund and Stevens (1990), The Loch was initially stocked with greenback cutthroat trout from the Estes Park fish hatchery. It was later stocked with about 16,000 Colorado River cutthroat trout from 1948 to 1955, with 42,000 rainbow trout in 1955. Currently, The Loch is dominated by greenback × rainbow hybrids. A few brook trout are also present. Inlet and outlet spawning habitat are available. Most of the hybrids observed in 1987 were fertile, but some sterile fish were also present. Unpublished National Park Service reports stated that brook trout represented 9% of the net sample in 1960. By 1987 brook trout represented only 6% of the net sample (Rosenlund and Stevens 1990).

Glass Lake today produces some of the largest brook trout in Rocky Mountain National Park, although the lake is dominated by cutthroat trout. Glass Lake was originally stocked near the turn of the century with greenback cutthroat trout, because nearly pure phenological specimens were found in 1987. Eighty-one thousand brook trout were stocked from 1922 to 1935, and 2,000 rainbow were added to this small lake in 1937. From 1938 to 1968 about 13,000 Colorado River cutthroat were stocked. A survey in 1960 observed no fish and supported fishing reports of no fish harvested from 1956 to 1959. The lake was stocked in 1962, 1965, and 1968 to revive this fishery. Since 1971, Glass Lake has been dominated by phenotypic greenback trout (Rosenlund and Stevens 1990).

While The Loch is currently dominated by rainbow × greenback cutthroat hybrids and Glass Lake by greenback cutthroat × Colorado cutthroat hybrids, Sky Pond is dominated by brook trout. Brook trout were stocked in 1935 and established a sustaining population despite introductions of cutthroat trout in 1939. Spawning is restricted to the Sky Pond outlet because of the lack of suitable habitat at the inlet. The 20 m of stream below the outlet contained numerous small brook trout near 5 cm in length in 1978.

References

Allan JD (1975) The distributional ecology and diversity of benthic insects in Cement Creek, Colorado. Ecology 56:1040–1053.

Allan JD (1982) Feeding habits and prey consumption of three setipalpian stoneflies. Ecology 63:26–34.

Bird DF, Kalff J (1986) Bacterial grazing by planktonic lake algae. Science 231:493–494.

Brook AJ, Woodward WB (1956) Some observations on the effects of water inflow and outflow on the plankton of small lakes. J. Anim. Ecol. 25:22–35.

Bushnell JH, Butler NM, Pennak RW (1982) Invertebrate communities and dynamics of alpine flowages. In: Halfpenny JC, ed. Ecological Studies in the Colorado Alpine: A Festschrift to John W. Marr, pp. 124–132. Occasional Paper No. 37, Institute for Arctic and Alpine Research, University of Colorado, Boulder.

Bushnell JH, Foster SQ, Wahle BM (1987) Annotated inventory of invertebrate populations of an alpine lake and stream chain in Colorado. Great Basin Nat. 47:500–511.

Canfield DE Jr., Bachman RW, Hoyer MV (1983) Freeze-out of salts in hard-water lakes. Limnol. Oceanogr. 28:970–977.

Canter HM, Lund JWG (1968) The importance of protozoa in controlling the abundance of planktonic algae in lakes. Proc. Linn. Soc. Lond. 179:203–219.

Carpenter SR, Kitchell JF (1988) Consumer control of lake productivity. BioScience 38:764–769.

Carpenter SR, Kitchell JF, Hodgson JR (1985) Cascading trophic interactions and lake productivity. BioScience 35:634–639.

Cather MR, Gaufin AR (1975) Life history and ecology of *Megarcys sognata* (Plecoptera: Nemouridae). Am. Midl. Nat. 95:464–471.

Cather MR, Gaufin AR (1976) Comparative ecology of three *Zapada* species of Mill Creek, Wasatch Mountains, Utah (Plecoptera: Nemouridae). Am. Midl. Nat. 95:464–471.

Dickman M (1969) Some effects of lake renewal on phytoplankton productivity and species composition. Limnol. Oceanogr. 14:660–666.

Dodds GS (1923) Mayflies from Colorado. Descriptions of certain species and notes on others. Trans. Am. Entomol. Soc. 49:93–114.

Dodds GS, Hisaw FL (1925) Ecological studies on aquatic insects. IV. Altitudinal range and zonation of mayflies, stoneflies and caddisflies in the Colorado Rockies. Ecology 6:380–390.

Elgmork K, Sæther OA (1970) Distribution of invertebrates in a high mountain brook in the Colorado Rocky Mountains. Univ. Colo. Stud. Ser. Biol. 31:1–55.

Ellsworth P (1983) Ecological seasonal cycles in a Colorado mountain pond. J. Freshwater Ecol. 2(3):225–237.

Hawkins CP, Sedell JR (1981) Longitudinal and seasonal changes in functional organization of macroinvertebrate communities in four Oregon streams. Ecology 62:387–397.

Hawkins CP, Murphy ML, Anderson NH (1982) Effects of canopy, substrate composition, and gradient on the structure of macroinvertebrate communities in Cascade Range streams of Oregon. Ecology 63:1840–1856.

Herrmann SJ (1978) Winter anomalies of seston, phytoplankton and cations in a Colorado (USA) alpine flowage lake. Int. Rev. Gesamten Hydrobiol. 63(6): 773–786.

Herrmann SJ, Ruiter DE, Unzicker JD (1986) Distribution and records of Colorado Trichoptera. Southwest. Nat. 31:421–457.

Hutchinson GE (1967) A Treatise on Limnology, Vol. 2. Wiley, New York.

Hynes HBN (1970) The Ecology of Running Waters. University of Toronto Press, Toronto, Canada.

Jensen S (1966) The Mayflies of Idaho (Ephemeroptera). M.S. Thesis, University of Utah, Salt Lake City.

Kalff J, Knoechel R (1978) Phytoplankton and their dynamics in oligotrophic and eutrophic lakes. Annu. Rev. Ecol. Syst. 9:475–495.

Keefer VM (1969) Seasonal Changes in Plankton and Seston Production in a Colorado Mountain Lake. Ph.D. Dissertation, University of Colorado, Boulder.

Keefer VM, Pennak RW (1977) Plankton and seston of a Colorado (USA) alpine lake: the winter anomaly and the inlet-outlet budget. Int. Rev. Gesamten Hydrobiol. 62(2):255–278.

Knight AW, Gaufin AR (1966) Altitudinal distribution of stoneflies (Plecoptera) in a Rocky Mountain drainage system. J. Kans. Entomol. Soc. 39:668–675.

Koob DD (1966) Parasitism of *Asterionella formosa* Hass. by a chytrid in two lakes of the Rawah wild area of Colorado. J. Phycol. 2:41–45.

Larsson P (1972) Distribution and estimation of standing crop of zooplankton in a mountain lake with fast renewal. Verh. Int. Verein. Limnol. 18:334–342.

Marr JW (1961) Ecosystems of the East Slope of the Front Range in Colorado. University of Colorado Studies, Series in Biology No. 8, University of Colorado Press, Boulder, Colorado.

McKnight D, Brenner M, Smith R, Baron J (1986) Seasonal changes in phytoplankton populations and related chemical and physical characteristics in lakes in Loch Vale, Rocky Mountain National Park, Colorado. U.S. Geological Survey Water-Resources Investigations Report 86-4101.

McKnight D, Miller C, Smith R, Baron J, Spaulding S (1988) Phytoplankton populations in lakes in Loch Vale, Rocky Mountain National Park, Colorado:

sensitivity to acidic conditions and nitrate enrichment. U.S. Geological Survey Water-Resources Investigations Report 88–4115.
McKnight D, Smith R, Bradbury JP, Baron J, Spaulding S (1990) Phytoplankton dynamics in three Rocky Mountain lakes. Arct. Alp. Res. 22:264–274.
Mecom JO (1972) Productivity and distribution of Trichoptera larvae in a Colorado mountain stream. Hydrobiologia 40:151–176.
Merrit RW, Cummins KW (1984) An Introduction to the Aquatic Insects of North America, 2d Ed. Kendall/Hunt, Dubuque, Iowa.
Merritt RW, Cummins KW, Burton TM (1984a) The role of aquatic insects in the processing and cycling of nutrients. In Resh VH, Rosenberg DM, eds. The Ecology of Aquatic Insects, pp. 134–163. Praeger, New York.
Merritt RW, Cummins KW, Resh VH (1984b) Collecting, sampling, and rearing methods for aquatic insects. In: Merritt RW, Cummins KW, eds. An Introduction to the Aquatic Insects of North America, 2d Ed. Kendall/Hunt, Dubuque, Iowa.
Morihara DK, McCafferty WP (1979) The *Baetis* larvae of North America (Ephemeroptera: Baetidae). Trans. Am. Entomol. Soc. 105:139–221.
Mutch RA, Pritchard G (1982) The importance of sampling and sorting techniques in the elucidation of the life cycle of *Zapada columbiana* (Nemouridae: Plecoptera). Can. J. Zool. 60:3394–3399.
Mutch RA, Pritchard G (1984) The life history of *Zapada columbiana* (Plecoptera: Nemouridae). Can. J. Zool. 62:1273–1281.
O'Connell MF, Andrews CW (1987) Plankton ecology in relation to flushing rate in four Newfoundland ponds. Int. Rev. Gesamten Hydrobiol. 72:487–515.
Pennak RW (1968) Field and experimental winter limnology of three Colorado mountain lakes. Ecology 49(3):505–520.
Pennak RW (1978) Freshwater Invertebrates of the United States, 2d Ed. Wiley, New York.
Pinder LCV (1986) Biology of freshwater Chironomidae. Annu. Rev. Entomol. 31:1–23.
Radar RB, Ward JV (1989) The influence of environmental predictability/disturbance characteristics on the structure of a guild of mountain stream insects. Oikos 54:107–116.
Reynolds CS (1984) The Ecology of Freshwater Phytoplankton. Cambridge University Press, Cambridge.
Richardson JW, Gaufin AR (1971) Food habits of some western stonefly nymphs. Trans. Am. Entomol. Soc. 97:91–121.
Rosenlund BD, Stevens DR (1990) Fisheries and Aquatic management: Rocky Mountain National Park 1988–1989. U.S. Fish and Wildlife Service, Colorado Fish and Wildlife Assistance Office, Golden, Colorado.
Ross HH (1956) Evolution and Classification of Mountain Caddisflies. University of Illinois Press, Urbana.
Sangren CD (1988) The ecology of chrysophyte flagellates: their growth and perennation strategies as freshwater phytoplankton. In Sangren CD, ed. Growth and Reproductive Strategies of Freshwater Phytoplankton. Cambridge University Press, Cambridge.
SAS Institute Inc. (1988) SAS Procedures Guide (release 6.03 edition). SAS Institute Inc., Cary, North Carolina.
Short RA, Ward JV (1980) Macroinvertebrates of a Colorado high mountain stream. Southwest. Nat. 25:23–32.
Short RA, Ward JV (1981) Trophic ecology of three winter stoneflies (Plecoptera). Am. Midl. Nat. 105:341–347.
Smith SD (1968) The Rhyacophila of the Salmon River drainage of Idaho with a special reference to larvae. Ann. Entomol. Soc. Am. 61:655–674.
Spaulding SA (1991) Phytoplankton Dynamics Under Ice-Cover in a Subalpine Lake M.S. Thesis, Colorado State University, Fort Collins.

Tukey JW (1977) Exploratory Data Analysis. Addison-Wesley, Reading, Massachusetts.
Vannote RL, Minshall GW, Cummins KW, Sedell JR, Cushing CE (1980) The river continuum concept. Can. J. Fish. Aquat. Sci. 37:130–137.
Ward JV (1982) Altitudinal zonation of Plecoptera in a Rocky Mountain stream. Aquat. Insects 4:105–110.
Ward JV (1986) Altitudinal zonation in a Rocky Mountain stream. Archiv für Hydrobiologie Monographische Beiträge, Supplementbänd 74:133–199.
Ward JV, Stanford JA (1982) Thermal responses in the evolutionary ecology of aquatic insects. Annu. Rev. Entomol. 27:97–117.
Wetzel RG (1983) Limnology. Saunders, Philadelphia.
Wiggins GB (1977) Larvae of the North American Caddisfly Genera (Trichoptera). University of Toronto Press, Toronto.
Wiggins GB, Richardson JS (1982) Revision and synopsis of the caddisfly genus *Dicosmoecus* (Trichoptera: Limnephilidae; Dicosmoecinae). Aquat. Insects 4:181–217.
Wright RT (1964) Dynamics of a phytoplankton community in an ice-covered lake. Limnol. Oceanogr. 9(2):163:173.

10. Biogeochemical Fluxes

Jill Baron

Chemical flux budgets are used widely today by goechemists and ecologists as a way to quantify earth and ecosystem process. Geochemical mass-balance studies, such as have been conducted by Mast for Loch Vale Watershed (LVWS) (see Chapter 6, this volume; Mast 1989; Mast et al. 1990), are used to identify mineral weathering reactions, mineral stability, and weathering rates within small watersheds (Garrels and Mckenzie 1967; Colman and Dethier 1986; Frogner 1990). Input-output budgets have been used to test hypotheses about ecological processes, such as succession and steady state (Vitousek and Reiners 1975) and disturbance (Likens et al. 1977; O'Neill et al. 1977; Vitousek and Melillo 1979). Comparison of watersheds with different states of soil and vegetation development can be used as a link between geochemistry and ecology to understand the importance of the biosphere on geological processes (Cleaves et al. 1970; Graustein et al. 1977; Likens et al. 1977; Antweiler and Drever 1982). Comparative studies have been valuable in increasing our knowledge of the fate of atmospherically derived pollutants; as a general conclusion, sequestering of metals, sulfur, and nitrogen increases with increasing proportions of soil and vegetation (Adriano and Havas 1989; Malanchuk and Nilson 1989; Dillon and Molot 1990).

Input-output budget studies are exercises in accounting. The flux of chemicals into an ecosystem is compared with the flux of chemicals out of an ecosystem, and conclusions may be drawn about biogeochemical behavior and processes from the difference. Velbel (1986) phrased it simply, and humorously, that mass-balance studies are based on the principle that "some of it plus the rest of it equals all of it." Where inputs are greater than outputs, materials are accumulated within an ecosystem; if inputs are less than outputs, mateials are lost from an ecosystem. Conservative behavior is displayed by chemicals whose inputs equal outputs (Likens et al. 1977). This simplicity is misleading, because there are many sources of uncertainty that go into creating a mass balance. The most important of these is the uncertainty associated with the hydrologic budget (Winter 1981), which was discussed in Chapter 3 (this volume). The differences between minimum and maximum values for inputs and outputs may be so large that real, but slight, contributions from or losses to the ecosystem may be masked. The sulfur mass balance for Loch Vale Watershed is an example of where this may occur; a slight internal mineral source of S to subalpine waters may be undetected because, within the bounds of uncertainly, SO_4 inputs balance SO_4 outputs. These are problems that must be solved through further investigation, but in the meantime mass balances can be used to understand general patterns. Used in conjunction with understanding of elemental cycles within the ecosystem, they provide a way to understand biogeochemical process through time and across ecosystems.

Annual Input-Output Budgets

There is variability in the major ion inputs (Table 10.1) that does not relate directly to variability in precipitation. Input variability may result from uncertainty in precipitation measurements, which we described in detail in Chapter 3 (this volume). It can also be due to real differences in atmospheric loading of major cations and anions from year to year, resulting from different storm tracks. Outputs of all ions except NH_4 and H varied with amount of discharge (Table 10.1). The highest runoff occurred in 1984 and 1986, corresponding to the greatest absolute losses of Ca, Mg, Na, K, NO_3, SO_4, and Cl. This supports the idea introduced in earlier chapters that on an annual basis the flux of most ions is hydrologically controlled. Biological and geochemical processes affect flux on smaller, seasonal scales.

The 5-year averge budgets show that there is a net loss from LVWS of base cations (see Table 10.1). Losses of Ca, Mg, Na, and K are between two- and fourfold greater than inputs from precipitation and can be accounted for by weathering of the bedrock minerals found within the drainage (see Chapter 6, this volume). There is also a net loss of SiO_2, Al, and HCO_3 from LVWS as a result of weathering (Chapter 6), because inputs of these ions in precipitation are negligible. There is a net gain of NH_4, H, and NO_3 to the watershed. Nearly all H (94%) that enters in precipitation is retained and is consumed by weathering and exchange reactions (Chapter 6; Mast et al. 1990). Of the wet-deposited NH_4, only 14% leaves the watershed in soluble form. Most of the wet-deposited NO_3 is flushed out of LVWS, but 14%

Table 10.1. Mass Balances (input-output budgets) of Major Ions 1984–1988, Loch Vale Watershed (kg ha^{-1})[a]

	1984	1985	1986	1987	1988	Average Annual	Ratio Out/In
Ca							
Input	2.06	3.99	3.46	1.32	2.29	2.62	
Output	10.36	8.38	10.13	8.04	8.55	9.09	
Out-In	8.30	4.39	6.67	6.72	6.27	6.47	3.46
Mg							
Input	0.39	0.77	0.55	0.23	0.25	0.44	
Output	1.81	1.26	1.70	1.39	1.33	1.50	
Out-In	1.42	0.49	1.15	1.15	1.09	1.06	3.41
Na							
Input	0.67	1.68	0.98	1.07	1.58	1.20	
Output	3.83	2.64	3.24	2.99	3.02	3.14	
Out-In	3.15	0.96	2.27	1.92	1.43	1.95	2.63
K							
Input	0.26	0.62	0.41	0.19	0.18	0.33	
Output	1.51	1.06	1.48	1.34	1.08	1.29	
Out-In	1.25	0.43	1.07	1.14	0.90	0.96	3.88
NH_4							
Input	1.65	1.81	2.06	1.29	0.88	1.54	
Output	0.49	0.20	0.16	0.13	0.10	0.10	
Out-In	−1.16	−1.61	−1.90	−1.16	−0.78	−1.32	0.14
H							
Input	0.18	0.16	0.21	0.14	0.13	0.16	
Output	0.009	0.008	0.013	0.007	0.008	0.009	
Out-In	−0.17	−0.15	−0.19	−0.13	−0.13	−0.15	0.06
NO_3							
Input	8.71	12.05	12.32	4.87	7.40	9.07	
Output	9.77	7.00	9.03	6.84	6.52	7.83	
Out-In	1.06	−5.06	−3.28	1.97	−0.88	−1.24	0.86
SO_4							
Input	9.68	12.37	13.28	6.55	8.33	10.04	
Output	12.00	9.72	12.17	9.98	10.58	10.89	
Out-In	2.33	−2.65	−1.10	3.43	2.25	0.85	1.08
Cl							
Input	1.18	2.07	1.43	0.81	1.01	1.30	
Output	1.04	0.85	1.03	0.89	0.97	0.96	
Out-In	−0.14	−1.22	−0.40	0.07	−0.04	−0.34	0.74
Precipitation (cm)	115	116	128	96	111		
Discharge ($m^3 * 10^6$)	5.8	4.3	6.1	4.7	4.7		

[a] Moderate precipitation estimates (Chapter 3) and mean annual water year concentrations were used to derive these values. Precipitation and discharge values presented in this table are measured values.

annually is retained. Because NH_4 is more easily assimilated than NO_3 by plants and microbial organisms (Syrett 1981; Dillon and Molot 1990), the greater annual accumulation of NH_4 is reasonable. Sulfate appears to behave conservatively, with 10.04 kg $SO_4\,ha^{-1}$ deposited annually, and 10.89 kg $SO_4\,ha^{-1}$ flushed out. Chloride, on the other hand, appears to be accruing slightly, because only 74% of the inputs are recovered in discharge. If 4-year budgets that exclude 1985 are used instead, Cl comes very close to having balanced inputs and outputs, while SO_4 then appears to have a watershed source. Sulfide-bearing minerals were not found in LVWS, but it seems more likely to this investigator that there is a mineral source of S than that chloride is undergoing immobilization within the watershed.

Interesting similarities and diferences appear when LVWS is compared with Hubbard Brook (Likens et al. 1977) and with the Sogndal control catchment of western Norway (Table 10.2) (Wright et al. 1988; Frogner 1990). Hubbard Brook, in the White Mountains of New Hampshire, is a forested catchment with extensive soil development underlain by metamorphosed sedimentary rocks. Precipitation at Hubbard Brook is acidic, with mean annual precipitation pH between 4.3 and 4.5. The Sogndal catchments have alpine vegetation and thin, acidic soils over gneiss. Precipitation at Sogndal (average, 98 cm yr^{-1}) is only slightly acidic (pH 4.9). All three sites export base cations, although proportionally more are lost from Hubbard Brook than from LVWS and proportionally less is lost from the Sogndal catchments. The greater export at Hubbard Brook could result from any of several factors, including greater annual moisture (130 cm yr^{-1}) distributed equally through the year, greater soil development, with proportionally greater surface area available for weathering and exchange (Drever 1988), and increased cations in soil soltion resulting from increased SO_4 deposition (Reuss and Johnson 1986). Frogner (1990) calculated cation denundation rates from Sogndal to be 150–200 $\mu eq\,ha^{-1}\,yr^{-1}$, which is half that which Mast calculated for LVWS

Table 10.2. Comparison of Outputs/Inputs for Loch Vale Watershed and Other Watersheds

	LVWS	Hubbard Brook	Sogndal
Ca	3.46	6.35	2.93
Mg	3.41	5.49	1.22
Na	2.63	4.57	1.07
K	3.88	2.11	0.48
NH_4	0.14	0.12	0.20
H	0.06	0.11	0.11
NO_3	0.86	0.85	0.11
SO_4	1.08	1.39	0.91
Cl	0.74	0.74	0.98
References	This study	Likens et al. 1977	Frogner 1990

(390 μeq ha^{-1} yr^{-1}), thus lower net export of base cations is in keeping with these estimates.

Sulfate shows nearly conservative behavior in LVWS and the Sogndal catchments and is being exported from Hubbard Brook. It appears that Cl at both Hubbard Brook and LVWS is being retained. Likens et al. (1977) suggested there may be an internal, but unknown, sink for Cl in the undisturbed forests, because a pulse of Cl occurred after clearcutting. In soils where atmospheric SO_4 deposition is declining or dry deposition of S is significant, SO_4 outputs can exceed the measured inputs (Galloway et al. 1983; Reuss and Johnson 1986). Both these circumstances have been documented for the eastern United States (Lindberg et al. 1990), in contrast to LVWS, where SO_4 deposition appears to be steady and dry deposition is low (see Chapter 4, this volume). Interestingly, both NO_3 and NH_4 behave similarly between Hubbard Brook and LVWS, in that most of the NH_4 is retained while most of the NO_3 flushes from both sites. There is a net N retention in both systems, of 1.9 kg N ha^{-1} yr^{-1} in LVWS, and 2.5 kg N ha^{-1} yr^{-1} in Hubbard Brook. There is a much greater net N retention in the Sogndal catchments (7.18 kg N ha^{-1} yr^{-1}), where only 20% of the NH_4 and 11% of the NO_3 from wet deposition is exported. Dillon and Molot (1990) explored NO_3 and NH_4 mass balances for a number of watersheds in eastern North America and concluded that NO_3, as a mobile anion, is readily leached. These authors suggest retention of NO_3 is a function of catchment grade, areal water discharge, and within lakes, lake depth, and water retention time. Ammonium, on the other hand, is much more likely to be immobilized by microbial transformation. The steep slopes of both LVWS (15%) and Hubbard Brook (20–30%) support the idea that water movement affects the flux of NO_3.

Seasonal Input-Output Budgets

Examination of monthly fluxes shows the overwhelming importance of spring snowmelt on the movement of all major ions except H and NH_4 (Figure 10.1A–I). The efflux patterns for major cations as well as anions are nearly indistinguishable, except for scale. The influx of base cations and major anions is nearly constant throughout the year. There are deposition peaks for most of these ions in April, and again in November. This may result from atmospheric transport of soil and agricultural materials off exposed bare soils at lower elevations to the west and east of LVWS. By April the snow is usually melted from the plains and plateaus, but vegetation has not provided substantial soil cover. The greatest deposition of NH_4 occurs in April, possibly related to precrop agricultural activity at lower elevations (Pacyna 1989). Animal feedlots and the use of fertilizers on agricultural fields are common along the plains east of the Front Range. Bare soil is again exposed in November, when a window exists between the harvest

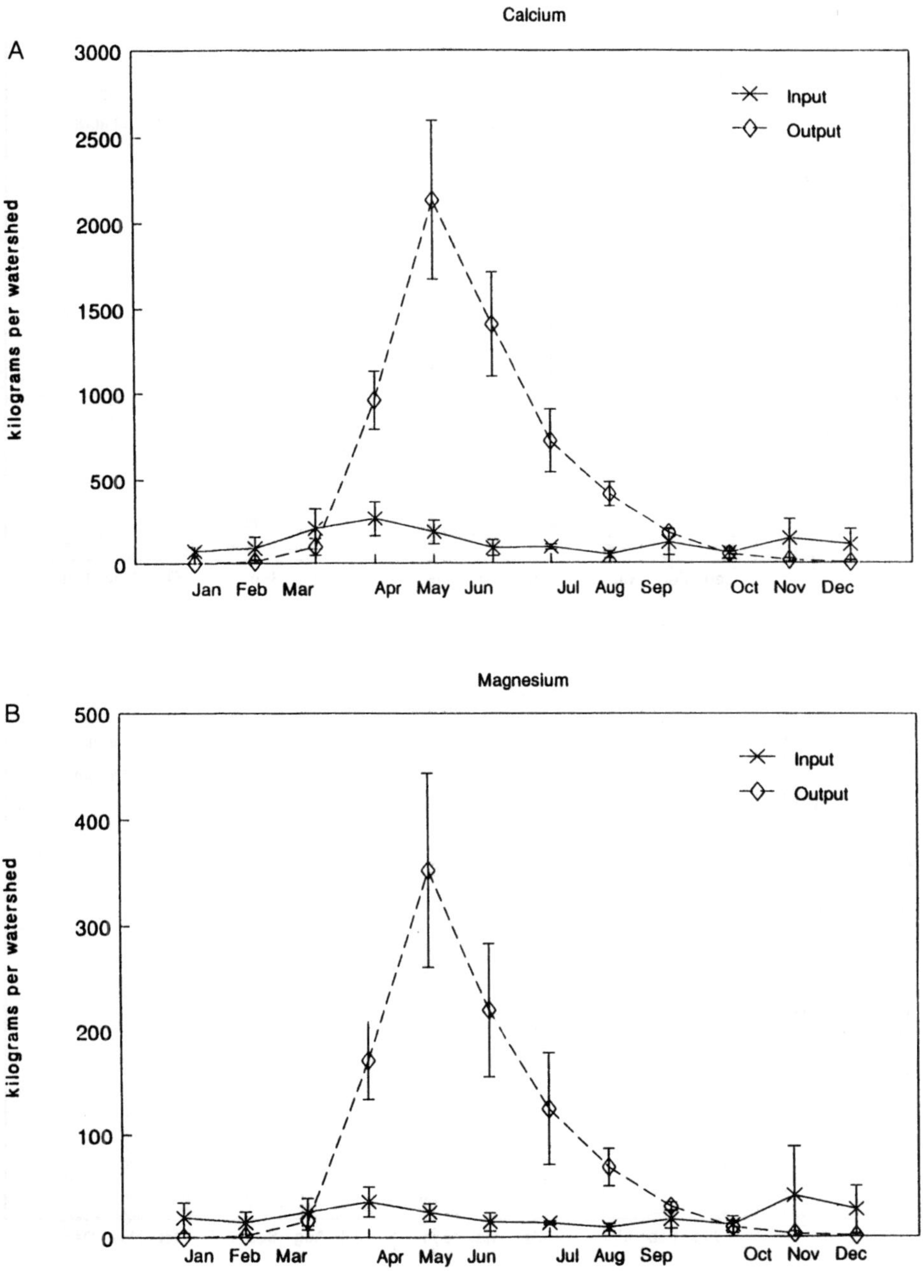

Figure 10.1(A–I). Monthly fluxes of major ions (kg per watershed area) through LVWS. Values are monthly averages for 1984–1988 in kg watershed^{-1}, error bars represent 1 S.D.; watershed area, 660 ha.

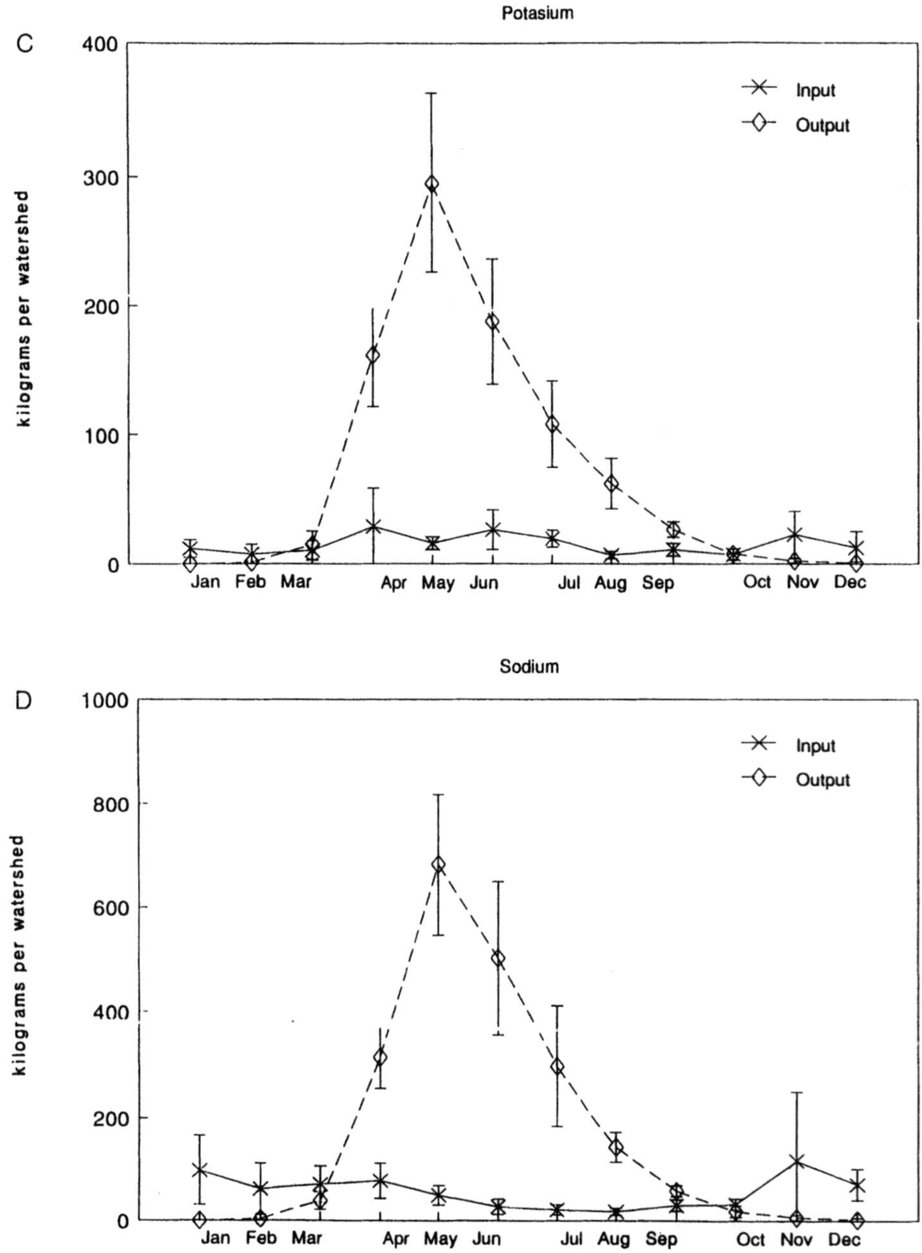

Figure 10.1 (*Continued*)

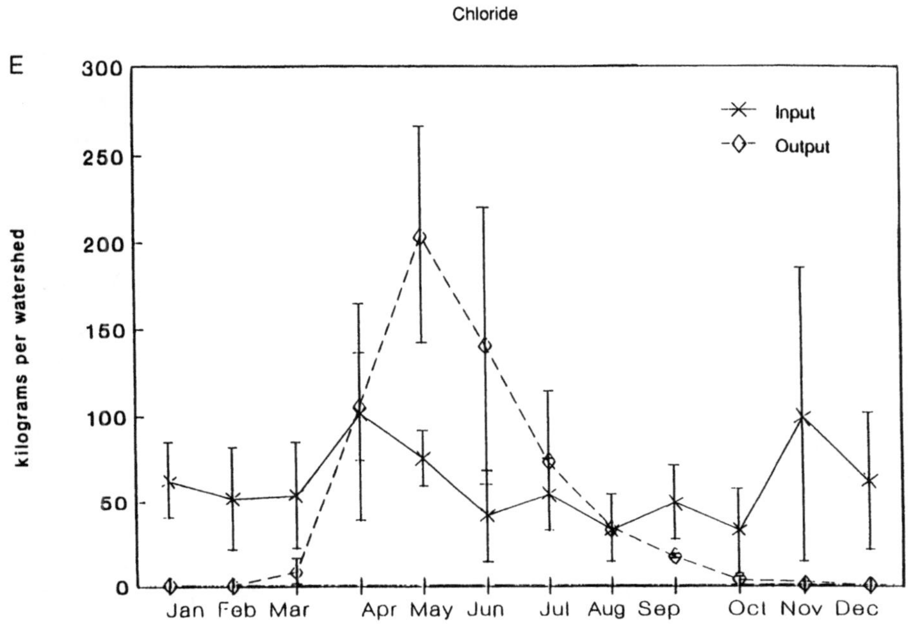

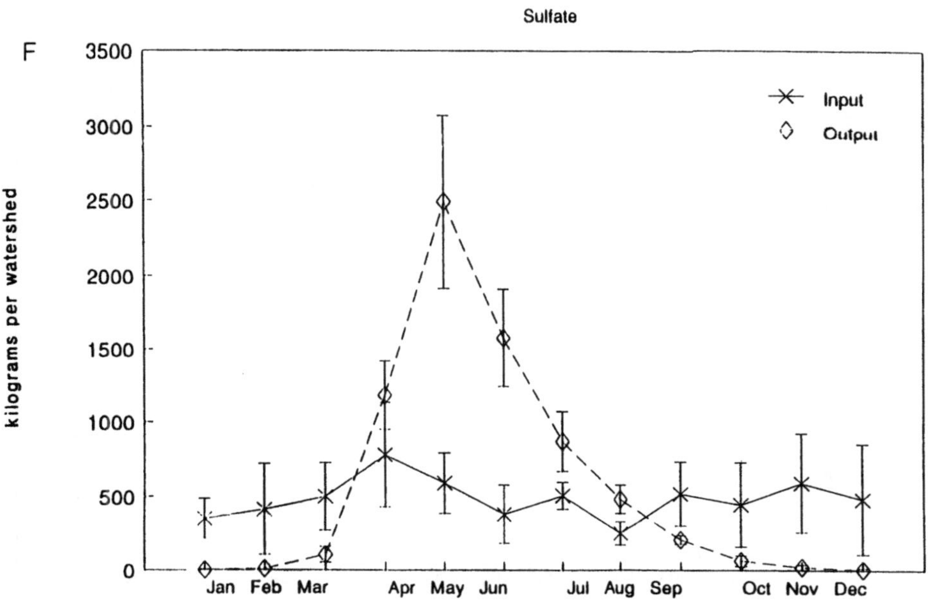

Figure 10.1 (*Continued*)

G

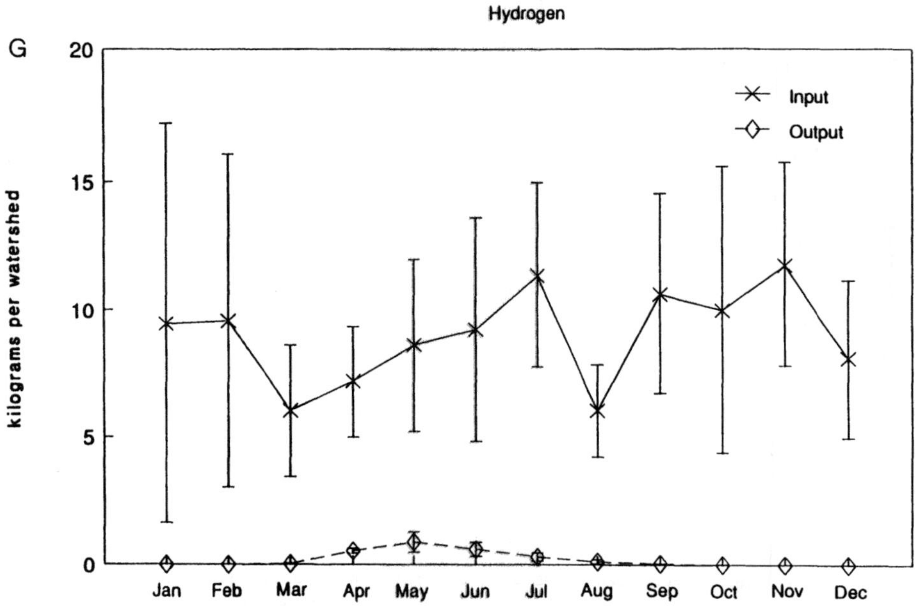

H

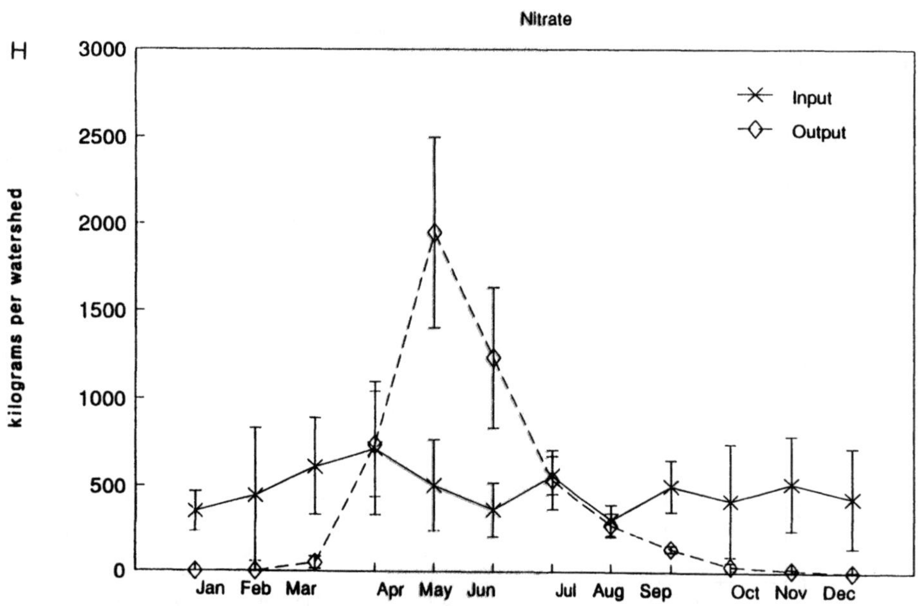

Figure 10.1 (*Continued*)

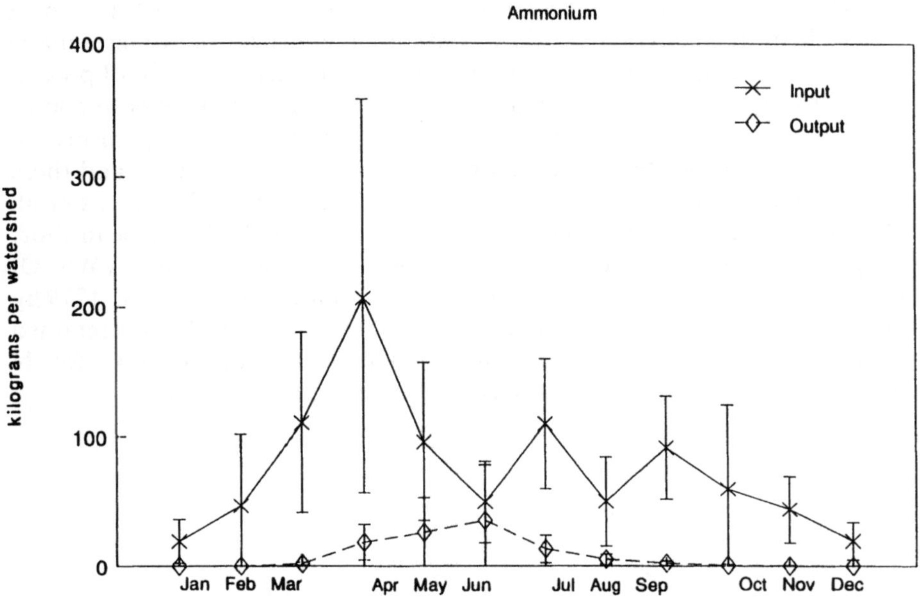

Figure 10.1 (*Continued*)

of crops and the onset of winter snow. Nitrate deposition is high in July, and SO_4, Cl, NH_4, and H also show somewhat higher July deposition than June or August. Summer upslope convective activity often transports air masses, which can be polluted, from east of the mountains to higher elevations (see Chapter 4).

Overall, the monthly budgets are distinguished by the *lack* of watershed biogeochemical control on fluxes, with the two exceptions of H and NH_4. The most plausible explanation for the observed monthly influxes relates to regional climatological phenomena. Hydrographic processes exert strong control over the observed monthly effluxes. Between October and March of each year influx is greater than effluxes, which corresponds to accumulation of atmospherically deposited materials in the snowpack.

The Nitrogen Cycle

"Some of it plus the rest of it" adequately explains the mass balances of all major ions deposited in precipitation and weathered from the bedrock except for H, NH_4, and NO_3. Most of the H from precipitation is consumed in weathering and exchange processes (see Chapter 6), and thus it is unavailable for transport downstream. There is a net consumption of N in

LVWS, in spite of the rapid hydrologic flux. Not nearly as much is known about the nitrogen cycle in LVWS as has been reported from other aquatic ecosystems such as Mirror Lake, New Hampshire (Likens 1985), but I present the following as a hypothesis that can be studied further in years to come.

Nitrogen is a major nutrient for terrestrial and aquatic organisms. In high-elevation ecosystems, the major source of N is precipitation, and these systems are typically nutrient limited by N (Coats et al. 1976; Likens et al. 1977; Keigley 1987). One could conclude that inorganic N should be in short supply, as a scarce resource, but that is not what is observed in LVWS. Of the 2145 kg N yr^{-1} that is deposited in precipitation on LVWS, 1279 kg N yr^{-1} is lost via the outflow (Figure 10.2). The remaining 866 kg N is retained within LVWS in terrestrial or aquatic biomass. Terrestrial sinks for N include plant and microbial biomass and other organic matter.

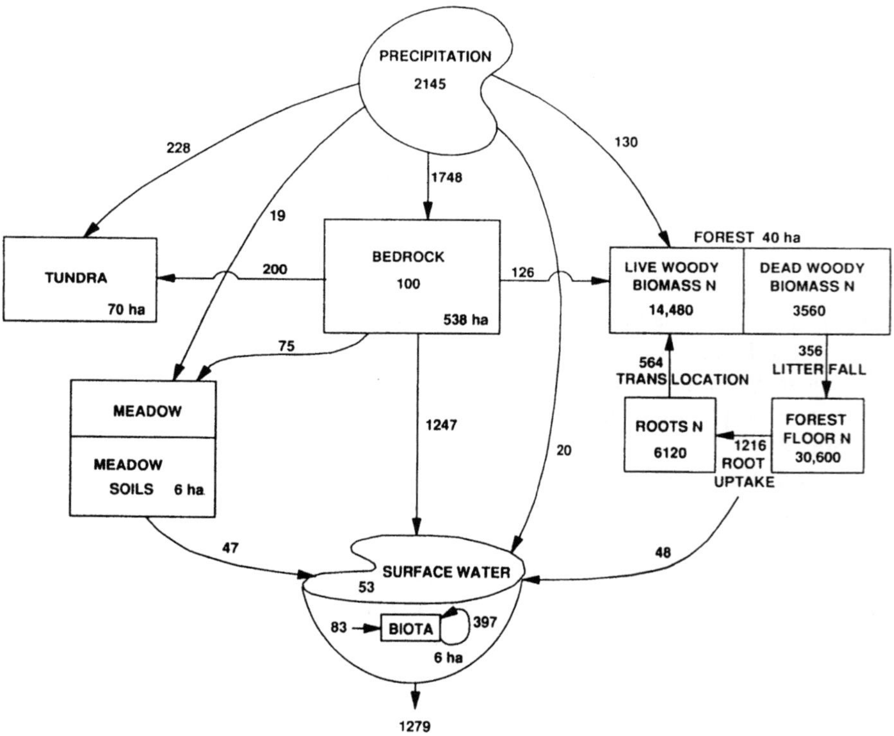

Figure 10.2. Annual nitrogen budget for LVWS based on current best estimates of sources and sinks. Values (numbers accompanying arrows) are kg N per watershed area (660 ha) per year. Ranges for inputs of N are based on different hydrologic scenarios, discussed in Chapter 3, as follows: minimum, 1845 kg N; maximum, 3339 kg N; ranges for outputs: minimum, 1234 kg N, maximum, 1466 kg N. Standard deviations for forest N cycle values are found in Chapter 5 (this volume).

The terrestrial ecosystem is divided into bedrock, tundra, meadow, and forest systems. These comprise 82%, 11%, 1%, and 5%, respectively, of the surface area of LVWS. A proportionate amount of N from precipitation was deposited on each system. Most of the N deposited on bedrock was assumed to flow off, although 100 kg N was left there to support lichens and small amounts of vegetation. Bedrock and tundra are concentrated at the higher elevations, so the largest proportion of runoff from bedrock was channeled into tundra. I assume all N inputs to tundra are retained within the alpine system, on the premise that N is in great demand (Chapin 1980; Keigley 1987) and that the alpine part of LVWS is physically disjunct from the rest of the watershed. Most of the tundra (70 ha) is located on top of Thatchtop, Taylor, and Powell Peaks, and is separated from the Icy Brook valley by 300 to 500 m cliffs. Tundra located below these cliffs is included with bedrock. Studies of arctic tundra have indicated that N is the most frequently limiting nutrient for plant production (Tieszen 1978; Shaver and Chapin 1980). Inputs of N to arctic tundra systems typically exceed known outputs by a factor of (Barsdate and Alexander 1975). The alpine tundra of LVWS, then, acts as a permanent sink for approximately 428 kg N per year. Nitrogen additions can be stored in tundra plants; higher leaf N concentrations were found after N amendments to tundra plots on Niwot Ridge, Colorado (Keigley 1987). Nitrogen can also be stored as organic matter within the soils. Gaseous losses from tundra are currently believed to be low (Giblin et al. 1991).

The forested ecosystem receives 256 kg N per year from direct precipitation and runoff from bedrock. Of this, 48 kg flushes from forested soils during snowmelt. Arthur (1990; see Chapter 5) has quantified the amount of N stored in woody biomass and forest floor. These figures show that aboveground woody biomass is accruing slightly each year. The largest pool of N is the forest floor, with 30,600 kg N; live woody biomass has the next largest pool, with 14,480 kg N.

Very little is known about LVWS meadows and their role in N cycling. Although there are only 6 ha of meadows in the watershed, an equal amount of N is flushed from meadows and forests. This is because the high water table in LVWS meadows causes saturated conditions during the summer as well as spring, and also because much of the meadow area is located adjacent to streams and lakes, allowing direct transport of N from meadows to surface waters. I assumed half the inputs to LVWS meadows from both precipitation and bedrock runoff were retained.

The pool of inorganic N in LVWS surface waters is approximately 53 kg. In the surface waters, N can be taken up by algae and passed through the food chain to the resident fish population. It can also be lost to the sediments when algal cells sink to the lake bottom or lost downstream as algae are flushed out of The Loch. Using an annual productivity of $45\,\mathrm{g\,C\,m^2\,yr^{-1}}$ (from Wetzel 1983 for a shallow alpine Austrian lake) and a C:N ratio of 5.7:1 (from Likens 1985), I obtained approximately $8\,\mathrm{g\,N\,m^2\,yr^{-1}}$, or 480 kg of algal N in LVWS. Of this amount, 83 kg N is taken up from inorganic N each year and 397 kg is recycled.

The N budget shown in Figure 10.2 suggests that approximately 438 kg N are being accumulated within the subalpine and aquatic systems of LVWS and that an equal amount, 428 kg N, is accumulating each year in alpine areas. If the tundra is truly disconnected from the rest of the watershed, and N deposited on tundra or flushed from bedrock remains there, then total inputs to the drainage part of LVWS are much lower than calculated from direct precipitation. This implies that the steep, snowmelt-dominated watershed is even less capable of retaining N than previously thought. This is in keeping with the conclusions of Dillon and Molot (1990), who found a strong correlation between catchment grade and mean NO_3 export.

References

Adriano DC, Havas M (1989) Acidic Precipitation, Vol. 1: Case Studies. Advances in Environmental Science, Springer-Verlag, New York.

Antweiler RC, Drever JI (1982) The weathering of late Tertiary volcanic ash: importance of organic solutes. Geochim. Cosmochim. Acta 47:623–629.

Arthur MA (1990) The Effects of Vegetation on Watershed Biogeochemistry at Loch Vale Watershed, Rocky Mountain National Park, Colorado. Ph.D. Dissertation, Cornell University.

Barsdate RJ, Alexander V (1975) Nitrogen balance in an arctic tundra. J. Environ. Qual. 4:111–117.

Chapin FS, III (1980) Nutrient allocation and responses to defoliation in tundra plants. Arct. Alp. Res. 12:553–563.

Cleaves ET, Godfrey AE, Bricker OP (1970) Geochemical balance of a small watershed and its geomorphic implications. Geol. Soc. Am. Bull. 81:3015–3032.

Coats RN, Leonard RL, Goldman CR (1976) Nitrogen uptake and release in a forested watershed, Lake Tahoe Basin, California. Ecology 57:995–1004.

Colman SM, Dethier DP (1986) Rates of Chemical Weathering of Rocks and Minerals. Academic Press, New York.

Dillon PJ, Molot LA (1990) The role of ammonium and nitrate retention in the acidification of lakes and forested catchments. Biogeochemistry 11:23–43.

Drever J (1988) The Geochemistry of Natural Waters. 2nd Ed. Prentice-Hall, New York.

Frogner T (1990) The effect of acid deposition on cation fluxes in artificially acidified catchments in western Norway. Geochim. Cosmochim. Acta 54:769–780.

Galloway JN, Norton SA, Church MR (1983) Freshwater acidification from atmospheric deposition of sulfuric acid: a conceptual model. Environ. Sci. Technol. 17:541A–545A.

Garrels RM, McKenzie FT (1967) Origin of the chemical compositions of some springs and lakes. In: Equilibrium Concepts in Natural Water Systems. Am. Chem. Soc. Adv. Chem. Ser. 67:222–242.

Giblin AE, Nadelhoffer KJ, Shaver GR, Laundre JA, McKerrow AJ. Biogeochemical diversity along a riverside toposequence in arctic Alaska. Ecol. Monogr. (in press).

Graustein WC, Cromack K, Jr., Sollins P (1977) Calcium oxalate: occurrence in soils and effect on nutrient and geochemical cycles. Science 198:1252–1254.

Keigley RL (1987) Effect of Experimental Treatments on *Kobresia myosuroides* with

Implications for the Potential Effect of Acid Deposition. Ph.D. Dissertation, University of Colorado, Boulder.

Lindberg SE, Page AL, Norton SA (1990) Acid Precipitation, Vol. 3: Sources, Deposition, and Canopy Interactions. Advances in Environmental Science, Springer-Verlag, New York.

Likens GE, ed (1985) An Ecosystem Approach to Aquatic Ecology: Mirror Lake and Its Environment. Springer-Verlag, New York.

Likens GE, Bormann FH, Pierce RS, Eaton JS, Johnson NM (1977) Biogeochemistry of a Forested Ecosystem. Springer-Verlag, New York.

Malanchuk JL, Nilsson J, eds. (1989) The role of nitrogen in the acidification of soils and surface waters. Miljørapport 1989:10, Nordic Council of Ministers 1989:92, Copenhagen, Denmark.

Mast MA (1989) A Laboratory and Field Study of Chemical Weathering with Special Reference to acid Deposition. Ph.D. Dissertation, University of Wyoming, Laramie.

Mast MA, Drever JI, Baron J (1990) Chemical weathering in the Loch Vale Watershed, Rocky Mountain National Park, Colorado. Water Resour. Res. 26:2971–2978.

O'Neill RV, Ausmus BS, Jackson DR, Van Hook RI, Van Voris P, Washburn C, Watson AP (1977) Monitoring terrestrial ecosystems by analysis of nutrient export. Water Air Soil Pollut. 8:271–277.

Pacyna JM (1989) Atmospheric emissions of nitrogen compounds. In: Malanchuk JL, Nilsson J, eds. The Role of Nitrogen in the Acidification of Soils and Surface Waters. Miljørapport 1989:10, Nordic Council of Ministers 1989:92, Copenhagen, Denmark.

Reuss JO, Johnson DW (1986) Acid deposition and the acidification of soils and surface waters. Ecological Studies 59, Springer-Verlag, New York.

Shaver GR, Chapin FS, III (1980) Response to fertilization by various plant growth forms in an Alaskan tundra: nutrient accumulation and growth. Ecology 61:662–675.

Syrett PJ (1981) Nitrogen metabolism of microalgae. In: Platt T, ed. Physiological bases of phytoplankton ecology. Can. Bull. Fish. Aquat. Sci 210:182–210.

Tieszen LL (1978) Vegetation and production ecology of an Alaskan arctic tundra. Springer-Verlag, New York.

Velbel MA (1986) Weathering and soil-forming processes. In: Swank WT, Crossley DA, Jr., eds. Forest Hydrology and Ecology at Coweeta, pp. 93–102. Ecological Studies 66, Springer-Verlag, New York.

Vitousek PM, Reiners WA (1975) Ecosystem succession and nutrient retention: a hypothesis. BioScience 25:376–381.

Vitousek PM, Melillo JM (1979) Nitrate losses from disturbed forests: patterns and mechanisms. For. Sci. 25:605–619.

Wetzel RG (1983) Limnology, 2d Ed. Saunders, Philadelphia.

Winter TC (1981) Uncertainties in estimating the water balances of lakes. Water Resour. Bull. 17:82–115.

Wright RF, Lotse E, Semb A (1988) Reversibility of acidification shown by whole-catchment experiments. Nature (London) 334:670–675.

11. Management Implications

Jill Baron

So What?

The federal land manager, whose days and budgets are often consumed by short-term deadlines, might question the value of highly specific ecological information such as we have gathered together here. A manager must balance many priorities, and the temptation is to emphasize the loudest, most persistent crisis instead of long-term, low-level monitoring. In the following pages, some examples show how information from a project such as this can contribute to improved natural resource management. The examples are grouped into four categories: (1) using understanding of natural processes to identify deviation from expected behavior; (2) using understanding of natural processes to predict and prepare for change; (3) maximizing educational and interpretive opportunities; and (4) using understanding of natural processes to minimize surprise. The value of categories (1) and (2) lies in the direct application of scientific results to resource management, while the value of categories (3) and (4) is in a political and social context.

Looking for Deviation from Expected Behavior

Years of observation and assessment have allowed us to develop a fairly good idea of what to expect in the hydrologic and chemical cycles and in the vegetation and algal dynamics of Loch Vale Watershed. For many parameters, there is both a quantitative assessment and a qualitative "feel" for natural fluctuations. We expect, for example, surface water alkalinity to fluctuate seasonally, ranging from 10 to 150 μeq L^{-1}; water clarity is good to very good; specific conductance is very low; and NO_3 concentrations between 1983 and 1990 averaged 16 μeq L^{-1}. The subalpine forest is currently composed of about equal basal areas of spruce and fir, although there are many more individuals of fir than spruce. Thus, the recruitment of fir is greater, but fir mortality is also greater. One of five trees is currently standing dead, with a ratio of fir to spruce of 2:1. Soils are very acidic and base saturation levels range between 10% and 40% in forested areas. These are some examples of the kinds of benchmarks we can refer to over the coming years as we monitor for change. If different patterns are observed, there is a base of information from which to question them.

Knowing how a natural system ought to behave is like expecting certain behavior from a healthy human being. When the signals differ from expected, scrutiny should intensify. In the case of a human being, the physician looks for evidence of illness or disease, based extensive knowledge of systems of the human body. She or he draws on experience plus all the accumulated medical knowledge available to diagnose and correct the illness. We who are managers must also have an accumulated body of knowledge to draw on to gauge the health of the ecosystems that are our responsibility.

Predicting and Preparing for Change

How much acidity is too much? Acid-neutralizing capacity in Loch Vale Watershed (LVWS) lakes averages about 60 μeq L^{-1} on a yearly basis, and weathering contributes about 200 μeq HCO_3 ha^{-1} yr^{-1}. Except for a brief period each spring when soils play a role, precipitation and bedrock supply essentially all base cations and alkalinity to LVWS surface waters. Because weathering in most of the drainage is a primarily physical process, we do not expect weathering rates to change given an increase in acidity from precipitation. Current precipitation averages 140 μeq H ha^{-1} yr^{-1}. An increase in H deposition of greater than 60 μeq H ha^{-1} yr^{-1} will deplete lake acid-neutralizing capacity, causing lakes to acidify. This converts to a deposition rate of 0.2 kg H ha^{-1} as compared to the current rate of 0.14 kg H ha^{-1}. Those who monitor precipitation must therefore be alert for increases in deposition acidity or new sources of emissions that could bring this increase. Potential new emission sources can be evaluated in light of their predicted effects to Rocky Mountain lakes and streams, and managers can influence the direction of the federal permitting process on the basis of combined knowledge of thresholds, current sources, and seasonal deposition cycles.

Spatial predictions are made easier because of in-depth knowledge about processes within LVWS. The representativeness of LVWS with respect to the entire population of Rocky Mountain lakes is rigorously defined by the 1985 Western Lake Survey (Eilers et al. 1986; Turk and Spahr 1989). This essentially defines the limits of extrapolation from LVWS to other systems and lends additional support to these extrapolations. LVWS represents the most sensitive extreme of Rocky Mountain lakes and therefore can be used as an early indicator of change.

Educational and Interpretive Opportunities

An Informed Public

An educated constituency is a powerful ally, which is reason enough to educate the public about natural ecosystems and threats to their integrity. National parks are not amusement parks, and their purpose is not solely to provide recreational opportunities. It is irresponsible to both a park and its visitors to portray the park as a pristine jewel of naturalness when there are real and potential threats. Many parks today have excellent natural history interpretive programs. These can be augmented with environmental education that places the park in a bigger context. Parks are affected by regional air and water pollution; parks may be affected by global-scale climate change; parks are valuable repositories of biological diversity and observatories for natural processes. Long-term data sets such as provided by the LVWS study can be used to illustrate many points about regional and national air and water quality or about other issues that might arise such as climate change.

Natural History Interpretation

The long-term records that are documented in a study such as this provide local examples for natural history education. Natural ecosystems are dynamic. They change over seasons or in response to disturbances. This is an important concept to convey to the public. Records like those initiated in LVWS will allow visitors to have a greater appreciation for ecological complexity.

Education by Example

The United States of America has the premier network of natural preserves on Earth. We can set an example of how to, rather than how not to, manage these resources for the future. A major role of the National Park Service is to protect and preserve resources for future generations, not only for our "enjoyment," but possibly for our very survival. This is becoming a major theme, not only among ecologists (Baron and Galvin 1990; Lubchenko et al. 1991), but among policymakers and economists (WCED 1987). The United

States has been instrumental in developing much of current world thinking in basic ecology; by using this knowledge to manage our natural resources, we can also become a world leader in applied ecology.

Minimizing Surprise

The LVWS research program was undertaken on the premise that interpretation of ecological observations leads to understanding of ecosystem processes. Increased understanding, in turn, can be used to detect and correct certain types of disturbances. Understanding, then, provides a position of strength from which a manager can influence the future integrity of the resources that are his responsibility. Long-term observations have been valuable to Lake Tahoe, where they were used to record increasing eutrophication from recreational development (Goldman and Horne 1983). It was the long-term record of precipitation and stream chemistry from Hubbard Brook that brought acidic deposition to the attention of North American scientists (Likens et al. 1980). In both examples the deviation from expected natural cycles of behavior was detected because of a priori understanding of natural variability. In both examples it was the scientific foundation that provided the basis for political and social change.

Why should managers of natural ecosystems keep long-term records of ecosystem parameters? Two reasons come to mind, and both minimize surprise (Holling 1978). Surprise type 1, or ecological surprise, occurs when natural systems operate in unanticipated ways. This type of surprise results from not understanding natural ecosystem variability. Type 1 surprises should be avoided at all costs, because of their potential to lead to ecological disaster. Monitoring of ecological parameters and regular examination of the newly collected information in comparison with the complete data records help to minimize the potential for ecological disaster caused by ignorance.

Surprise type 2, or managerial surprise, occurs when there is inadequate communication between those who do the research and monitoring and those who do the resource managing. Surprise of this type can also lead to ecological disaster through mismanagement rather than ignorance. Type 2 surprises can lead to political embarrassment; managers of national lands are very much in the public eye. The antidote to type 2 surprise is, of course, better communication between researcher and manager. Trends in natural resource variability can become regular discussions at staff meetings. Resource fluctuations should be scrutinized as closely as are park monies. Just as educated publics become strong allies, so do educated colleagues. And the ebb and flow of nature should be inherently more interesting than the ebb and flow of funds to a manager of wild lands.

Humankind is entering a period when human-caused global-scale changes will affect local ecosystem dynamics. Scientists have barely begun to outline what the consequences will be to natural ecosystems as we know them. We are breaking new ground by trying to prepare for these changes. There is no question but that knowledge of the past and present will help us address the future. As we move into the twenty-first century, let us use every tool and increase understanding at every level to protect and preserve our natural lands.

References

Baron J, Galvin KA (1990) Future directions of ecosystem science: toward an understanding of the global biological environment. BioScience 40:640–642.

Eilers JM, Kanciruk P, McCord RA, Overton WS, Hook L, Blick DJ, Brakke DF, Kellar P, Silverstein ME, Landers DH (1986) Characteristics of Lakes in the Western United States: Vol. II: Data Compendium for Selected Physical and Chemical Variables. EPA-600/3-86/054B, U.S. Environmental Protection Agency, Washington, D.C.

Goldman CR, Horne AJ (1983) Limnology. McGraw-Hill, New York.

Holling CS, ed. (1978) Adaptive Environmental Assessment and Management. Wiley, London.

Likens GE, Bormann FH, Eaton JS (1980) Variations in precipitation and streamwater chemistry at the Hubbard Brook Experimental Forest during 1964 to 1977. In: Hutchinson TC, Havas M, eds., Effects of Acid Precipitation on Terrestrial Ecosystems, pp. 443–464. NATO Conference Series I: Ecology. Plenum Press, New York.

Lubchenko J, Olson AM, Brubaker LB, Carpenter SR, Holland MM, Hubbell SP, Levin SA, MacMahon JA, Matson PA, Melillo, JM, Mooney HA, Peterson CH, Pulliam HR, Real LA, Regal PJ, Risser PG (1991) The sustainable biosphere initiative: an ecological research agenda. A report from the Ecological Society of America. Ecology 72: 371–412.

Turk JT, Sphar NE (1989) Chemistry of Rocky Mountain Lakes. In: Adriano DC, Havas M, eds., Acidic Precipitation, Vol. I: Case Studies, pp. 181–208. Advances in Environmental Science, Springer-Verlag, New York.

World Commission of Environment and Development (WCED) (1987) Our Common Future. Oxford University Press, New York.

Index

U

V

W

Y

Z

Ecological Studies

Volume 74
Inorganic Contaminants in the Vadose Zone (1989)
Edited by B. Bar-Yosef, N.J. Barnow, and J. Goldshmid

Volume 75
The Grazing Land Ecosystems of the African Sahel (1989)
By H.N. Le Houérou

Volume 76
Vascular Plants as Epiphytes: Evolution and Ecophysiology (1989)
Edited by U. Lüttge

Volume 77
Air Pollution and Forest Decline: A Study of Spruce *(Picea abies)* on Acid Soils (1989)
Edited by E.-D. Schulze, O.L. Lange, and R. Oren

Volume 78
Agroecology: Researching the Ecological Basis for Sustainable Agriculture (1990)
Edited by S.R. Gliessman

Volume 79
Remote Sensing of Biosphere Functioning (1990)
Edited by R.J. Hobbs and H.A. Mooney

Volume 80
Plant Biology of the Basin and Range (1990)
Edited by C.B. Osmond, L.F. Pitelka, and G.M. Hidy

Volume 81
Nitrogen in Terrestrial Ecosystems (1990)
By C.O. Tamm

Volume 82
Quantitative Methods in Landscape Ecology (1990)
Edited by M.G. Turner and R.H. Gardner

Volume 83
The Rivers of Florida (1990)
Edited by R.J. Livingston

Volume 84
Fire in the Tropical Biota: Ecosystem Processes and Global Challenges (1990)
Edited by J.G. Goldammer

Volume 85
The Mosaic-Cycle Concept of Ecosystems (1991)
Edited by H. Remmert

Volume 86
Ecological Heterogeneity (1991)
Edited by J. Kolasa and S.T.A. Pickett

Volume 87
Horses and Grasses: The Nutritional Ecology of Equids, and Their Impact on the Camargue (1991)
By P. Duncan

Volume 88
Pinnipeds and El Nino: Responses to Environmental Stress (1991)
Edited by F. Trillmich and K.A. Ono

Volume 89
Plantago: A Multidisciplinary Study (1992)
Edited by P.J.C. Kuiper and M. Bos

Volume 90
Biogeochemistry of a Subalpine Ecosystem: Loch Vale Watershed (1992)
Edited by Jill Baron

MIX
Papier aus verantwortungsvollen Quellen
Paper from responsible sources
FSC® C105338

If you have any concerns about our products,
you can contact us on
ProductSafety@springernature.com

In case Publisher is established outside the EU,
the EU authorized representative is:
Springer Nature Customer Service Center GmbH
Europaplatz 3, 69115 Heidelberg, Germany

Printed by Libri Plureos GmbH
in Hamburg, Germany